Membrane Fluidity in Biology

Volume 2

General Principles

Membrane Fluidity in Biology

Volume 2
General Principles

EDITED BY

ROLAND C. ALOIA

Departments of Anesthesiology and Biochemistry
Loma Linda University
School of Medicine
and
Anesthesia Service
Pettis Memorial Veterans Hospital
Loma Linda, California

1983

ACADEMIC PRESS

A Subsidiary of Harcourt Brace Jovanovich, Publishers

New York London
Paris San Diego San Francisco São Paulo Sydney Tokyo Toronto

ACADEMIC PRESS, INC.
111 Fifth Avenue, New York, New York 10003

United Kingdom Edition published by
ACADEMIC PRESS, INC. (LONDON) LTD.
24/28 Oval Road, London NW1 7DX

Library of Congress Cataloging in Publication Data

Main entry under title:

Membrane fluidity in biology.

Bibliography: p.
Includes index.
Contents: v. 2. General Principles --
1. Membranes (Biology)--2. Membranes (Biology)--
Mechanical properties. I. Aloia, Roland C.
QH601.M4664 1982 574.87'5 82-11535
ISBN 0-12-053002-3

PRINTED IN THE UNITED STATES OF AMERICA

83 84 85 86 9 8 7 6 5 4 3 2 1

Contents

7. *Ionotropic Effects on Phospholipid Membranes: Calcium/Magnesium Specificity in Binding, Fluidity, and Fusion*

Nejat Düzgüneş and Demetrios Papahadjopoulos

8. *The Effect of the Proton and of Monvalent Cations on Membrane Fluidity*

Hansjörg Eibl

9. *Membrane Fluidity and Cytoplasmic Viscosity*

Alec D. Keith and Andrea M. Mastro

Contributors

Numbers in parentheses indicate the pages on which the authors' contributions begin.

Joan M. Boggs (89), Biochemistry Department, The Hospital for Sick Children, Toronto, Ontario M5G 1X8, and Department of Clinical Biochemistry, University of Toronto, Toronto, Ontario M5G 1L5, Canada

Dennis Chapman (5), Department of Biochemistry and Chemistry, Royal Free Hospital School of Medicine, University of London, London NW3 2PF, England

Frank S. Davis (1), Department of Biological Chemistry, University of Illinois Medical Center, Chicago, Illinois 60612

Nejat Düzgüneş (187), Cancer Research Institute, University of California, School of Medicine, San Francisco, California 94143

Hansjörg Eibl (217), Max-Planck-Institut für Biophysikalische Chemie, D-3400 Göttingen-Nikolausberg, Federal Republic of Germany

Chris W. M. Grant (131), Department of Biochemistry, University of Western Ontario, London, Ontario N6A 5C1, Canada

George M. Helmkamp, Jr. (151), Department of Biochemistry, College of Health Sciences and Hospital, School of Medicine, The University of Kansas Medical Center, Kansas City, Kansas 66103

Alec D. Keith (237), The Pennsylvania State University, Biophysics Program, University Park, Pennsylvania 16802

William E. M. Lands (1), Department of Biological Chemistry, University of Illinois Medical Center, Chicago, Illinois 60612

Anthony G. Lee (43), Department of Biochemistry, University of Southampton, Southampton SO9 3TU, United Kingdom

Andrea M. Mastro (237), The Pennsylvania State University, Microbiology Program, University Park, Pennsylvania 16802

Demetrios Papahadjopoulos (187), Department of Pharmacology, Cancer Research Institute, University of California, School of Medicine, San Francisco, California 94143

Preface

Oh the comfort we feel when the image we see
Fits the model we've drawn in our books to a "T"!
It's easy to say "artifacts? there are none!"
For the image and model agree—we are done!
K. A. Platt-Aloia

There have been thousands of studies and numerous publications on various aspects of membrane structure–function relationships. However, it has been apparent for many years that a major treatise focusing on the tenets and facets of membrane fluidity was needed. This set of volumes, entitled *Membrane Fluidity in Biology,* is intended to be such a treatise. The contributors to these volumes examine the many membrane properties influenced by alterations in membrane lipid composition and/or other organizational parameters encompassed by the term *fluidity.* The treatise will be a comprehensive source that examines the precepts of membrane fluidity and conceptualizes the significance of fluidity changes in both normal and pathological cellular states. Each volume presents a state-of-the-art review and should serve as a valuable reference source for all scientists whose research involves cellular membrane function.

The first volume of this treatise, *Concepts of Membrane Structure,* examined the fundamental concepts of currently popular membrane models. The contributors to Volume 1 presented new ideas about membrane structural organization and the architectural arrangement of molecular components of cell membranes. They provided insight into the complexity and diversity of cell membrane structure and function as well as a sound conceptual framework for evaluating the principles of membrane fluidity to be discussed in subsequent volumes.

The present volume is devoted to an elaboration of the basic tenets of membrane fluidity. The contributors, all experts in their respective fields, discuss such topics as lateral phase separations and phase transitions, hydrophobic and electrostatic effects of membrane lipid–protein interactions, isothermal phase transitions and the effects of ionic factors, and the influence of such components as cholesterol, phospholipids, fatty acids, and cellular water on the parameters of membrane fluidity. Each of these topics is elaborated in great detail and clarity to provide a unique insight into the factors that influence the thermal molecular motions of membrane components and hence cellular membrane function.

Because such activities as membrane transport, enzyme kinetics, and receptor function are modulated by the physical state of the membrane lipids and proteins, a thorough comprehension of the molecular aspects of membrane fluidity is necessary for evaluating the arcane aspects of membrane-related cellular activities. This volume covers a broad spectrum of subjects related to membrane fluidity and should therefore be essential reading for all scientists and researchers concerned with the molecular principles of cellular and organelle function. Furthermore, this volume provides an appropriate background for Volume 3, *Cellular Aspects and Disease Processes,* and for subsequent discussion of the relationship of membrane fluidity to environmental parameters, drugs, anesthetics, and other exogenous agents.

I wish to express my thanks to the Department of Anesthesiology of Loma Linda University and the Anesthesia Service at the Pettis Memorial Veterans Hospital. I am indebted to Drs. George Rouser, Gene Kretchevsky, and William W. Thomson for kindling my interest in membranes and for their continued support. I am grateful to Gizete Babcock, Helen Mayfield, and Penelope Winkler for their dedicated secretarial support, and to Judy Daviau, Mary Ann Meyer, and Shelley Tucker for their help with the index.

Roland C. Aloia
1983

Contents of Volume 1

Chapter 1

Definitions, Explanations, and an Overview of Membrane Fluidity

William E. M. Lands and Frank S. Davis

Research on membranes and the fluidity of their components has involved a variety of instrumental techniques. Occasionally, the various techniques provide measurements for which the specific underlying molecular interactions are complex. Although no single measurement can fully describe the broad general concept of *membrane fluidity,* all descriptions have one common aspect; they are based on interactions that are in part related to the density of the liquid. This chapter reviews some of the density-related molecular events that are common to all the different measures of membrane fluidity.

Fluidity (ϕ) is the property of a liquid that describes its ease of movement. Viscosity (η), the inverse of fluidity, is the property describing the resistance of a fluid to movement. Viscosity has been defined as the force per unit area

$$\text{fluidity} = \phi = 1/\eta = 1/\text{viscosity} \tag{1}$$

that must be applied to a moving plate separated from a fixed plate by 1 cm of fluid such that the moving plate maintains a constant velocity of 1 cm/sec (see Hildebrand, 1972). The standard unit of viscosity, the *poise,* has the dimensions of gram/second/centimeter.

The above definition expresses viscosity as the absorption of momentum by a liquid in terms of mass/time/distance. Energy is expended to maintain the velocity of the moving plate as momentum is transferred to the liquid between the plates. The absorption of momentum by the liquid provides a quantitative measure of the viscosity η.

$$\eta = \text{absorption of momentum/cross-sectional area} \tag{2}$$

Transfer of momentum from an individual molecule to its neighbor pro-

ISBN 0-12-053002-3

vides the basis for viscosity in a liquid. These adjacent molecules also absorb and transfer momentum, so that displacing one molecule in a highly interactive (viscous) liquid requires that a great many molecules be moved.

Molecular momentum has several forms: rotational, vibrational, and translational; in membranes, each form of motion has a component perpendicular to the plane of the membrane and a component parallel to the membrane. Because we can describe six major forms of momentum transfer in membranes, there are six corresponding forms of viscosity to consider.

The density of a liquid influences the rate of transfer of all forms of momentum and thereby provides a common factor for all forms of viscosity. A dense liquid will have a greater frequency of momentum-transferring collisions than a less dense liquid.

$$\text{momentum transferred/time cross-sectional area} = [\tfrac{1}{3}\, nm(dx/dt)^2]/V \tag{3a}$$

Here, n equals the number of molecules in a volume V, each with mass m and velocity dx/dt, which pass through a given area in time t. This expression represents the flux of momentum in the fluid, and with density (ρ) equal to nm/V, Eq. (3a) may be rewritten as

$$\text{flux of momentum} = \tfrac{1}{3}\, \rho\, (dx/dt)^2. \tag{3b}$$

This expression emphasizes that the transfer of momentum in a liquid is directly proportional to the density of the liquid.

The volume occupied by 1 mol of liquid (the molal volume) is also directly related to density, and it usually increases with increasing temperature. The coefficient of expansion β is

$$V = \beta\, \Delta T + V_0 \tag{4}$$

useful in predicting the expansion of a particular liquid with temperature. V_0 can be regarded as the volume occupied by 1 mol of liquid at the transition from solid to liquid phase. In this state, the molecules rotate and vibrate in the liquid matrix, but do not move laterally because of crowding by their neighbors. V_0 is found by extrapolation from plots of fluidity at different temperatures vs the molal volume V at each temperature (Hildebrand, 1972).

Hildebrand (1971; 1972) modified the earlier (1913) concept of Batschinski to establish a simple quantitative relationship between fluidity and excess molar volume. The term $(V-V_0)$ is the excess molal volume and represents

$$\phi = B\, (V-V_0)/V_0 \tag{5}$$

the degree of expansion of a liquid beyond that of the solid–liquid transition state. The B term represents the contribution of molecular spatial features to fluidity. The value for this coefficient B may be expected to differ significantly for the different types of molecular motion. Thus one technique used to examine the transfer of molecular momentum in a fluid could determine a type of fluidity based on spatial orientations (movement) of molecules that are very different from those measured by another technique. This difference can be expected even though the same molal excess volume is common to both measurements.

When considering how fatty acids affect membrane fluidity, we can consider their effect on membrane density (Holub and Lands, 1975). As an example, the unsaturated acyl chains have a greater cross-sectional area than do the saturated acids (Demel *et al.*, 1972), resulting in lower densities and a higher reported fluidity. This effect is expressed by the excess molal volume in Hildebrand's equation. In addition, the low phase transition temperatures observed for unsaturated acyl esters result in their greater degree of expansion (and thus lower densities) at physiological temperatures. This interpretation of the contribution of fatty acids to the volume of a membrane helps explain their contribution to the fluidity.

It would be helpful if the different methods of measuring membrane fluidity were well characterized in regard to the types of molecular motion they measure, and some progress has been made in this area. Watts (1981) has summarized how NMR and ESR analyses measure different time scales of molecular motion in membranes. Lakowicz (1979) has noted that fluorescence depolarization measurements of the rotational motion of a molecular probe may be a useful method for comparing density effects on steady state anisotropy. Many general discussions of fluidity related to differental scanning calorimetry or other measures of transition temperature may be seen as reflecting in part the density of the lipid microenvironment irrespective of molecular motion. Recognition of this common factor in all fluidity phenomena can help us interpret all of the diverse instrumental methods of analysis that are presently being applied to describe membrane behavior.

As we discover how different lipids contribute to the density of a membrane, we can develop an improved understanding of the role of lipids in determining membrane fluidity. Despite the progress noted above, a challenge remains for us all to discover the extent to which each different instrumental technique (electron spin resonance, fluorescence depolarization, differential scanning microcalorimetry, diffusivity, etc.) measures each of the six different types of fluidity. We must also describe how different fats, hydrocarbons, sterols, and other molecules of interest interact to affect membrane density, fluidity, and function. Finally, it seems likely that certain molecular motions will have greater significance in one phenomenon than another. Thus, we need to learn which form of fluidity is influencing or limiting each particular membrane-related phenomenon in which we are interested.

References

Batschinski, A. J. (1913). *Z. Phys. Chem. (Leipzig)* **84,** 643.
Demel, R. A., Geurts Van Kessel, W. S. M., and Van Deenen, L. L. M. (1972). *Biochim. Biophys. Acta* **266,** 26–40.
Hildebrand, J. H. (1971). *Science* **174,** 490–493.
Hildebrand, J. H. (1972). *Proc. Natl. Acad. Sci.* **69,** 3428–3431.
Holub, B. J., and Lands, W. E. M. (1975). *Can. J. Biochem.* **53,** 1262–1267.
Lakowicz, J. R., Prendergast, F. G., and Hogan, D. (1979). *Biochemistry* **18,** 508–519.
Watts, A. (1981). *Nature (London)* **294,** 512–513.

Chapter **2**

Biomembrane Fluidity: The Concept and Its Development

Dennis Chapman

The Concept of Biomembrane Fluidity

In 1966, Chapman *et al.* postulated a *fluidity concept* for biomembranes that proposed "the particular distribution of fatty acyl residues which occurs with

Membrane Fluidity in Biology, Vol. 2
General Principles

ISBN 0-12-053002-3

a particular biomembrane is present so as to provide the appropriate membrane fluidity for a particular environmental temperature to match the required diffusion rate or rate of metabolic processes required for the tissues." It was also suggested that "there appear to be biosynthetic feedback mechanisms by which a cell will attempt to retain a constant membrane fluidity" (Chapman *et al.*, 1966), and examples were given of poikilothermic organisms that altered their membrane fluidity to match different environmental temperatures.

This fluidity concept was based upon a range of studies. An important contribution to the concept of biomembrane fluidity was a study of the detailed molecular nature of the major thermotropic phase transition of long-chain amphiphilic molecules, delineated by IR spectroscopic studies (Chapman, 1958). Studies (1964–1966) of pure phospholipids utilizing such techniques as infrared spectroscopy and nuclear magnetic resonance (NMR) spectroscopy pointed to (and emphasized) the extent of molecular mobility that can be associated with phospholipid molecules. This is particularly valid when these molecules exceed a certain critical transition temperature (Byrne and Chapman, 1964; Chapman *et al.*, 1966, 1967). The reduction of line width that occurs at the critical transition temperature even with pure anhydrous phospholipids is shown in Fig. 1.

These and later studies showed the molecular details underlying the membrane fluidity concept. Above a phase transition temperature (T_c characteristic for a particular lipid of given chain length:

1. The lipid chains show flexing and twisting of the methylene (i.e., CH_2) groups and a marked increase of rotational isomers
2. The oscillations and rotational disorder of the methylene groups are most marked at the methyl end of the lipid chains (Chapman and Salsbury, 1966)
3. In addition to the chain motion, other parts of the molecule, such as the polar groups of the lecithin molecules, exhibit a marked increase in mobility (Veksli *et al.*, 1969)
4. Lipid self-diffusion occurs when sufficient water is present to weaken any ionic linkages between the polar groups (Penkett *et al.*, 1968)
5. The transition temperature from ordered to fluid "melted" chains is related to the chain length and degree of unsaturation of the lipid (Chapman *et al.*, 1967)
6. The fluidity of the biomembrane can be modulated by molecules which penetrate the lipid bilayer such as cholesterol (Chapman and Penkett, 1966; Ladbrooke *et al.*, 1968).

Since these early studies, many workers have confirmed, extended, and quantified the mobility, order, and diffusion of lipid molecules in model and natural biomembranes. As a result of many more studies, the concept of

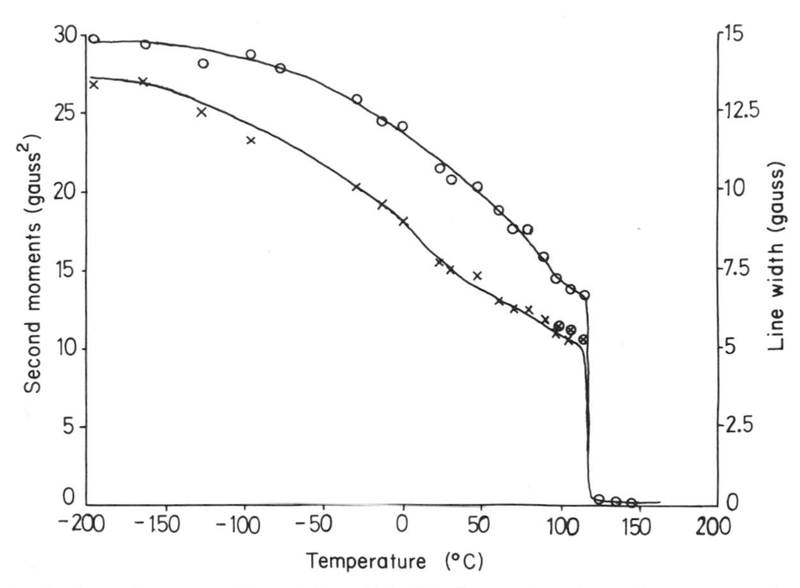

Fig. 1. Second moment (x) and linewidth (○) data as functions of temperature for 2,3-dimyristoyl-DL-phosphatidylethanolamine. The points shown thus (⊗) correspond to the second moment of the broad components only. From Chapman and Salsbury (1966).

fluidity has become more sophisticated. Furthermore, once it was appreciated that a particular biomembrane had its own characteristic fluidity, varying methods for measuring this were developed. This led directly to the application of ESR and fluorescent probe methods. Later studies have attempted to relate certain disease conditions to modifications of biomembrane fluidity.

The concept of biomembrane fluidity is now an integral part of popular models for biomembrane structure and function. A recent summary of abstracts (183) associated with oncology shows the continued appreciation of the importance and relevance of biomembrane fluidity (see Cooper, 1981).

Fluidity and Lipid Dynamics

Lipid Order Parameters

On the basis of calorimetric studies, Chapman *et al.* (1969) suggested that the fluid character of the bilayer matrix of biomembranes is different from that of a simple paraffinic melt. [This contrasted with the view that the interior lipid chains were in a random chaotic state (see Luzzati, 1968).] The ΔH values associated with the gel to liquid crystal transition are lower, in fact, than the ΔH values for the melting of pure hydrocarbons. This is also true for the entropy change in the process. The incremental ΔS per CH_2 group is only about 1 entropy unit (e.u.) for the gel to liquid–crystal transition of bilayers, but it is almost twice as large for the melting of simple paraffins (Phillips *et al.*, 1969). These thermodynamic results showed that the hydrocarbon chains in the bilayer core are not as disordered as they are in a pure liquid hydrocarbon.

Although the early studies with proton NMR (Chapman and Salsbury, 1966; Veksli *et al.*, 1969) were valuable in emphasizing the mobility of the lipid molecules, another useful innovation was the introduction of nonperturbing deuterium probes for deuterium magnetic resonance studies, by Oldfield *et al.* (1971). This has been particularly valuable for studying the details of the lipid dynamics within model and natural biomembranes. It has been utilized (Seelig and Seelig, 1974; Seelig and Seelig, 1980) to examine the *order parameter* of the lipid chains and to show that in the region of constant order parameter gauche conformations can occur only in complementary pairs leaving the hydrocarbon chains essentially parallel to each other. The deuterium probe results differ from those obtained using spin-labeled molecules (Hubbell and McConnell, 1971), for example, the spin labels detect a continuous decrease of the order parameter, whereas the deuterium probe shows that the order parameter remains approximately constant for the first nine segments.

The number of order profiles reported in the literature is still small. ^{2}H-NMR order profiles using selectively deuterated phospholipids have been established now for 1,2-dipalmitoyl-*sn*-glycero-3-phosphocholine (DPPC) (Seelig and Seelig, 1974, 1975), 1,3-dipalmitoyl-*sn*-glycero-3-phosphocholine (Seelig and Seelig, 1980), 1,2-dimyristoyl-*sn*-glycero-3-phosphocholine (Oldfield *et al.*, 1978a,b), 1-palmitoyl-2-oleoyl-*sn*-glycero-3-phosphocholine (POPC) (Seelig and Seelig, 1977; Seelig and Waespe-Šfrčevic, 1978), and 1,2-dipalmitoyl-*sn*-glycero-3-phosphoserine (DPPS) (Seelig and Browning, 1978; Browning and Seelig, 1980). In addition, egg yolk lecithin with perdeuterated palmitic acid acyl chains intercalated physically (Stockton *et al.*, 1976), perdeuterated 1,2-dipalmitoyl-*sn*-glycero-3-phosphocholine (Davis *et al.*, 1979), and a glycolipid (Skarjune and Oldfield, 1979) have been studied. Although limited in number, these order profiles comprise a fairly representative collection of different lipid classes, including saturated and unsaturated fatty acyl chains as well as different polar groups.

A comparison of the rotational correlation times with the deuterium order parameter as a function of the labeled segment is given in Fig. 2. The shapes of the correlation time and order parameter are similar to correlation times,

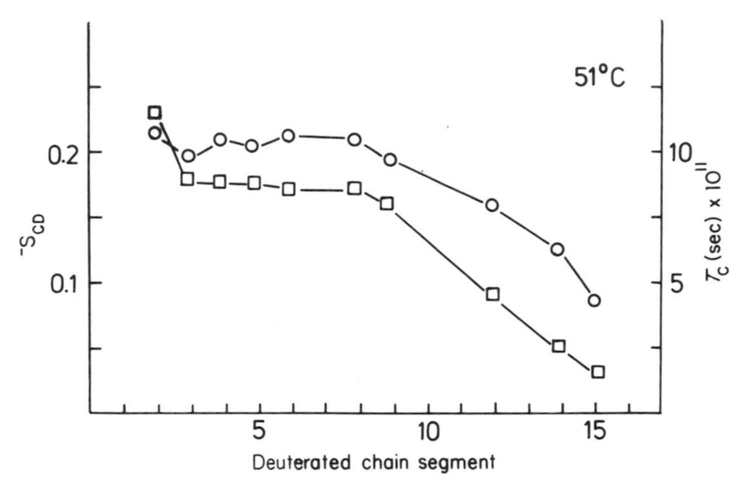

Fig. 2. A comparison of the rotational correlation times (T_{υ}, □) determined from the T_1 data and the deuterium order parameter ($-S_{CD}$,○) as a function of segment position. Reproduced with permission from Brown *et al.* (1979).

ranging from about 8×10^{-11} sec for the plateau region to about 3×10^{-11} sec for the C-15 methylene segment (Brown *et al.*, 1979).

The deuterium magnetic resonance technique has also been applied to the study of the lipid dynamics of natural biomembranes. This was first carried out by Oldfield *et al.* (1972a) with *Acholeplasma laidlawii* cells and later by Stockton *et al.* (1977) with selectively deuterated palmitic acid. Quite recently, similar studies have also been made with *Escherichia coli* membranes. The fatty acyl chains in elaidate 9,10-$^2H_2$-enriched *E. coli* inner and outer membranes are less ordered by 10–20% than in the corresponding model systems (Gally *et al.*, 1980).

This characteristic order parameter signature of model membranes is carried over into biological membranes. The agreement between the order profile of *Acholeplasma laidlawii* and those of the pure phospholipid membranes is striking. The incorporation of a cis double bond promotes larger changes in the order profile than, for example, the introduction of a net negative charge in the polar head group (DPPS) or the incorporation of proteins into the membrane. The divergence of the POPC order profile between carbon atoms 5 to 9 can be explained by a specific stiffening effect of the cis double bond (Seelig and Seelig, 1977).

Measurements of the deuterium quadrupole splittings ($\Delta\nu_Q$) furnishes information about the time-averaged orientation of the segments involved. In contrast, measurement of the deuterium NMR relaxation times gives information on the *rate* of segmental motion. The correlation between chain *order* and chain mobility is not yet well understood, but the distinction between time-averaged structural parameters (such as relaxation times, correlation times, and microviscosity) refers to membranes in general and is independent of the specific technique employed (Seelig and Seelig, 1980).

Lipid Diffusion

Quantitative studies of lipid diffusion in model bilayers were made by Träuble and Sackmann (1972) and Devaux and McConnell (1972); values reported were of the order $D = 1.8 \times 10^{-8}$ cm²/sec. Triplet probes have also been used (Naqvi *et al.*, 1974) to measure the diffusion coefficient of lipids in the fluid state; a value of 1.6×10^{-8} cm²/sec was derived by this method. Similar studies have been carried out using a variety of methods including NMR spectroscopy, fluorescence photobleaching, and spin label photochemical methods.

A technique of pattern photobleaching has been used to measure the diffusion of fluorescent-labeled phospholipids (Smith and McConnell, 1978). In this technique, the light from an argon-ion laser is passed through a microscope and a Runci ruling onto a sample under the objective of the microscope that is in the form of phospholipid multilayers. The fluorescent molecules are photobleached in a striped pattern. The time-independent intensity of the stripes is recorded photographically and used to determine the diffusion coefficients below and above the T_c phase transition temperature. Values of 10^{-10} cm²/sec below T_c and $>10^{-8}$ cm²/sec above this temperature were obtained. Studies of the antibodies bound to lipid haptens in model biomembranes have been observed by the pattern photobleaching method to diffuse as rapidly as the lipids themselves (B. A. Smith *et al.*, 1979); these ranged in value from 10^{-11} to 10^{-8} cm²/sec.

Correlation with Monolayer Properties

Prior to the introduction of the concept of fluidity in 1966, some attempts had been made to relate the monolayer properties at the air–water interface with biomembrane structure. However, these attempts were rather unsatisfactory, primarily because there was no clear concept of the molecular details of the monolayer. For instance, what is the molecular organization or dynamics underlying the "expanded" state of a monolayer?

The situation became clarified when Chapman *et al.* (1966) and Phillips and Chapman (1968) pointed out that a correlation exists between the lipid monolayer properties at the air–water interface of lipids and the properties of the lipid bilayers in aqueous dispersions. The "condensed monolayer" corresponds to the crystalline or gel phase and the expanded state to the "fluid" or melted state which occurs above the lipid transition temperature. Similar thermotropic phase changes occur with the monolayers (Fig. 3) as do with lipid bilayers.

All monolayer states are possible with the saturated lecithin and phosphatidylethanolamine homologs (Phillips and Chapman, 1968). It is apparent that if the hydrocarbon chains are sufficiently long, condensed monolayers are formed; with shorter chains, liquid-expanded films occur. Figure 3 shows the pressure/area (π/A) curves for a series of 1,2-diacylphosphatidylcholines. These two limiting states are sufficiently well defined that at any particular temperature only one of the homologs studied exhibits the transition state. The data indicate that variations in hydrocarbon chain length which do not give rise to change in monolayer state do not have a significant effect on the π/A curves. Temperature changes can also give rise

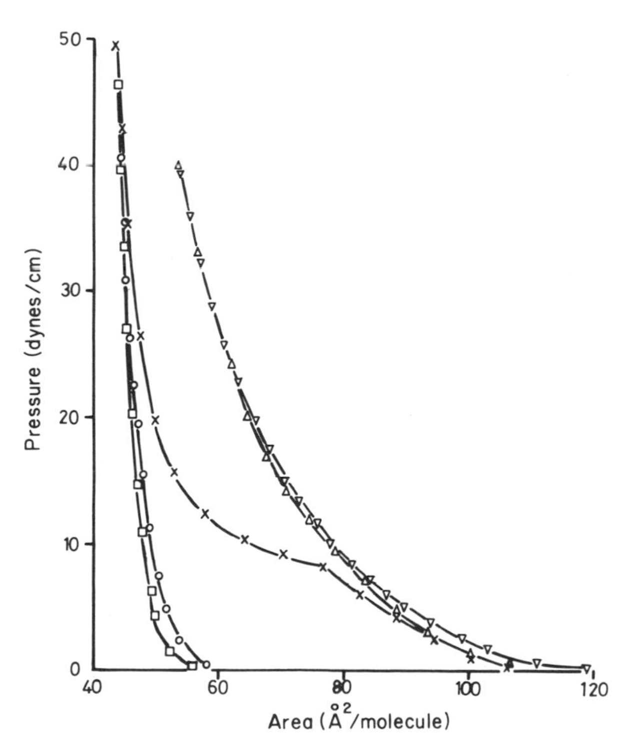

Fig. 3. Pressure–area curves for saturated 1,2-diacylphosphatidylcholines on 0.1 *M* NaCl at 22°C. C_{22}, □; C_{18}, ○; C_{16}, × C_{14}, △; C_{10}, ▽. From Phillips and Chapman (1968).

to the condensed and expanded states for a monolayer of a single homolog. Obviously, a sufficiently low temperature causes the film to become completely condensed, whereas at higher temperatures it is fully expanded. Monolayers in the two limiting states are more or less invariant with temperature, and it is the sensitivity of the phase transition to temperature that leads to the variety of isotherms.

The molecules in a completely condensed phosphatidylethanolamine monolayer are much more closely packed than are those in the equivalent lecithin monolayer. This also correlates with the lipid bilayer behavior. The lecithins have a lower transition temperature in a bilayer structure than does the equivalent chain length phosphatidylethanolamine. This presumably arises from steric factors associated with the large polar groups on the lecithin molecules. Marčelja (1974) has used this correlation between the monolayers and lipid–water dispersions to calculate the lateral pressure in the hydrocarbon chain region. He estimates this to be approximately 20 dynes/cm for each half of the bilayer.

Phase Separation

Most biomembranes contain a range of lipid classes and a variety of acyl chain lengths and degrees of unsaturation. This led Ladbrooke and Chapman (1969) and Phillips *et al.* (1970) to carry out calorimetric studies of mixtures of lipids within model biomembrane structures.

The first studies on phase separation of lipid–water systems were discussed by Ladbrooke and Chapman (1969), who reported studies of binary mixtures of lecithins using calorimetry (Phillips *et al.*, 1970). These authors examined mixtures of distearoyl and dipalmitoyl lecithin (DSL–DPL) and also distearoyl lecithin and dimyristoyl lecithin (DSL–DML). With the DSL–DPL mixtures, the phase diagram shows that a continuous series of solid solutions are formed below the T_c line. It was concluded that compound formation does not occur, and that in this pair of molecules which have only a small difference in chain length, co-crystallization occurs.

With the system DSL–DML, monotectic behavior was observed with limited solid solution formation. The difference in chain length is already too great for co-crystallization to occur, so that as the system is cooled migration of lecithin molecules occurs within the bilayer to give crystalline regions corresponding to the two compounds (Ladbrooke and Chapman, 1969).

Examination of a series of fully saturated lecithins with dioleoyl lecithin gave similar results, with phase separation of the individual components taking place (Phillips *et al.*, 1970). Later calorimetric studies were reported by Clowes *et al.* (1971) on mixed lecithin–cerebroside systems and on

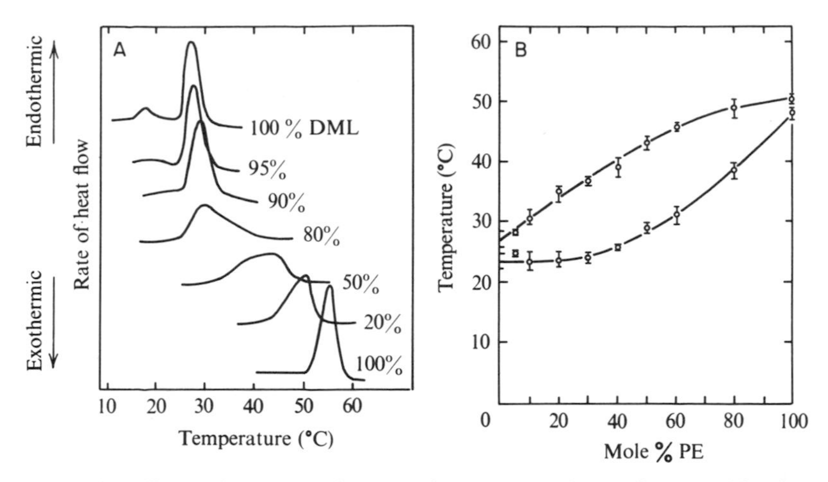

Fig. 4. (A) Differential scanning calorimetry heating curves for 1,2-dimyristoyl lecithin–1,2-dimyristoylphosphatidyl-ethanolamine–water mixtures. (B) Phase diagram of the 1,2-dimyristoyl lecithin–1,2-dimyristoylphosphatidylethanolamine–water mixtures. Upper curve, onset (cooling) temperature; lower curve, T_c heating. From Chapman *et al.* (1974).

lecithin–phosphatidylethanolamine mixtures (reviewed by Oldfield and Chapman, 1972; Chapman *et al.*, 1974). The lecithin–phosphatidylethanolamine systems of the same chain length give a wide melting range with some separation of the different lipid classes. Some calorimetric curves and a phase diagram for mixtures of 1,2-dimyristoyllecithin and 1,2-dimyristoylphosphatidylethanolamine are shown in Fig. 4.

The use of spin labels such as TEMPO to prove phase separation of mixed lipid systems was reported by Shimshick and McConnell (1973) on similar lipid mixtures. Other phase separation properties, observed by Ito and Ohnishi (1974), indicate that lipid phase separation can occur in phosphatidic acid–lecithin membranes because of the effects of Ca^{2+}. Butler *et al.* (1974) have shown that the probe stearic acid spin label tends to migrate to the more fluid lipid phase in multiphase systems. This confirms the earlier conclusions of Oldfield *et al.* (1972b); measurements of membrane fluidity in heterogeneous systems are not necessarily representative of the entire membrane.

Metal Ion and pH Effects

For some years, metal ion interactions have been known to affect the thermotropic phase transition of soap systems. The thermotropic phase transition of stearic acid occurs at 114°C with the sodium salt and at 170°C with the potassium salt. These phase transitions can be linked to the monolayer characteristics.

Early studies of stearic acid monolayers (Harkins and Anderson, 1937; Shanes and Gershfeld, 1960) showed that interaction with Ca^{2+} caused an increase in surface pressure (i.e., condensation) and also decreased the permeability to water. The same effect has been observed with phosphatidylserine monolayers (Rojas and Tobias, 1965), but Na^{+} and K^{+} addition gave no such condensation. More extensive studies showed that a variety of acidic phospholipid monolayers undergo an increase in surface potential and decrease in surface pressure on addition of Ca^{2+} and otherbivalent cations (Papahadjopoulos, 1968). Phosphatidylserine is found to be more selective than phosphatidic acid, but in both systems the order of cation effectiveness is

$$Ca^{2+} > Ba^{2+} > Mg^{2+}$$

The formation of linear polymeric complexes was proposed to account for these findings.

Cationic charge has been observed to be important in some bilayer studies of phosphatidylserine (Ohki, 1969). Black membranes formed in the presence of Ca^{2+} ions are more stable with a higher electrical resistance than those formed in the presence of Na^{+} only. The concentrations of cationic species required to produce charge reversal in phosphatidylserine dispersions have been determined, with association constants for the species formed (Barton, 1968). These results agree well with those obtained previously (Blaustein, 1967) with the exception of uranyl cation UO_2^{2+}. Studies with this cation (Chapman *et al.*, 1974) indicate that it causes the thermotropic phase transition temperature of lecithins to increase. Two main phase transitions were observed, corresponding to the presence of complexed and uncomplexed lipid. When the titration is complete only the higher melting transition remains. The studies by Chapman *et al.* (1974) indicate that the interaction between cations and phosphatidylserine causes greater shifts of transition temperature than is observed with lecithin molecules. All the cations studied shifted the phase transition temperature of the phospholipids to higher values.

The precise nature of the interaction between ions and phospholipids is still undefined. There is some evidence that charge neutralization is the prime interaction of charged phospholipids with divalent cations (Verkleij *et al.*, 1974; Träuble and Eibl, 1974). Divalent cations were found to increase the transition temperature; the monovalent cations fluidize the bilayer. Some authors believe that the primary effect of the cation on lecithins may be on the aqueous portion of the lipid bilayer (Gottlieb and Eanes, 1972; Ehrström *et al.*, 1973; Godin and Ng, 1974). A recent study of an extensive range of salts with lecithin bilayers (Chapman *et al.*, 1977b) indicates that the anion present has a very large effect in determining the state of fluidity of the bilayer; the results obtained were explained best by a thermodynamic treatment based on relative association constants. Verkleij *et al.* (1974) have shown that the thermotropic phase transition of a synthetic phosphatidylglycerol is influenced by pH, Ca^{2+}, and a basic protein of myelin. Träuble (1972) has shown that pH can affect lipid transition temperatures, particularly lipids such as the phosphatidylethanolamines and phosphatidic acids. Träuble (1976) also pointed out that surface charges tend to fluidize or expand lipid biomembranes. Fluidization of a biomembrane can be induced by an increase either in pH or in salt concentration. Rigidification can be achieved by a decrease in pH or (when the lipid is fully ionized) by an increase in ionic strength.

Fluidity and Its Modulation by Cholesterol

Condensation and Fluidity

For some years it was known that cholesterol could affect and apparently condense monolayers (at the air–water interface) with certain unsaturated phospholipids. The meaning of this was, however, obscure and controversial, some workers believing that a cis double bond was essential for this, invoking unusual structures and complexes between the lipid and cholesterol. Chapman *et al.* showed, however, that phospholipids containing trans double bonds and even saturated phospholipids could exhibit these effects. (The introduction of the concept of biomembrane fluidity, and the correlation and understanding that this provided for monolayer studies; proved to be particularly valuable for understanding and rationalizing the action of cholesterol molecules).

Proton NMR studies clearly indicated that cholesterol molecules affected the fluid lipid chains within a model biomembrane system (Chapman and Penkett, 1966). Studies using ESR probes (Barratt *et al.*, 1969; Hubbell and McConnell, 1971), and fluorescent probes have confirmed the interpretation of these experiments. Studies using deuterium NMR have recently examined model biomembranes containing various amounts of cholesterol (Rice *et al.*, 1979). Addition of cholesterol to the sample at the equimolar level (about 33 wt %) (see Fig. 5) increases quadrupole splitting from 3.6 to 7.8 kHz,

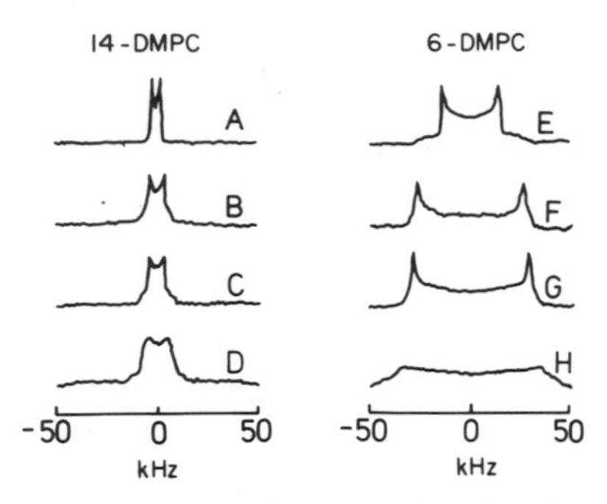

Fig. 5. Deuterium NMR spectra obtained by the quadrupole echo Fourier transform method at 35 mHz (corresponding to a magnetic field strength of 5.2T) of l-myristoyl-2-[14,14,14-$^2H_3$]myristoyl-*sn*-glycero-3-phosphocholine [6,6-$^2H_2$]DMPC bilayers in the absence and presence of cholesterol as a function of temperature. Trace A, pure [$^3H_2$]DMPC, 30°C, 100 kHz effective spectral width, 0.54 sec recycle time, 2048 data points $\tau_1 = \tau_2 = 50$ μsec, 7 μsec 90° pulse widths, 20,000 scans, and 150-Hz line broadening. Trace B, [$^3H_2$]DMPC–50 mol % cholesterol; other conditions as described in trace A, except 10,000 scans. Trace C, [$^3H_2$]DMPC–50 mol % cholesterol, 18°C; other conditions as described in trace A, except 10,000 scans. Trace D, pure [$^3H_2$]DMPC, 18°C; other conditions as described in trace A, except $\tau_1 = \tau_2 = 70$ μsec and 15,000 scans. Trace E, pure [6,6-$^2H_2$]DMPC, 30°C and other conditions as described in trace A, except 0.054 sec recycle time, 10,000 scans and $\tau_1 = \tau_2 = 40$ μsec. Trace F, [6,6-$^2H_2$]DMPC–50 mol % cholesterol, 30°C, other conditions as described in trace A, except, 0.100 sec recycle time and 100,000 scans. Trace G, [6,6-$^2H_2$]DMPC–50 mol % cholesterol, 18°C; other conditions as described in trace A, except 0.110 sec recycle time and 200,000 scans. Trace H, pure [6,6-$^2H_2$]DMPC, 21°C; other conditions as described in trace A, except 0.110 sec recycle time, 5,000 scans, 4096 data points, and $\tau_1 = \tau_2 = 40$ μsec. All samples were hand dispersions in deuterium-depleted water. Sample size was about 200 μliter and samples typically contained 20–50 mg of phospholipid. After Rice *et al.* (1979).

corresponding to an increase in molecular order parameter (from S_{mol} = 0.18 to S_{mol} = 0.41). Cooling the sample to a temperature some 5°C below that of the gel to liquid crystal phase transition temperature (T_c = 23°C) has little effect on the quadrupole splitting; this is consistent with previous data (Ladbrooke *et al.*, 1968; Oldfield *et al.*, 1971). Cooling the pure lipid to the same temperature, however, results in hydrocarbon chain crystallization into the rigid crystalline gel phase (Chapman *et al.*, 1967), and a broad, rather featureless spectrum with $\Delta\nu_Q \sim 14.0$ kHz is observed (Oldfield *et al.*, 1978a). Analysis of this result in terms of a molecular order parameter is not possible, because the details of the motion of the rest of the hydrocarbon chain are unclear.

Similar ordering effects of cholesterol are seen when a C_6-labeled phospholipid, 1-myristoyl-2-(6,6-dideuteromyristoyl)-*sn*-glycero-3-phospho-choline [6,6-$^2H_2$]DMPC, is used. A ^2H-NMR spectrum of [6,6-$^2H_2$]DMPC in excess water at about 30°C is shown in Fig. 5. The quadrupole splitting

$\Delta\nu_Q$ is 29.0 kHz (which corresponds to an $S_{mol} = 0.45$). The region of the lipid bilayer closer to the polar head group is more ordered than the terminal methyl region ($S_{mol} = 0.18$). Addition of cholesterol at the 33 wt % (1:1 mol ratio) level increases $\Delta\nu_Q$ to 54.5 kHz (Fig. 4F) which corresponds to an $S_{mol} = 0.85$. The relative increase in order parameter on addition of cholesterol becomes smaller toward the top of the hydrocarbon chain. Cooling the sample to 18°C has little effect on $\Delta\nu_Q$. However, when samples of pure [6,6-$^2H_2$]DMPC were cooled to about 18°C, it was found that there is a very large line broadening, and at lower temperatures there is an increase in $\Delta\nu_Q$ to the rigid lattice value of 127.0 kHz. (At temperatures close to but below T_c, the line shapes are complex, indicating that the rates of motion of the hydrocarbon chain about its long axis are decreasing but that the angular fluctuations are rather similar to those above T_c). At lower temperatures, however ($\simeq$ 10°C), chain motion ceases (Rice *et al.*, 1979).

Another feature of the inclusion of cholesterol in biomembrane structures is that the presence of large amounts of cholesterol prevents lipid chain crystallization and thus removes phase transition characteristics. This was shown by Ladbrooke *et al.* (1968), who described studies on lecithin–cholesterol–water interactions by differential scanning calorimetry (DSC) and X-ray diffraction. The 1,2-dipalmitoyl-L-phosphatidylcholine (DPPC)–cholesterol–water system was studied as a function of both temperature and concentration of components. This particular lecithin was used because it exhibits the thermotropic phase change in the presence of water at a convenient temperature (41°C). The addition of cholesterol to the lecithin in water lowers the transition temperature between the gel and the lamellar fluid crystalline phase, and decreases the heat absorbed at the main transition. No transition is observed with an equimolar ratio of lecithin with cholesterol in water. This ratio corresponds to the maximum amount of cholesterol that can be introduced into the lipid bilayer before cholesterol precipitation occurs.

X-ray evidence (Ladbrooke *et al.*, 1968) indicates that a lamellar arrangement occurs, and that at 50% cholesterol an additional long spacing pattern occurs as a result of the separation of crystalline cholesterol. These results may be interpreted in terms of penetration of the lipid bilayer by cholesterol. In the lamellae of aqueous lecithin, the chains are hexagonally packed and tilted at 58°. It can be envisioned that penetration will be facilitated when the chains are vertical. This causes an increase in the X-ray long spacing. At concentrations of cholesterol greater than 7.5%, the long spacing decreases. Above this critical concentration, a reduction of the cohesive forces between the chains occurs, producing chain fluidization.

Below the lipid T_c transition temperature, calorimetric studies show that the main lipid endotherm is removed as the amount of cholesterol is increased. The presence of the cholesterol molecules modulates the lipid fluidity above and below the transition temperature of the lipid. The first studies of Ladbrooke *et al.* (1968) suggested that the enthalpy was totally removed at 50 mol %; later studies (Hinz and Sturtevant, 1972) suggested that this occurred at 33 mol %. The latter conclusions led to the concept that cholesterol existed as a 2:1 lipid–cholesterol complex.

Very recent studies using sensitive scanning calorimeters confirm (Mabrey *et al.*, 1978) in the early conclusion that the enthalpy is in fact removed at 50 mol % of cholesterol/lipid as originally conceived.

Theoretical Models

Engelman and Rothman (1972) suggested that each cholesterol molecule is separated from a nearest neighbor cholesterol by a single layer of lipid chains, which are thereby removed from the cooperative phase transition. They arranged the structure so that ΔH became zero at ≈ 0.33, in agreement with the results of Hinz and Sturtevant (1972). This result is now acknowledged to be incorrect (Mabry *et al.*, 1978) because of instrumental difficulties (see above). Forslind and Kjellander (1975) constructed a structural model of the phosphatidylcholine (PC)–cholesterol–water system for $T > T_c$. They were concerned with neither a phase diagram nor thermodynamic quantities, but they did obtain good agreement with measured values of the lamellar repeat distances as a function of c. Martin and Yeagle (1978) extended the model of Engelman and Rothman (1972) to allow for the possibility that each cholesterol could be surrounded by its own "annulus" of lipid chains, and to allow also for cholesterol dimer formation both with their annulus model and the chain-sharing model of Engelman and Rothman. They predicted that phase boundaries could occur at $c \approx 0.22$, $c \approx 0.31$–0.35, and $c \approx 0.47$. Scott and Cherng (1978) performed a Monte Carlo simulation to study the effect on hydrocarbon chain-order parameters of a rigid cylinder (representing the cholesterol molecule) immersed to various depths in the lipid layer. They considered eight 10-link chains (each link representing a C—C bond) surrounding the cylinder. Depending on the depth of penetration, the order parameters can be increased or decreased. They considered only excluded volume effects (hard core repulsion) and not van der Waals forces. Pink and Carroll (1978) considered the PC chains or cholesterol molecules occupied the sites of a triangular lattice, approximating the chain states by two states; an all-trans state and a melted state, as well as steric repulsions represented by an effective pressure (Marčelja, 1974). They did not calculate a phase diagram but they did study the dependence of ΔH on c, predicting the shape that has now been measured. Subsequently,

Pink and Chapman (1979) used the same model to calculate a phase diagram and various thermodynamic quantities. They confirmed the dependence of ΔH on c, and calculated that a phase boundary existed at $c \approx 0.2$ for a model of DPPC–cholesterol bilayers, as is now widely accepted. The region $0 \leq c \leq 0.2$ was found to be generally a single phase region for $T < T_c$ (see above). They also showed that the average area per chain increased (decreased) as c increased for $T < T_c$ ($T > T_c$) and that the area per chain adjacent to a cholesterol was greater (smaller) than the average for $T < T_c$ ($T > T_c$).

Fluidity and Protein Dynamics

The fact that many biological membranes are built upon a fluid lipid bilayer leads to the related phenomenon that in some biomembranes the intrinsic proteins are able to exhibit both rotational and lateral diffusion.

Rotational Diffusion

Early attempts to study the rotational diffusion of proteins in lipid bilayers were made using the technique of fluorescence depolarisation (Tao, 1971; Wahl *et al.*, 1971). However, although this method had previously been used to study the rotational diffusion of protein free in solution (Weber, 1952), it proved to have limited use for studying the proteins in membranes.

Later work on the protein rhodopsin showed that this molecule did rotate within the membrane, but in a much slower time domain than the initial studies had investigated. This was made possible by using the properties of the intrinsic chromophore, retinal, present in this protein (Cone, 1972). Following absorption of light, this protein undergoes a series of changes which result in changes in its absorption spectrum. By following these polarized absorbance changes, information about the protein dynamics could be obtained.

About the same time as the investigations in rhodopsin were performed, Naqvi *et al.* (1973) performed similar studies on the protein bacteriorhopsin. This unusual protein is found as the sole protein component in the patches of purple membrane present in certain halobacteria. Like rhodopsin, bacteriorhodopsin is capable of absorbing light and consequently undergoing a series of photochemical changes. However, whereas rhodopsin normally bleaches on exposure to light, the photochemistry of bacteriorhodopsin is such that it undergoes a cyclical process resulting in the regeneration of the ground state. Because this cyclical sequence of events occurs in a few milliseconds, it can also be utilized to measure slow rotational motion. In contrast to rhodopsin, however, bacteriorhodopsin was found to be immobile when present in the purple membrane. This is consistent with what is known about the structure of this specialized membrane. The proteins have been shown to exist in a hexagonal lattice with strong protein–protein interaction (Henderson, 1975). Such an ordered structure is unlikely to allow significant mobility of the protein constituents.

The successful studies on rhodopsin and bacteriorhodopsin both relied on the presence of a naturally occurring chromophore. Most proteins, however, do not possess such a suitable chromophore. It was, therefore, necessary to develop a suitable artificial chromophore which would have had an excited state lifetime of the order of milliseconds. Naqvi *et al.* (1973) suggested that molecules such as eosin, which have a long-lived triplet state, might be used for this purpose. These workers then demonstrated that it was indeed possible to measure the rotational mobility of proteins free in solution using these triplet probes.

Since these early studies, other developments have included the synthesis of other triplet probes that can be covalently linked to the protein being examined (Cherry *et al.*, 1976) and the preparation of probes other than eosin derivatives which have higher triplet yields (Moore and Garland, 1979). Methods have been described for detecting the triplet state by means of the phosphorescence emitted as the triplet state decays back to the ground state (Garland and Moore, 1979). Using this method of phosphorescence depolarization, it has become technically possible to measure protein rotation on samples as small as 6.25 n*M*. These developments in the technique have permitted the study of many different systems including the Band 3 anion transport protein, sarcoplasmic reticulum ATPase, the acetylcholine receptor, the concanavalin A receptor in lymphocytes, cytochrome P450, cytochrome *c* oxidase, and cytochrome b_5 (Cherry *et al.*, 1976; Hoffmann *et al.*, 1979; Austin *et al.*, 1979; Junge and Devaux, 1975; Vaz *et al.*, 1979).

Other methods have been developed for measuring rotational mobility of proteins, such as the saturation-transfer electron spin resonance (ESR) technique. Using this method, studies on rhodopsin (Baroin *et al.*, 1977) have yielded a rotational correlation time of 20 μsec at 20°C, which is in reasonable agreement with the earlier studies of Cone (1972). Other workers have used saturation-transfer ESR to measure the rotational motion of the (Ca^{2+}, Mg^{2+})-ATPase from sarcoplasmic reticulum. However, the results obtained show a sizeable variation; one report gave the value of t_2 as 60 μsec at 4°C (Thomas and Hidalgo, 1978), and another gave the value of t_2 at 2°C as 800 μsec (Kirino *et al.*, 1978). However, both groups of workers report a break in the plot of rotational mobility as a function of temperature at around 15°C. Thomas and Hidalgo (1978) interpret this as caused by a lipid effect, whereas Kirino *et al.* (1978) suggest it results from a protein conformational change.

Saturation-transfer ESR has also been applied to the study of the rotational motion of the acetylcholine receptors for the electric organ of *Torpedo marmorata* (Rousselet and Devaux, 1977) and also to the study of muscle proteins (Thomas *et al.*, 1975).

Lateral Diffusion

Proteins in some biomembranes are also free to diffuse laterally within the lipid matrix. However, in some biomembranes, the cytoskeleton can restrict such diffusion. Several methods exist for measuring lateral diffusion of membrane components, but the method commonly employed today is the technique of fluorescence photobleaching recovery (FPR). In this method, a pulse of light is used to bleach a suitable chromophore on the cell surface, thus creating an asymmetrical distribution of chromophores. By measuring the time required for the chromophores to become symmetrically distributed once more, the rate of lateral diffusion can be deduced. Once again, rhodopsin was the first protein to be investigated in this manner (Poo and Cone, 1974; Liebman and Entine, 1974) because of its intrinsic properties. With this particular example, the distribution of rhodopsin molecules was monitored spectrophotometrically. However, in studies on other proteins it is necessary to attach a suitable probe molecule to the protein under study. Typically, fluorescent probes such as rhodamine or fluorescein are used. These are covalently linked to either a lectin or antibodies, which enables the fluorescent probe to be directed to specific receptors on the cell surface. Once bound to the cell, the system is viewed with a fluorescence microscope. An intense flash from a laser is used to bleach the chromophore in a small area of the cell surface. The laser, now much attenuated, is then used to excite any fluorescence from the bleached area. Initially no fluorescence is observed, as all the chromophores have been bleached; however, the fluorescence intensity increases with time as the unbleached chromophores diffuse back into the bleached area from the rest of the cell surface. From the rate of fluorescence recovery, the lateral diffusion coefficients can be calculated.

An experimental difficulty associated with the FPR method is the problem of keeping the cell position constant, especially when trying to measure very low diffusion coefficients. A variation of the technique termed periodic pat-

TABLE I

Lateral Diffusion Constants for Protein in Biomembranes

Membrane	Species	Method	Temperature (°C)	D (cm^2/s)
Erythrocyte	Protein (?)	Fluorescent label	20–23	3×10^{-12}
Heterokaryons	Antigen	Fluorescent label	40	2.5×10^{-10}
Retinal rod outer segment (frog)	Rhodopsin	Adsorption	37	5×10^{-9}
Fibroblast	Protein	Fluorescent label	23	2.6×10^{-10}
Salmonella	Protein (?)	Ferritin label	25	3×10^{-13}
typhimurium	Protein (?)	Fluorescent label	40	$1–2 \times 10^{-9}$
Cultured muscle	Protein (?)	Gold particle	37	$<2 \times 10^{-10}$
fiber				
3T3 mouse				
fibroblast				

[a]Adapted from Jain and White (1977).

tern photobleaching (L. M. Smith *et al.*, 1979) may be useful in overcoming this difficulty. With this method, a periodic pattern of parallel stripes is bleached into the surface of the membrane; fluorescence photomicrographs of the cell are made at intervals of time after photobleaching. Diffusion of the fluorescent molecules causes a decay in the contrast of the pattern with time. Fourier analysis of the photographs can then be used to quantify the rate of decay of the contrast and yield information about the diffusion coefficient. The values of the lateral diffusion coefficient obtained for the various proteins are 10^{-9}–10^{-13}cm^2/sec (Table I). It is not certain whether all or only a part of the membrane (for example, the defect structures and regions of mismatch) is available for lateral diffusion.

Bretscher (1980) has commented on the fact that the values obtained for lateral diffusion of phospholipids and for rhodopsin are very different from those determined for a number of proteins in membranes and cell systems. Thus rhodopsin has a diffusion coefficient of 4×10^{-9}cm^2/sec at 20°C, but values from other systems give values 10–10^4-fold lower.

Redistribution of Proteins

The fluid nature of the lipid matrix allows redistribution of proteins in the plane of the membrane and can lead to patch formation, which is a passive discrete clustering of macromolecules into two-dimensional aggregates. Thus interaction with multivalent antibodies induces aggregation of the surface antigens into patches. Aggregation effects are also observed when lectins bind to cell surfaces. Reversible patch formation or particle aggregation has been observed under a variety of conditions such as pH change (da Silva, 1972), temperature shift (Speth and Wunderlich, 1973), addition of anesthetics (Poste *et al.*, 1975), proteolysis, and glycerol treatment. Aggregation induced by these various processes is rapid, reversible, and inhibited by glutaraldehyde fixation. It is possible that some of these effects are caused by altered cytoplasmic structures. For example, the sensitivity of polymorphonuclear leucocytes for agglutination with concanavalin A (Con-A) is thought to result from the presence of Con-A receptors in patches in the cell membrane. After treatment with colchicine or vinblastine, however, the agglutination is much reduced, suggesting that microtubules maintain some fixed distribution of membrane constituents (Berlin and Ukena, 1972). In cells interacting with each other in tissues, there is an uneven distribution of plasma membrane proteins. In addition to being restricted to only one surface of the cell, some of these proteins are apparently confined to certain regions, such as gap junctions and synapses. Selective and nonrandom localization of membrane proteins and morphological specialization (in folds and microvilli) would also be expected in membranes of intestinal, kidney, liver, and other epithelial cells.

The redistribution of cross-linked membrane components resulting in polar segregation away from the nucleus is termed capping. During cap formation, the cross-linked molecules are segregated from other membrane components by an active process that is probably dependent on microfilaments and microtubules.

Intrinsic Protein–Lipid Interactions

We have seen that cholesterol molecules can modulate the fluidity of the lipid bilayer. It is therefore of interest to see how intrinsic protein can perturb the lipid dynamics within the lipid bilayer. Unfortunately, considerable confusion has developed in recent years with regard to our understanding of intrinsic protein–lipid interactions, and even the recent review literature (1980) contains contradictory statements. Apparently, part of the confusion centers on an overly ready acceptance of conclusions based on evidence from one physical technique, and part relates to attempts to link together a number of different concepts that in fact may not be related.

Concepts combining biochemical effects, solvent extraction effects, and physical perturbing effects have all been linked together. Such concepts as bound lipid, specific lipid, minimum lipid, and the number of lipid molecules that can surround the circumference of the hydrophobic core have all at times been linked with the degree of dynamic perturbation that an intrinsic protein may cause to neighboring lipids.

Time-Scale Effects

Chapman *et al.* (1979) suggested three important points to consider in the perturbation of lipid dynamics by intrinsic proteins.

1. The time scale appropriate to the particular physical technique used to study the protein–lipid perturbation. This can be appreciated when we realize that measured on one time scale, for example 10^{-8} sec, using ESR spin-labeled molecules, a molecule may appear to be rigid; on another time scale, for example, 10^{-5} sec, using ^{2}H-NMR methods, the same molecule could appear to be mobile. When attempts are made to relate some measured perturbation effects with enzymic effects, another time scale must be considered, that is, the time interval over which the enzymic conformational effect occurs. An example of this situation is shown by studies of reconstituted cytochrome oxidase (Kang *et al.*, 1979).

ESR spin label and ^{2}H-NMR results for three lipid states, a pure liquid–crystalline lecithin bilayer, *boundary lipid*, and lipid in the crystalline gel state are presented in Fig. 6. As viewed by ESR, the rigid gel state of lecithin below T_c is similar in motional properties to the rigid or ordered boundary lipid of cytochrome oxidase. However, as viewed by ^{2}H-NMR of a terminal methyl-labeled lecithin, boundary lipid is even more disordered than the lipid above its T_c transition temperature (Kang *et al.*, 1979).

2. The concentration of the protein and its arrangement within the lipid bilayer which is being examined. When we look at the plane of the lipid bilayer, it is immediately apparent that as the concentration of the protein in the lipid bilayer increases the number of multiple contacts of each lipid with proteins also become important. At low protein concentration, single contacts of lipid with the protein prevail.

3. Whether, in the system considered, the lipid is above or below its T_c transition temperature. Lipids below the T_c transition temperature squeeze the proteins out of the cyrstalline lipid lattice when they crystallize. The shape and size of the protein causes packing faults in the lattice, and patches are formed of high protein-to-lipid content, that is, eutectic mixtures occur. These regions, which have been demonstrated by freeze-fracture electron microscopy, are distinct from the remaining crystalline lipid regions. This is illustrated for bacteriorhodopsin reconstituted in lipid model biomembranes in the freeze-fracture pictures of Fig. 7.

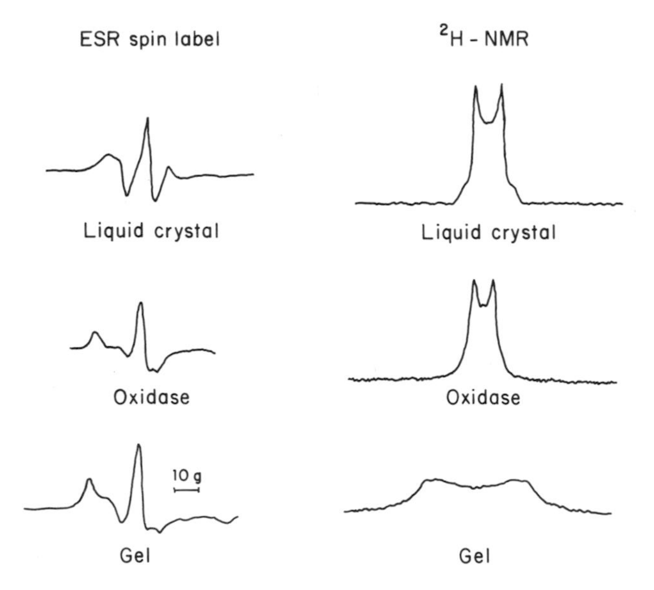

Fig. 6. A comparison between ESR and ^{2}H-NMR view of hydrocarbon chain order in gel, fluid, and cytochrome oxidase lipid (boundary lipid). From Oldfield *et al.* (1978b).

There is a further point which also requires consideration. This is the degree to which an intrinsic molecule must be lipid solvated in order to retain the structural integrity of the lipid bilayer matrix.

ESR experiments with cytochrome oxidase and sarcoplasmic reticulum ATPase (time scale $\sim 10^{-8}$ sec have shown the existence of *immobile components* in the corresponding spectra. It has been proposed that the ESR immobile component is an indication of a special lipid shell (called the boundary layer lipid or the *annulus lipid*) separate and distinct from the bulk lipid. The annulus lipid is said to control enzyme activities of intrinsic proteins and to rigorously exclude cholesterol.

Other workers have pointed out a number of observations which cast doubt on this interpretation of the observed ESR immobile component and of the concept of special annulus lipid.

1. It has been shown that the observation of such an ESR immobile component is not restricted to proteins which may require and possess captive or tightly bound lipids. It is indeed observed in gramicidin A–lipid–water systems (Chapman *et al.*, 1977a). The gramicidin A molecule is a relatively simple polypeptide.
2. In ESR experiments of rhodopsin using spin-labeled molecules that are attached to the protein but penetrate and sense the boundary layer, there is little effect on the mobility of the probe (Davoust *et al.*, 1980). This

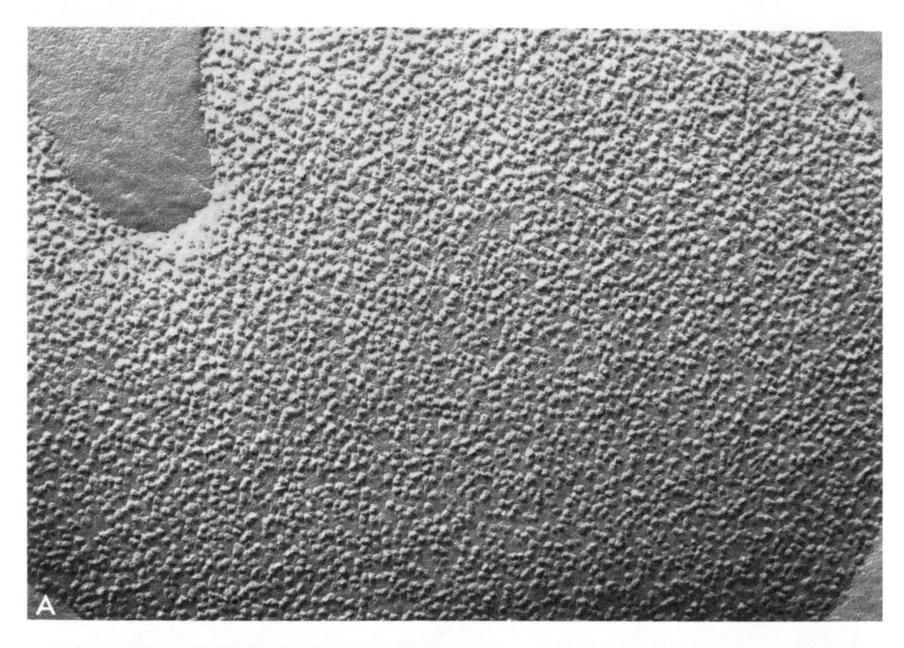

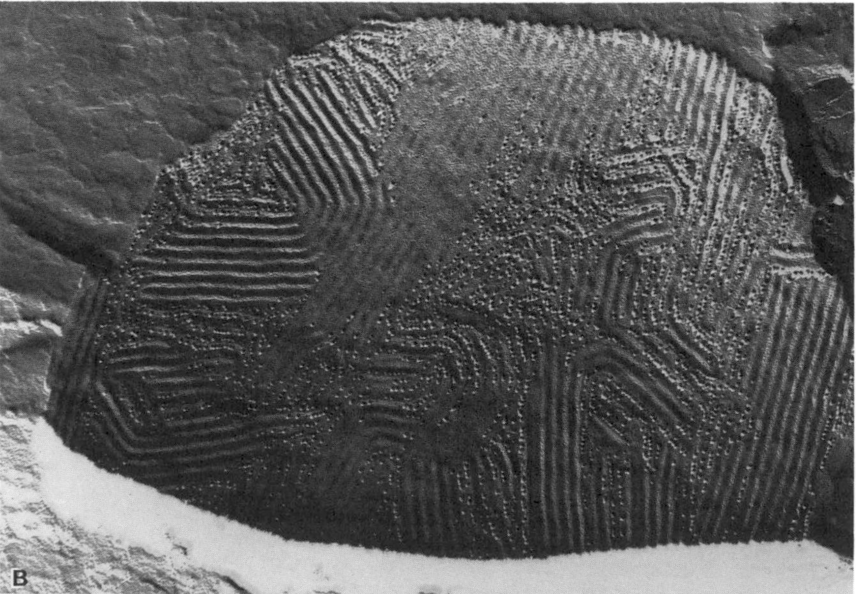

Fig. 7. Freeze-fracture micrographs of dimyristoylphosphatidylcholine recombinants, quenched from different temperatures. Sample (A) has a molar ratio of 31:1 and was quenched from 30°C (66,000×). Sample (B) has a molar ratio of 54:1 and was quenched from 17°C (29,000×).

shows that a high degree of fluidity exists in the boundary layer phospholipids. However, when the membranes are delipidated, an immobile ESR component is observed. These workers further conclude that lipid-depleted membranes cannot be used to characterize the viscosity of the boundary layer of native membranes.

3. Nuclear magnetic resonance studies using either ^{1}H, ^{19}F, or ^{2}H nuclei (time scale, 1–0.01 msec) on various biomembranes or reconstituted systems do not show the occurrence of two types of lipid. This is the case with ^{1}H-NMR studies of rhodopsin in disc membranes, ^{19}F-NMR studies of *E. coli* membranes, and ^{2}H-NMR studies of the cytochrome oxidase and sarcoplasmic reticulum systems. Thus a continuity between the bulk lipid phase and the boundary layer lipids and ready diffusion between these lipids takes place, as indicated by NMR experiments. The ^{2}H-NMR spectra of reconstituted sarcoplasmic reticulum ATPase are shown in Fig. 8. Although cholesterol increases the quadrupole splitting (compare with Fig. 5), the ATPase protein causes a slight decrease of quadrupole splitting.

The general observations in *all* ^{2}H-NMR reconstitution studies reported so far is that only *one* homogeneous lipid environment is present *above* T_c, even when a substantial amount of protein is present. The ^{2}H-NMR experiments give no indication for a strong, long-lived interaction between the membrane protein and the lipid. Instead, data can be explained by a relatively rapid exchange between those lipids in contact with the protein and those further away from it. This exchange must be fast on the ^{2}H-NMR time scale (exchange rate $> 10^4$ Hz) in order to produce a single component ^{2}H-NMR spectrum, but slow compared to the ESR time scale (exchange rate $<$ 10^7 Hz) to account for the two-component spin label spectrum.

^{2}H-NMR is more sensitive to structural and dynamic changes than spin label EPR, and although the method does not detect a specific boundary layer of lipids, a closer inspection of the ^{2}H-NMR spectra of reconstituted membranes reveals three interesting features. (i) Membranes with protein exhibit a small but finite decrease (10–25%) in the quadrupole splitting compared to pure lipid samples, (ii) the deuterium T_1 relaxation times are shorter by about 20–30% in reconstituted membranes, and (iii) the apparent linewidth of the ^{2}H- and ^{31}P-NMR spectra increases in protein-containing samples (Seelig and Seelig, 1980).

The reduction in the deuterium quadrupole splitting has been ascribed to a disordering effect of the protein interface. In most membrane models present in the literature, the membrane proteins are drawn as smooth cylinders or rotational ellipsoids. Even if the protein backbone is arranged in an α-helical configuration, the protrusion of amino acid side chains will lead to an uneven shape of the protein surface.

The observed disordering effect does not necessarily imply an increase in

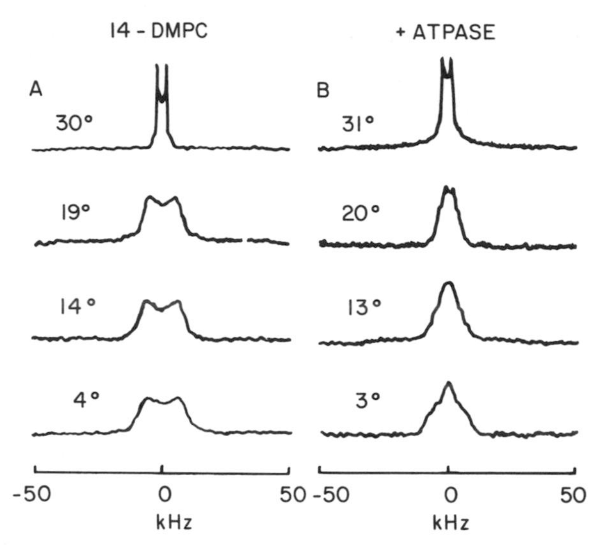

Fig. 8. Deuterium NMR spectra of [$^{3}H_2$]DMPC, [$^{3}H_2$]DPPC, and their ATPase (sarcoplasmic reticulum ATP phosphohydrolase, E.C. 3.6.1.6.) complexes as a function of temperature. (A) Pure [$^{3}H_2$]DMPC, 100 kHz effective spectral width, 0.54 sec recycle time, 2048 data points, $\tau_1 = \tau_2 = 50$ μsec, 90° pulse widths, 15,000–20,000 scans, 150 Hz line broadening, at the temperatures indicated. (B) [$^{3}H_2$]DMPC–ATPase (41:1) and other conditions as described in (A), except that $\tau_1 = \tau_2 = 90$ μsec, 40,000–80,000 scans. The protein–lipid complex samples contained about 10 mg of phospholipid which was 25% ^{2}H-labeled. After Rice *et al.* (1979).

the configurational space available to the fatty acyl chains. In fact, it appears more probable that the total number of chain configurations is lowered, and that the statistical probability of more distorted chain conformations increases at the same time (Seelig and Seelig, 1980).

From an increase in spatial disorder it cannot be concluded that the membrane is also more fluid. On the contrary, deuterium T_1 relaxation time measurements suggest a decrease in the rate of segment reorientation in the presence of protein. Such deuterium T_1 measurements have been performed with cytochrome *c* oxidase and reconstituted sarcoplasmic reticulum and some representative results are summarized in Table II. The addition of protein decreases the relaxation time in both cases. Above T_c, the motion still falls into the fast correlation time regime as shown by the longer T_1 relaxation times at higher temperatures. It is suggested that shorter T_1 relaxation times are therefore equivalent to an increase in the microviscosity. This conclusion is supported by ^{13}C-NMR experiments with reconstituted sarcoplasmic reticulum (Stoffel *et al.*, 1977). The T_1 relaxation rates of ^{13}C-labeled lipids decrease continuously with increasing protein concentration in the membrane.

TABLE II

Deuterium T_1 Relaxation Times (at 46.03 MHz) of Reconstituted Membranes[a]

System	Temperature (°C)	T_1 (ms) Reconstituted membrane	T_1 (ms) Pure lipid bilayer
Cytochrome *c* oxidase[b]	5	6.7	7.9
lipid-to-protein ratio	15	8.8	10.9
~0.175 (wt/wt)	28	11.9	15.5
Sarcoplasmic reticulum[c]	24	11.1	13.9
lipid-to-protein ratio			
~0.33 (wt/wt)			

[a]Data from Seelig and Seelig, 1980.

[b]Reconstituted with 1,2-di(9,10-^{2}Hz)oleoyl-*sn*-glycero-3-phosphocholine.

[c]The lipid employed is 1,2-di(9,10-^{2}Hz)elaidoyl-*sn*-glycero-3-phosphocholine.

Fluorescent Probes

Fluorescent probes have been used to study the fluidity of model and natural biomembranes. However, a clear interpretation of steady state properties is difficult to obtain, particularly when intrinsic proteins are present. This is illustrated by recent fluorescent probe studies of reconstituted systems (Hoffmann *et al.*, 1981). A study of the polarization of the probe (in reconstituted systems of an intrinsic polypeptide gramicidin A or of various intrinsic proteins as the concentration of the intrinsic molecule increases) shows that the value of the polarization P reaches a limited value. If it is argued that the increase of polarization is associated with an increase of lipid order parameter, this is in contradistinction to the deuterium NMR studies which indicate a reduction of order parameter with protein concentration. The fluorescent probe is a long rigid molecule, unlike the flexible lipid molecule. Empirically, each of the curves of polarization with protein concentration has been observed to fit a simple exponential equation. The value of the exponent is, in each equation, related to the number of probe molecules or lipids that surround the intrinsic molecule. It may be that a dominant influence on the probe motion is its interaction with the intrinsic proteins rather than the microviscosity or order parameter of the lipid chains.

A careful theoretical analysis using probability theory shows that the occurrence of such exponential curves is also consistent with the occurrence of a random arrangement of intrinsic molecules (see Fig. 9 for a comparison of theoretical and experimental values). The probe molecule is markedly affected by the presence of the intrinsic molecule $P_1 = e^{-Mx}$ where M is the number of DPH molecules that can be accommodated around the intrinsic molecule in half of the lipid bilayer and x is related to the concentration of intrinsic molecules in the lipid bilayer matrix. The dependence of $P(M,x)$ on x reflects the fact that as the concentration of polypeptide or protein increases, the probability of (for example) protein–protein contacts also increase, but the number of lipid molecules and probe molecules which can contact the protein decrease (see Fig. 8). Thus the present available evidence is that *there is no fixed stoichiometry of lipid to integral protein, instead the number of lipid molecules that contact the protein varies with the protein concentration in the lipid bilayer.*

Cortijo *et al.* (1982) have used infrared (IR) spectroscopy to study lipid phase transitions, lipid–cholesterol interactions and lipid–protein interactions.

The temperature shifts for the frequencies of the hydrocarbon chain methylene symmetric and asymmetric stretching vibrations provide a convenient probe for monitoring the phase transitions of lecithins dispersed in water. The frequencies and widths of these bands have been related to specific molecular properties, that is, a shift in frequency to the introduction of gauche conformers and changes in band width to variations in the rate of librational motions of the chains. The way in which the abrupt endothermic lipid phase transition is indicated by the shift of the asymmetric methylene band at the appropriate T_c value (Fig. 10) is reassuring for the application of this technique.

The effect observed upon incorporation of cholesterol into the lipid bilayers provides results which are in general accord with those obtained by the use of a range of physical techniques (Ladbrooke *et al.*, 1968: Oldfield and Chapman, 1972), that is, there is an increase in the number of gauche conformers below T_c (less order) and a decrease in this number above T_c (more order). This effect increases with greater amounts of cholesterol in the bilayer. At very high cholesterol concentrations (Fig. 10) almost no change occurs with temperature in the relative population of gauche and trans conformers of the lipid chains. (The abrupt lipid phase transition is essentially smeared out and removed.)

The effect of gramicidin A, bacteriorhodopsin, and ATPase on the lipid chain conformation below its T_c value is similar to what occurs with cholesterol regarding the general disordering effect produced by the incorporation of all these molecules into the lipid bilayer (Figs. 10, 11, and 12).

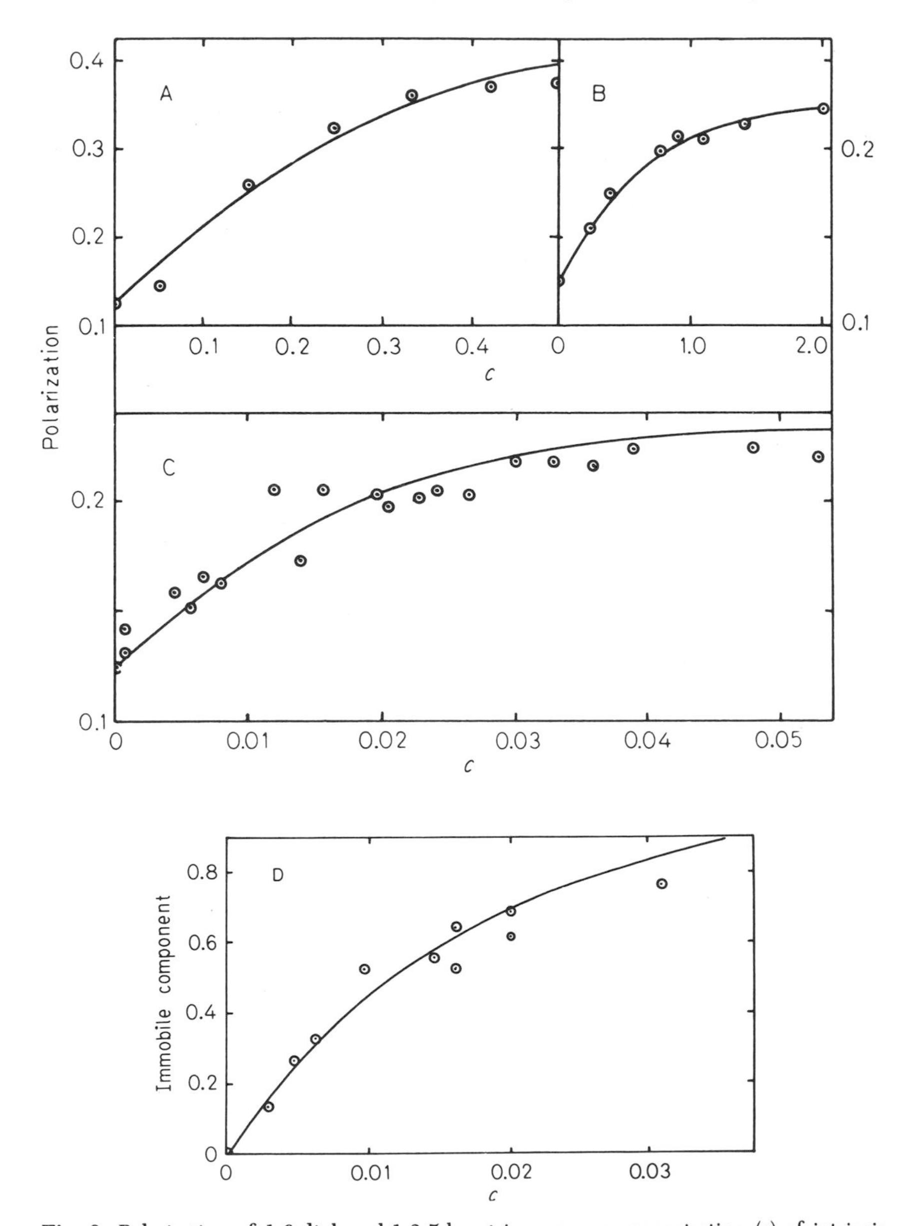

Fig. 9. Polarization of 1,6-diphenyl-1,3,5-hexatriene versus concentration (c) of intrinsic molecules in DMPC vesicles at 36°C. The solid lines are calculated values, and circles are data. (A) DMPC + cholesterol, (B) DMPC + gramicidin A, (C) DMPC + cytochrome oxidase (after Hoffmann *et al.*, 1981). (D) The immobile component from ESR in dimyristoylglycerophosphocholine + cytochrome oxidase using a nitroxide probe. The solid lines are the calculations. $1 - p_1\ (M,x)$ for $M = 60$. From Hoffmann *et al.* (1981).

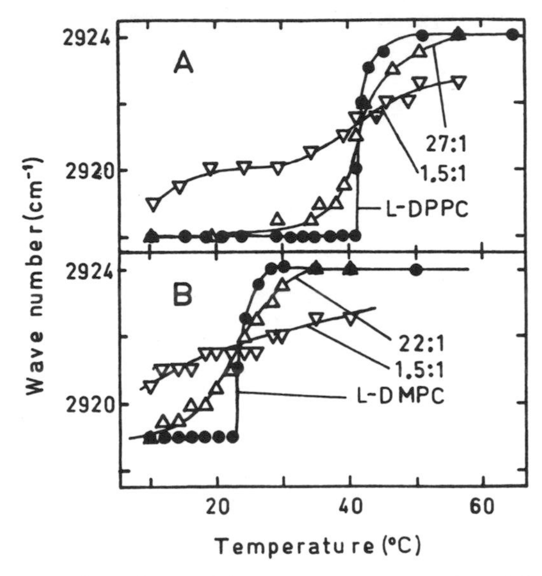

Fig. 10. Temperature dependence of the maximum wave number of the CH_2 asymmetric stretching vibrations in (A) L–DPPC/cholesterol and (B) L–DMPC/cholesterol at the molar ratios indicated. The temperature dependence for the pure lipids (●) is also given. After Cortijo *et al.* (1981).

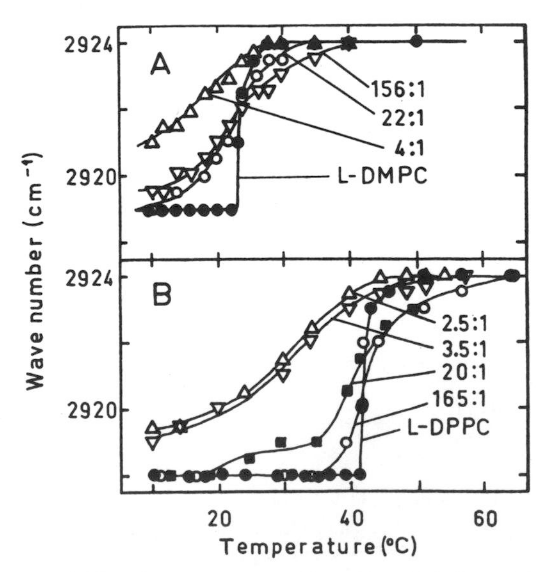

Fig. 11. Temperature dependence of the maximum wave number of the CH_2 asymmetric stretching vibrations in (A) L–DMPC/gramicidin A and (B) L–DPPC/gramicidin A at the molar ratios indicated. The temperature dependence for the pure lipids (●) is also given. After Cortijo *et al.* (1982).

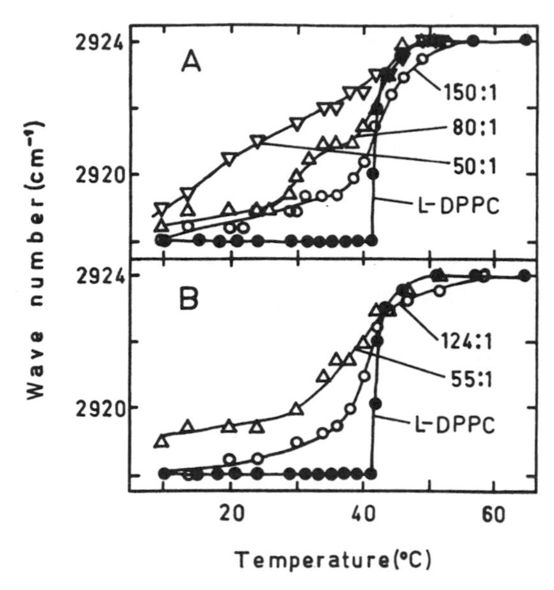

Fig. 12. Temperature dependence of the maximum wave number of the CH_2 asymmetric stretching vibrations in (A) L–DPPC/Ca^{2+}-ATPase and (B) L–DPPC/bacteriorhodopsin at the molar ratios indicated. The temperature dependence for the pure lipids (●) is also given After Cortijo *et al.* (1982).

There are, however, differences which can be discerned from the effect that occurs with cholesterol. Thus at high concentrations of intrinsic protein within the lipid bilayer the lipid phase transition is not so effectively removed. Furthermore, in the case of the Ca^{2+}-ATPase incorporated into the L–DPPC-lipid bilayer at a lipid/protein ratio of 80:1 there is a definite indication of a new transition centered around 30°C. This is in accord with recent results on reconstituted Ca^{2+}-ATPase systems (Hoffmann *et al.*, 1980). This shows then that when lipid crystallization occurs it leads to protein segregation. There are then two regions of lipid, that is, the pure lipid from which the protein has been segregated and high protein–lipid patches. Two lipid melting transitions then occur, one at around 30°C and the other at 41°C. The first corresponds to the melting of the high protein–lipid patches and the second to the melting of the remaining lipid. As the protein content incorporated into the lipid bilayer is increased so does the extent of the remaining pure lipid decrease. This causes the melting transition centered around 41°C also to become smeared and broadened. Protein segregation has also been observed with bacteriorhodopsin within the lipid bilayer after lipid crystallization, but a second distinct melting transition is not apparent (Fig. 12). In this case the high protein–lipid patches contain crystalline arrangements of protein with lipid chains so tightly squeezed between the proteins that a lower melting transition does not occur. A broadening of the transition centered around 41°C does occur as the protein concentration within the lipid bilayer increases (Fig. 12).

At temperatures above the lipid T_c value the observed effects are more complex than those observed with cholesterol. At high lipid/protein ratios the presence of gramicidin or the intrinsic proteins in the bilayer decrease the average number of gauche isomers (Figs. 11 and 12): for example, gramicidin at 20:1, the Ca^{2+}-ATPase at 150:1 and bacteriorhodopsin at 124:1, lipid/protein ratios. (This ordering effect can be abolished by sufficient increase in temperature. It is perhaps understandable that a small decrease in the entropy of the lipid chains produced by a relatively small amount of the lipid-soluble proteins can be reversed by the general tendency of the system to increase the entropy with an increased temperature.) The small ordering effect produced by the incorporation of the gramicidin polypeptide into the lipid bilayer at temperatures a few degrees above T_c has also been observed by means of Raman spectroscopy (Chapman *et al.*, 1977).

When higher concentrations of gramicidin or either of the intrinsic proteins are incorporated into the lipid bilayer no ordering effect is observed, that is, the same level of gauche conformers occurs as that of the pure lipid alone (see gramicidin 3.5:1, Ca^{2+}-ATPase 80:1, and bacteriorhodopsin 55:1, Figs. 11 and 12). The addition of these intrinsic proteins or polypeptide in increasing concentrations produces consecutively (a few degrees above T_c) first an ordering and later a disordering effect on the lipid chain.

Recent NMR spectroscopic studies of the effects of intrinsic proteins such as cytochrome oxidase and Ca^{2+}-ATPase upon lipid chain conformation using specifically labeled deuterium chains and deuterium NMR spectroscopy has led to the view that these intrinsic proteins cause an increase in lipid chain disorder. This disordering effect of intrinsic proteins has been related to the roughness of the protein surface. It has been contrasted with the effect of cholesterol, which induces static order in the lipid chains (see Seelig and Seelig, 1980; Kang *et al.*, 1979; Rice *et al.*, 1979). These authors do not report that an *ordering* effect occurs at lower intrinsic protein concentrations.

Studies of gramicidin A–lipid model membranes using deuterium NMR spectroscopy and deuterated methyl groups do indicate that an ordering effect occurs at low concentrations of gramicidin and that a disordering effect occurs at higher polypeptide concentrations (Rice and Oldfield, 1979). This effect is similar to what is observed by studies using infrared spectroscopy.

Rice and Oldfield (1979) explain their data on gramicidin by suggesting that the lipid chains adjacent to the polypeptide are constrained in a twisted configuration within the crevices of the surface of the molecule. They propose that this gramicidin–lipid complex then presents a smooth cholesterol-like surface to the remainder of the lipids which it then orders.

Pink *et al.* (1982) give a different interpretation of the same NMR data. They consider three populations of lipid can occur within the lipid bilayer which will vary with the protein concentration: (1) those not adjacent to any intrinsic protein, "free" lipid; (2) those adjacent to a protein; (3) those touched or trapped between two or three proteins. These authors conclude that the NMR data are satisfied by the varying populations of these different lipid environments where the methyl groups of adjacent lipids are slightly more statically ordered than those of free lipids, and where the methyl groups of "trapped" lipids are more statically disordered than those of the "free" lipid.

The deuterium NMR data on natural and model biomembranes can be influenced by vesicle tumbling or lipid exchange processes. The IR spectrum is not influenced by these factors. It provides clear independent evidence of the effect of intrinsic proteins and polypeptides upon the *average* number of gauche isomers of the lipid chains.

Calorimetric Studies

When reconstituted protein–lipid systems are cooled to temperatures below the main lipid T_c transition temperature, on addition of protein the transition gradually broadens and the transition enthalpy decreases. At very high protein concentrations, the phase transition may be completely undetectable (Curatolo *et al.*, 1977; Chapman *et al.*, 1977; Van Zoelen *et al.*, 1978; Mombers *et al.*, 1979). A broadening of the phase transition has also been confirmed by other methods as, for example, fluorescence spectroscopy (Gomez-Fernandez *et al.*, 1980; Heyn, 1979).

Some workers have interpreted this observed reduction in enthalpy with protein concentration as a measure of the boundary lipid associated with the protein. This assumes that each individual protein remains separate (as it is above the lipid T_c transition temperature) during the lipid crystallization process. Boggs and Moscarello (1978) have attempted to determine how the amount of boundary lipid depends on fatty acid chain length with lecithin vesicles containing the hydrophobic protein from myelin proteolipid. These authors used scanning calorimetry and determined the enthalpy of the transition as a function of protein/lipid molar ratio. They suggest that 21–25 molecules correspond to boundary lipid for fatty acid chain lengths of 14–18 carbons. The boundary lipid was, however, only 16 molecules per molecule of lipophilin for lipids containing fatty acids of chain length 12, or for a molecule with a trans double bond. [Their calculations suggest that 70 phospholipids would surround the circumference of the lipophilin protein (if both chains touch the protein).]

However, when the temperature of a reconstituted system is lowered below the lipid T_c transition temperature, the lipid chains crystallize, proteins are squeezed out and patches of high protein content are formed (see Fig. 7). This has significance for a number of situations. The crystallization process produces two regions, one including the crystalline lipid relatively free of protein, and the other consisting of patches of high protein–lipid content. As more protein is included to the lipid system, the aggregated patch increases in size at the expense of the remaining crystalline region. This is why lipid–water systems show a broadening of the main melting transition and a lowering of the enthalpy as protein is incorporated. The decreased enthalpy value may correspond to an average perturbed lipid.

Where the intrinsic protein remains isolated, it is possible that the en-

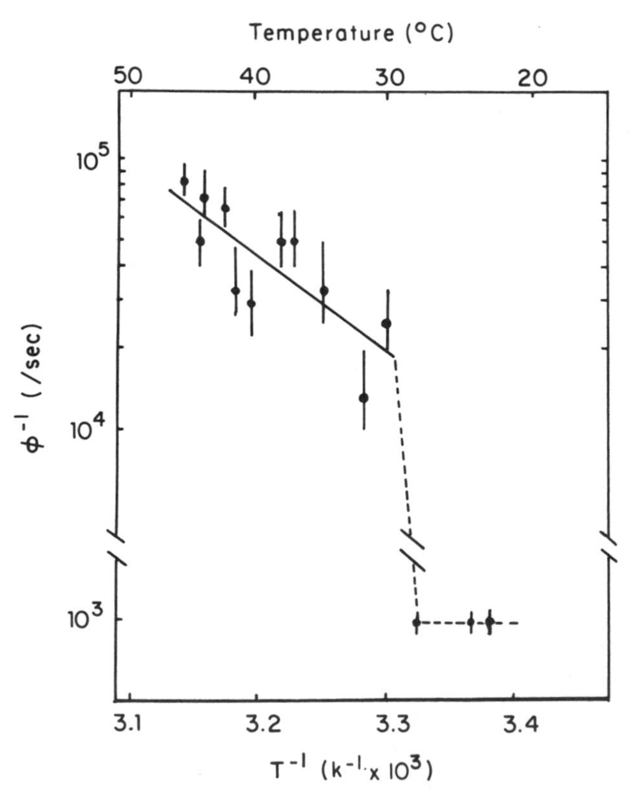

Fig. 13. (A) Arrhenius plot of rotational motion parameter for a Ca^{2+}-ATPase recombinant having a lipid/protein molar ratio of 86:1. The line is drawn by linear regression; from the slope an activation energy of 67 kJ/mol (16 kcal/mol) is calculated. From Hoffmann *et al.* (1980).

thalpy reduction might provide information about the number of lipid molecules which surround the intrinsic hydrophobic segment of the protein; on the other hand, more than one layer might be perturbed. It is not a measure of how much lipid is bound to the protein.

In some cases, the proteins in the patches do not remain isolated. This appears to be the case with reconstituted purple membrane protein in which a structure is formed similar to that which occurs in the membrane itself. In other cases, the high protein-to-lipid patches are associated with hexagonally packed lipid that can be detected by X-ray diffraction (Hoffmann *et al.*, 1980). A marked increase of protein rotation can occur below the lipid T_c transition temperature, because a melting of the protein–lipid patches can be determined by the melting of the lipids. This can mean that protein rotation increases dramatically (Hoffmann *et al.*, 1980), as enzymatic activity sometimes does about 10°C below the transition temperature T_c of the pure lipid (e.g., at 30°C with ATPase incorporated into dipalmitoyl–lecithin–water systems (T_c = 41°C). The rapid increase of rotation of the ATPase protein at 30°C observed using a triplet probe is shown in Fig. 13.

Theoretical Models

A number of theoretical studies have estimated the extent to which an intrinsic protein may affect the surrounding lipid in a lipid bilayer structure. Marčelja (1976) has published a microscopic mean field model of order in lipid bilayers based on chain conformation; Schroeder (1977) described a method of incorporating lipid–protein interactions into a preexisting mean field treatment of lipid bilayers. The protein acted formally as an external field on the lipids. In both these studies, the attractive lipid-mediated interaction between two identical proteins was demonstrated.

Jähnig (1977) also has developed a microscopic mean field bilayer model based on chain conformation. Owicki *et al.* (1978) studied the order perturbation as a function of temperature and lateral pressure using Landau–de Gennes theory and a variational procedure. They conclude that for a given lateral pressure there is a temperature dependence, and that the greater amount of boundary layer is present at the lipid T_c transition temperature.

Pink and Chapman (1979), using a lattice model, have examined the lipid systems in which the proteins interact only via van der Waals interactions and systems in which the proteins have bound or attached lipids on their circumference. These calculations have been used to examine the melting temperatures of eutectic protein–lipid patches.

Restriction of Movement and Fluidity of Biomembranes

There are a number of ways in which selective restriction of membrane component dynamics can be accomplished. Here we will discuss two approaches.

Photooxidation of Proteins

The triplet probes commonly used to measure the rotational diffusion of membrane proteins are very potent agents for causing photooxidation of biological materials. Under normal circumstances, these undesirable side-effects can be avoided by keeping labeled samples in the dark. In addition, because the samples for flash photolysis are always kept under anaerobic conditions during the course of the measurements, normally no damage to the system under study is observed. We have recently attempted (Restall *et al.*, 1981) to make use of this property of photooxidation to study the role of the thiol groups in the (Mg^{2+}, Ca^{2+})-ATPase from sarcoplasmic reticulum. Our recent studies have shown that after labeling the ATPase protein and illuminating the sample under anaerobic conditions, a steady decrease in the ATPase activity is observed. Associated with this loss of activity, we have established, both from freeze-fracture electron microscopy and from measurements of the protein dynamics, that aggregation of the proteins is occurring. Although the effects of oxidation on lipid composition and fluidity have been extensively studied, the effects of oxidation on membrane proteins have received little attention. These studies so far have all utilized the photo-oxidative properties of a triplet probe covalently attached to the protein. For future work we hope to use other oxidizing agents, which may give greater selectivity in what is affected. In addition, it is useful to examine the state of protein oxidation in certain pathogenic conditions, notably the protoporphyrias, which may be associated with photosensitivity and a function of tissue aging.

Hydrogenation and Polymerization

Two new approaches to markedly affect biomembrane fluidity and lead to a restriction of dynamics in the membrane utilize specific interaction with the lipid components of the membranes. This can be accomplished by techniques such as hydrogenation (Chapman and Quinn, 1976), and more recently, polymerization (Johnston *et al.*, 1980), which has been achieved in monolayer and multilayer structures. This new approach causes extensive cross-linking of the lipid chains to occur by a polymerization process triggered with UV radiation. Extensive polymerization has also been produced in certain natural biomembranes (e.g., *Acholeplasma laidlawii*) merely upon irradiation with UV radiation (Leaver *et al.*, 1982).

Conclusions

The concept of membrane fluidity has been and continues to be useful because of its emphasis on the dynamic character of biomembrane structures. This is quite distinct from the previous static structure proposed and emphasized on the basis of electron microscope studies. It has revolutionized our thinking and our understanding of many biomembrane functions, and may prove valuable for understanding certain disease conditions. Some of these implications are discussed in other chapters in this book.

Acknowledgments

I wish to thank the Wellcome Trust and the Medical Research Council for financial support and my many research colleagues for helpful discussions and collaboration during our studies of biomembrane structure and function.

References

Austin, R. H., Chan, S. S., and Jovin, T. M. (1979). *Proc. Natl. Acad. Sci. U.S.A.* **76,** 5650–5654.

Baroin, A., Thomas, D. D., Osborne, B., and Devaux, P. F. (1977). *Biochem. Biophys. Res. Commun.* **78,** 442–447.

Barratt, M. D., Green, D. K., and Chapman, D. (1969). *Chem. Phys. Lipids* **3,** 140–144.

Barton, P. G. (168). *J. Biol. Chem.* **243,** 3884–3890.

Berlin, R. D., and Ukena, T. E. (1972). *Nature (London)* **238,** 120–122.

Blaustein, M. P. (1967). *Biochim. Biophys. Acta* **135,** 653–668.

Boggs, J. M., and Moscarello, M. A. (1978). *Biochemistry* **17,** 5374–5379.
Bretscher, M. S. (1980). *TIBS* **5,** 6–7.
Brown, M. F., Seelig, J., and Häberlen, U. (1979). *J. Chem. Phys.* **70,** 5045–5053.
Browning, J. L., and Seelig, J. (1980). *Biochemistry* **19,** 1262–1270.
Butler, K. W., Tattrie, N. H., and Smith, I. C. P. (1974). *Biochim. Biophys. Acta* **363,** 351–360.
Byrne, P., and Chapman, D. (1964). *Nature (London)* **202,** 987–988.
Chapman, D. (1958). *J. Chem. Soc.* **152,** 784–789.
Chapman, D., and Penkett, S. A. (1966). *Nature (London)* **211,** 1304–1305.
Chapman, D., and Quinn, P. J. (1976). *Proc. Natl. Acad. Sci. U.S.A.* **73,** 3971–3975.
Chapman, D., and Salsbury, N. K. (1966). *Trans. Faraday Soc.* **62,** 2607–2621.
Chapman, D., Byrne, P., and Shipley, G. G. (1966). *Proc. R. Soc. London, Ser. A* **290,** 115–142.
Chapman, D., Williams, R. M., and Ladbrooke, B. D. (1967). *Chem. Phys. Lipids* **1,** 445–475.
Chapman, D., Keough, K., and Urbina, J. (1974). *J. Biol. Chem.* **249,** 2512–2521.
Chapman, D., Cornell, B. A., Eliasz, A. W., and Perry, A. (1977a). *J. Mol. Biol.* **113,** 517–538.
Chapman, D., Kingston, B., Peel, W. E., and Lilley, T. H. (1977b). *Biochim. Biophys. Acta* **464,** 260–275.
Chapman, D., Gomez-Fernandez, J. C., and Goni, F. M. (1979). *FEBS Lett.* **98,** 211–223.
Cherry, R. J., Cogoli, A., Oppliger, M., Schneider, G., and Parish, G. R. (1976). *Nature (London)* **263,** 389–393.
Clowes, A., Cherry, R. J., and Chapman, D. (1971) *Biochim. Biophys. Acta* **249,** 301–307.
Cortijo, M., Alonso, A., Gomez-Fernandez, J. C., and Chapman, D. (1982). *J. Mol Biol* **157,** 597–618.
Cone, R. A. (1972). *Nature (London)* **236,** 39–43.
Cooper, R. (1981). "Membrane Fluidity and Cell Surface Receptor Mobility." U.S. Dept. of Health.
Curatolo, W., Sakura, J. D., Small, D. M., and Shipley, G. G. (1977). *Biochemistry* **16,** 2313–2319.
da Silva, P. R. (1972). *J. Cell Biol.* **53,** 777–787.
Davis, J. H., Nichol, C. P., Weeks, G., and Bloom, M. (1979). *Biochemistry* **18,** 2103–2112.
Davoust, J. A., Bienvenue, P. Fellmann, P., and Devaux, P. F. (1980). *Biochim. Biophys. Acta*, **596,** 28–42.
Devaux, P. F., and McConnell, H. M. (1972). *J. Am. Chem. Soc.* **94,** 4475–4481.
Ehrström, M., Eriksonn, L. E. G., Israelachvili, J., and Ehrenberg, A. (1973). *Biochem. Biophys. Res. Commun.* **55,** 396–402.
Engelman, D. M., and Rothman, J. E. (1972). *J. Biol. Chem.* **247,** 3694–3697.
Forslind, E., and Kjellander, R. (1975). *J. Theor. Biol.* **51,** 97–109.
Gally, H. U., Pluschke, G., Overath, P., and Seelig, J. (1980). *Biochemistry* **19,** 1638–1643.
Garland, P. B., and Moore, C. H. (1979). *Biochem. J.* **183,** 561–572.
Godin, D. V., and Ng, T. W. (1974). *Mol. Pharmacol.* **9,** 802–819.
Gomez-Fernandez, J. C., Goni, F. M., Bach, D., Restall, C., and Chapman, D. (1980). *Biochim. Biophys. Acta* **598,** 502–516.
Gottlieb, M. H., and Eanes, E. D. (1972). *Biophys. J.* **12,** 1533–1548.
Harkins, W. D., and Anderson, T. F. (1937). *J. Am. Chem. Soc.* **59,** 2189–2197.
Henderson, R. (1975). *J. Mol. Biol.* **93,** 123–138.
Heyn, M. P. (1979). *FEBS Lett.* **108,** 359–364.
Hinz, H. J., and Sturtevant, J. M. (1972). *J. Biol. Chem.* **247,** 3697–3700.
Hoffmann, W., Sarzala, M. G., and Chapman, D. (1979). *Proc. Natl. Acad. Sci. U.S.A.* **76,** 3860–3864.

Hoffmann, W., Sarzala, M. G., Gomez-Fernandez, J. C., Goni, F. M., Restall, C. J., Chapman, D., Heppeler, G., and Kreutz, W. (1980). *J. Mol. Biol.* **141,** 119–132.
Hoffmann, W., Pink, D. A., Restall, C., and Chapman, D. (1981). *Eur. J. Biochem.* **114,** 585–589.
Hubbell, W. L., and McConnell, H. M. (1971). *J. Am. Chem. Soc.* **93,** 314–326.
Ito, T., and Ohnishi, S. (1974). *Biochim. Biophys. Acta* **352,** 29–37.
Jähnig, F. (1977). Dissertation, Max-Planck-Institut für Biophysikalische Chemie, Göttingen-Nikolausberg, FRG.
Jain, M. K. and White, H. B. (1977). *Adv. Lipid Res.* **15,** 1–60.
Johnston, D. S., Sanghera, S., Pons, M., and Chapman, D. (1980). *Biochim. Biophys. Acta* **602,** 57–69.
Junge, W., and Devaux, D. (1975). *Biochim. Biophys. Acta* **408,** 200–214.
Kang, S., Gutowsky, H. S., Hsung, J. C., Jacobs, R., King, T. E., Rice, D., and Oldfield, E. (1979). *Biochemistry* **18,** 3257–3267.
Kirino, Y., Ohkuma, T., and Shimizu, H. (1978). *J. Biochem. (Tokyo)* **84,** 111–115.
Ladbrooke, B. D., and Chapman, D. (1969). *Chem. Phys. Lipids* **3,** 304–367.
Ladbrooke, B. D., Williams, R. M., and Chapman, D. (1968). *Biochim. Biophys. Acta* **150,** 333–340.
Liebman, P. A., and Entine, G. (1974). *Science* **185,** 457–469.
Luzzati, V. (1968). *In* Biological Membranes (D. Chapman, ed.), p. 71. New York Academic Press.
Mabrey, S., Mateo, P. L., and Sturtevant, J. M. (1978). *Biochemistry* **17,** 2464–2468.
Marčelja, S. (1974). *Biochim. Biophys. Acta* **367,** 165–176.
Marčelja, S. (1976). *Biochim. Biophys. Acta* **455,** 1–7.
Martin, R. B., and Yeagle, P. L. (1978). *Lipids* **13,** 594–597.
Mombers, C., Verkleij, A. J., de Gier, J., and Van Deenan, L. L. M. (1979). *Biochim. Biophys. Acta* **551,** 271–281.
Moore, C. H., and Garland, P. B. (1979). *Biochem. Soc. Trans.* **7,** 945–946.
Naqvi, R. K., Gonzalez-Rodriguez, J., Cherry, R. J., and Chapman, D. (1973). *Nature (London)* **245,** 249–251.
Naqvi, R. K., Behr, J. P., and Chapman, D. (1974). *Chem. Phys. Lett.* **26,** 440–444.
Ohki, S. (1979). *Biophys. J.* **9,** 1195–1205.
Oldfield, E., and Chapman, D. (1972). *FEBS Lett.* **23** (3), 285–297.
Oldfield, E., Chapman, D., and Derbyshire, W. (1971). *FEBS. Lett.* **16,** 102.
Oldfield, E., Chapman, D., and Derbyshire, W. (1972a). *Chem. Phys. Lipids* **9,** 69–81.
Oldfield, E., Keough, K., and Chapman, D. (1972b). *FEBS Lett.* **20,** 344–346.
Oldfield, E., Gilmore, R., Glaser, M., Gutowsky, H. S., Hshung, J. C., Kang, S. Y., King, T. E., Meadows, M., and Rice, D. (1978a). *Proc. Natl. Acad. Sci. U.S.A.* **75,** 4657–4660.
Oldfield, E., Meadows, M., Rice, D., and Jacobs, R. (1978b). *Biochemistry* **17,** 2727–2740.
Owicki, J. C., Springgate, M. W., and McConnell, H. M. (1978). *Proc. Natl. Acad. Sci. U.S.A.* **75,** 1616–1619.
Papahadjopoulos, D. (1968). *Biochim. Biophys. Acta* **311,** 330–348.
Penkett, S. A., Flook, A. G., and Chapman, D. (1968). *Chem. Phys. Lipids* **2,** 273–290.
Phillips, M. C., and Chapman, D. (1968). *Biochim. Biophys. Acta* **163,** 301–313.
Phillips, M. C., Williams, R. M., and Chapman, D. (1969). *Chem. Phys. Lipids* **3,** 234–244.
Phillips, M. C., Ladbrooke, B. D., and Chapman, D. (1970). *Biochim. Biophys. Acta* **193,** 35–44.
Pink, D. A., and Carroll, C. E. (1978). *Phys. Lett. A* **66A,** 157–160.
Pink, D. A., and Chapman, D. (1979). *Proc. Natl. Acad. Sci. U.S.A.* **76,** 1542–1546.
Pink, D. A., Georgallis, A., and Chapman, D. (1981). *Biochemistry* **20,** 7152–7157.

Poo, M. M., and Cone, R. A. (1974). *Nature (London)* **247,** 438–441.
Poste, G. D., Papahadjopoulos, D., Jacobson, K., and Vail, W. J. (1975). *Nature (London)* **253,** 552–554.
Restall, C., Arrondo, J. L. R., Elliot, D. A., Jaskowska, A., Weber, W. V., and Chapman, D. (1981). *Biochim. Biophys. Acta* **670,** 433–440.
Rice, D. M., Meadows, M. D., Scheinman, A. O., Goni, F. M., Gomez-Fernandez, J. C., Moscarello, M. A., Chapman, D., and Oldfield, E. (1979). *Biochemistry* **18,** 5893–5903.
Rice, D. M. and Oldfield, E. (1979). *Biochemistry* **18,** 3272–3279.
Rojas, E., and Tobias, J. M. (1965). *Biochim. Biophys. Acta* **94,** 394–404.
Rousselet, A., and Devaux, P. F. (1977). *Biochem. Biophys. Res. Commun.* **78,** 448–454.
Schroeder, H. (1977). *J. Chem. Phys.* **67,** 1617–1619.
Scott, H. L., and Cherng, S. L. (1978). *Biochim. Biophys. Acta* **510,** 209–215.
Seelig, A., and Seelig, J. (1974). *Biochemistry* **13,** 4839–4845.
Seelig, A., and Seelig, J. (1975). *Biochim. Biophys. Acta* **406,** 1–5.
Seelig, A., and Seelig, J. (1977). *Biochemistry* **16,** 45–50.
Seelig, J., and Browning, J. L. (1978). *FEBS Lett.* **92,** 41–44.
Seelig, J., and Seelig, A. (1980). *Q. Rev. Biophys.* **13,** 19–61.
Seelig, J., and Waespe-Šarčevic, N. (1978). *Biochemistry* **17,** 3310–3315.
Shanes, A. M., and Gershfeld, N. L. (1960). *J. Gen. Physiol.* **44,** 345–363.
Shimshick, E. J., and McConnell, H. M. (1973). *Biochemistry* **12,** 2351–2360.
Skarjune, R., and Oldfield, E. (1979). *Biochim. Biophys. Acta* **556,** 208–218.
Smith, B. A., and McConnell, H. M. (1978). *Proc. Natl. Acad. Sci. U.S.A.* **75,** 2759–2763.
Smith, B. A., Clark, W. R., and McConnell, H. M. (1979). *Proc. Natl. Acad. Sci. U.S.A.* **76,** 5641–5644.
Smith, L. M., Smith, B. A., and McConnell, H. M. (1979). *Biochemistry* **18,** 2256–2259.
Speth, V., and Wunderlich, F. (1973). *Biochim. Biophys. Acta* **291,** 621–628.
Stockton, G. W., Polnaszek, C. F., Tulloch, A. P., Hasan, F., and Smith, I. C. P. (1976). *Biochemistry* **15,** 954–966.
Stockton, G. W., Johnson, K. G., Butler, K., Tulloch, A. P., Boulanger, Y., Smith, I. C. P., Davis, J. H., and Bloom, M. (1977). *Nature (London)* **269,** 268.
Stoffel, W., Zierenberg, O., and Scheefers, H. (1977). *Hoppe-Seyler's Z. Physiol. Chem.* **358,** 865–882.
Tao, T. (1971). *Biochem. J.* **122,** 54p.
Thomas, D. D., and Hidalgo, C. (1978). *Proc. Natl. Acad. Sci. U.S.A.* **75,** 5488–5492.
Thomas, D. D., Seidel, J. C., Hyde, J. S., and Gergley, J. (1975). *Proc. Natl. Acad. Sci. U.S.A.* **72,** 1729–1733.
Träuble, H. (1972). *Biomembranes* **3,** 197.
Träuble, H. (1976). *Nobel Found. Symp.* **34,** 509–550.
Träuble, H., and Eibl, H. (1974). *Proc. Natl. Acad. Sci. U.S.A.* **71,** 214–219.
Träuble, H., and Sackmann, E. (1972). *J. Am. Chem. Soc.* **94,** 4499–4510.
Van Zoelen, E. J. J., Van Dijck, P. W. M., de Kruijff, B., Verkleij, A. J., and Van Deenen, L. L. M. (1978). *Biochim. Biophys. Acta* **514,** 9–24.
Vaz, W. L. C., Austin, R. H., and Vogel, H. (1979). *Biophys. J.* **26,** 415–426.
Veksli, Z., Salsbury, N. J., and Chapman, D. (1969). *Biochim. Biophys. Acta* **183,** 434–446.
Verkleij, A. J., de Kruijff, B., Ververgaert, P. H. J. T., Tocanne, J. F., and Van Deenen, L. L. M. (1974). *Biochim. Biophys. Acta* **339,** 432–437.
Wahl, P., Kasai, M., and Changeux, J.-P. (1971). *Eur. J. Biochem.* **18,** 332–341.
Weber, G. (1952). *Biochem. J.* **51,** 145–155.

Chapter **3**

Lipid Phase Transitions and Mixtures

Anthony G. Lee

Introduction

The study of membranes takes the biochemist into very unfamiliar territory—the nature and theory of mixtures. Mixtures are generally avoided in biochemistry as far as is possible, but in the study of membranes they are not avoidable. The essential nature of the membrane is that of a complex mixture of lipid and protein, so that an understanding of the interactions between these components is an essential step in describing the function of the membrane. A study of lipid–protein interaction is particularly important because a cell has the capability of varying the lipid composition of its membrane, and thus changing the environment and so the activity of proteins in its

ISBN 0-12-053002-3

membrane; there is no comparable mechanism for altering the activity of a water-soluble protein.

The most successful approach to the problem of membrane structure and function is that of reconstitution. By understanding the properties of single lipids, lipid mixtures, and finally lipid–protein mixtures, we will approach an understanding of the intact biological membrane. Because the basis for an understanding of mixtures is in classical thermodynamics, a great deal of attention has been directed toward the study of the thermodynamic properties of lipids. In this chapter, the study of the thermodynamics of simple lipids will be related to real membranes.

First-Order Phase Transitions

The physical state of phospholipids depends markedly on temperature. When heated, phospholipids undergo an endothermic transition at a temperature well below the true melting point. At this temperature, a change of state from the crystalline (or gel) state to the liquid–crystalline state occurs, which is associated with increased conformational freedom for the lipid fatty acyl chains. The fatty acyl chains are said to have "melted."

In the gel phase below the transition temperature, phospholipids adopt a bilayer structure in which the fatty acyl chains are packed in highly ordered, all-trans arrays. In the liquid–crystalline phase, X-ray diffraction studies show that the lipids still hold a bilayer structure but the lipid fatty acyl chains are considerably more disordered than in the gel phase. The disorder is due to rotation about C—C bonds, with the appearance of gauche rotational isomers (Lee, 1975).

Although the transition between gel and liquid–crystalline phases can be described as melting, this is somewhat misleading. The changes in both volume and entropy of a lipid bilayer at the phase transition are less than the corresponding changes for fatty acids or alkanes. The differences arise from the obvious differences in structure between an isotropic liquid and an anisotropic lipid bilayer. In the bilayer, one end of the fatty acyl chain is attached to the glycerol backbone of the lipid, and one would expect this to reduce the extent of the disordering that occurs in a bilayer.

The transition between gel and liquid–crystalline phases is usually very sharp, as is the melting of a normal liquid (Fig. 1). If the molecules in the bilayer were completely isolated, then it would be possible for rotational isomerization to occur in the fatty acyl chains of one lipid independently of

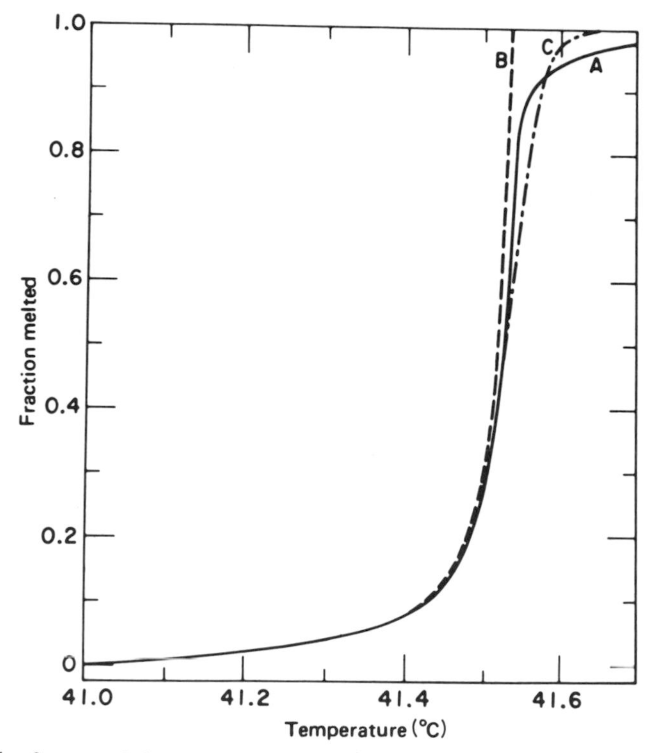

Fig. 1. The fraction of dipalmitoyl phosphatidylcholine that has melted as a function of temperature. Curve A, the fraction melted; Curves B and C, theoretical calculations assuming various values for impurity levels. From Albon and Sturtevant (1978).

what its neighbors were doing. The disorder in each chain could then gradually increase with increasing temperature. However, lipids in bilayers are not isolated, and rotation in one chain will cause it to bump into neighboring chains. The appearance of disorder in the lipid fatty acyl chains therefore must be cooperative. Disordering or melting of one lipid creates the space allowing another lipid to disorder or melt. The cooperativity appears because the tendency of a particular molecule to change from one phase to another is dependent on the state of the other molecules around it.

The sharpness of the phase transition observed for highly purified lipid (Fig. 1) suggests that it can be reasonably described as a first-order transition; therefore, the thermodynamic relationships derived for other first-order transitions, such as the melting of a normal liquid, will be applicable. The effect of pressure on the transition temperature accurately fits the Clapeyron equation, which is also consistent with a first-order transition (Liu and Kay, 1977). The pure first-order transition is, however, an abstraction.

First, by definition, a first-order transition occurs discontinuously at a single temperature. This implies that the number of molecules taking part in the transition (the cooperative unit) is infinitely large. In practice, because of the presence of physical imperfections in the bilayer that can act as nuclei for melting in local regions, the cooperative unit will be smaller than the whole bilayer, and the transition will have a finite width. Second, when melting is discussed as an equilibrium between solid and liquid, it is assumed that the regions of solid and liquid are so large that contributions from their surfaces can be neglected. For lipid bilayers, this will not be true. In the bilayer, both liquid–crystalline and gel phases could coexist. Molecules in the "walls" separating the domains of lipid in each of the two phases in the hybrid bilayers will be in states of unusual packing, which will make the transition to some extent indeterminate.

In practice, unless great care is taken in purifying the lipid, the transition will not be as sharp as that shown in Fig. 1. This is almost certainly due to the presence of impurities. Indeed, the transition is often detected spectroscopically, in which case an impurity, the spectroscopic probe, is deliberately added to the system. As described on page 67 the presence of an impurity will often lead to a broadening of the transition; the transition however remains first order. Unfortunately, the finite width of the transition has led to its interpretation as non–first order. In particular, many attempts have been made to calculate a "cooperative unit" for the transition from the width of the transition, with the size of the cooperative unit decreasing with increasing transition width. Such an attempt is misplaced.

In the small, single-shelled lipid vesicles that can be formed by sonication, the effects of temperature are rather different from those for the normal bilayer. In general, it is agreed that the transition between liquid–crystalline and gel becomes very broad. This can be understood in terms of the difficulty of packing lipid in the gel phase into the tightly-curved vesicles. Blaurock and Gamble (1979) have suggested that such vesicles in the gel state are polygonal with small facets of planar packing, rather than being smoothly curved. The cooperative unit for the transition in such vesicles would then be very small, and a broad transition would result.

Although the thermodynamic approach allows us to calculate a number of interesting properties of the bilayer (see page 67) it does not, of course, supply any understanding of what the molecules in the bilayer are actually doing. For that, a statistical approach is necessary. Unfortunately, the lipid bilayer is too complicated for a full statistical mechanical calculation, and simplifications must be made; there is considerable disagreement about which simplifications are appropriate and which are not.

In general, the state of the bilayer is determined by the balance between the entropy associated with the rotameric disorder of the fatty acyl chains and the attractive van der Waals forces between the chains, and attractions between head groups. At low temperatures, the lipid fatty acyl chains will generally be in an all trans form. With increasing temperature, rotameric disorder will appear in the chains; this is described most simply in terms of gauche rotamers about C—C bonds. With increasing disorder, the packing density of the chains will decrease, leading to decreased van der Waals interactions between the chains. It is clearly not possible to calculate the van der Waals interactions between all the possible rotational states of fatty acyl chains in bilayers, and some sort of approximating calculation is necessary. One common approach is the mean field approximation; the interactions between individual chains are represented by an average interaction, ignoring the fact that the individual behaviors and interactions of the chains can be widely distributed about the average. The theory necessarily ignores any short-range order that might exist. As the present state of theoretical approaches to the lipid phase transition has been reviewed recently by Nagle (1980), no further discussion will be given here.

Before proceeding to look at the phase transitions in lipid bilayers in detail, it will be sensible to look at what is known about the structures of such bilayers.

Structures of Lipid Phases

Lamellar Phase

A wide variety of phases can be observed in lipid–water systems, but the most important for biological membranes are the lamellar phases formed at high water concentrations. (Other phases are considered on page 54.) Unfortunately, the most detailed structural information has come from studies of lipid crystals containing only a few molecules of solvation. However, it appears that the structures adopted in these crystals are very similar to those adopted in fully hydrated lipid bilayers. We will therefore first review the crystal structure data, and then consider the evidence that this is of relevance to the hydrated bilayer.

Information about the conformation of lipid head groups has been deduced from X-ray crystal structures of phosphatidylcholines, phosphatidylethanolamines, and cerebroside (Hitchcock *et al.*, 1974; Pascher and Sundell, 1977; Pearson and Pascher, 1979; Hauser *et al.*, 1980). The conformations of the phosphorylcholine and phosphorylethanolamine groups are very similar in crystals of glycerylphosphorylcholine, lysophosphatidyl-

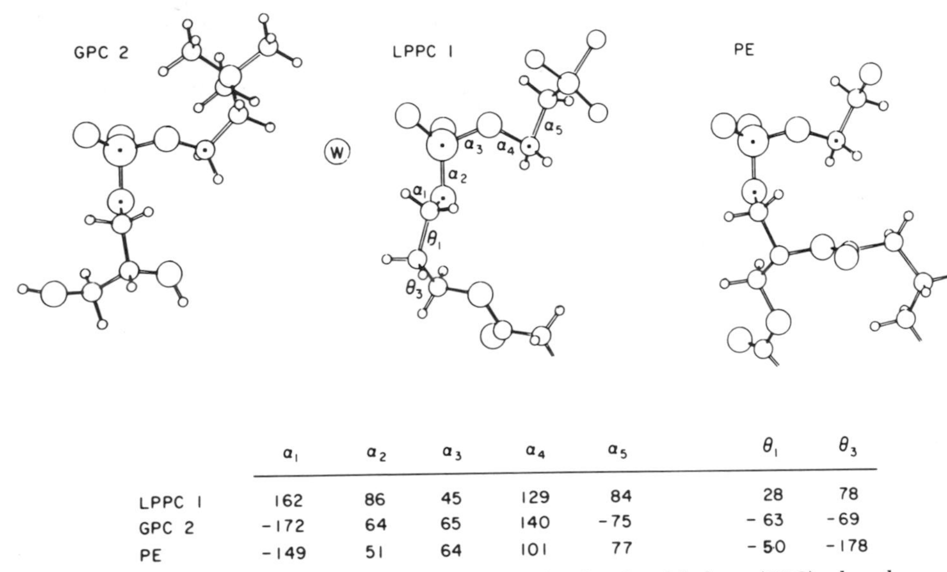

	α_1	α_2	α_3	α_4	α_5	θ_1	θ_3
LPPC I	162	86	45	129	84	28	78
GPC 2	-172	64	65	140	-75	-63	-69
PE	-149	51	64	101	77	-50	-178

Fig. 2. Polar group conformation of glyceryl phosphorylcholine (GPC), lysophosphatidylcholine (LPPC), and dilauroyl phosphatidylethanolamine. With permission from Hauser *et al.* (1980). Copyright by Academic Press Inc. (London) Ltd.

choline, and dilauroylphosphatidylethanolamine (Fig. 2). These similar conformations reflect the tendency of the ammonium nitrogen to fold back towards the phosphate groups so as to minimize the distance between the groups of opposite charge. Intramolecular forces are clearly important in defining the preferred head group conformation.

In dilauroylphosphatidylethanolamine, the molecules occupy an area of 39 Å^2 in the bilayer phase, with the polar groups parallel to the bilayer surface. In the structure, the NH_3^+ group forms hydrogen bonds 2.8 Å long with the unesterified phosphate oxygens of adjacent molecules. The fatty acyl chains are packed parallel and untilted with respect to the bilayer structure (Fig. 3). The conformation of the glycerol diester group is such that the initial part of the 2-fatty acyl chain extends parallel to the bilayer surface.

The structure of dimyristoylphosphatidylcholine dihydrate is more complex (Fig. 4). The molecular area occupied by the lipids is the same as that for dilauroyl phosphatidylethanolamine (39 Å^2), but this is too small for the bulky phosphorylcholine group. The phosphatidylcholine molecules therefore pack mutually displaced in the direction perpendicular to the bilayer surface. This means that the effective surface area per molecule is kept small, and the head groups remain almost parallel to the bilayer surface (Hauser *et al.*, 1980). This results in a ripple-like structure, which may be

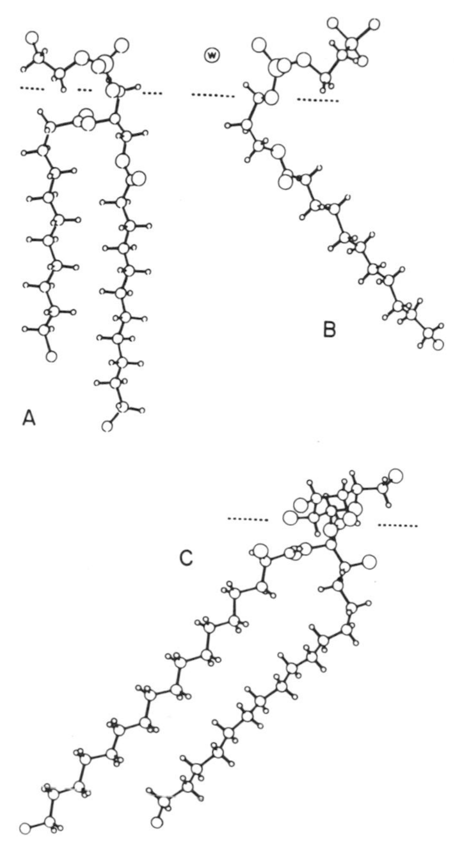

Fig. 3. Conformation of membrane lipids as found in crystals. A, dilauroyl phosphatidylethanolamine, B, lyso phosphatidylcholine, and C, cerebroside. With permission from Hauser *et al.* (1980). Copyright by Academic Press Inc. (London) Ltd.

related to the ripples seen in freeze-fracture electron microscopy (see later).

The head group conformation of the two lipid arrangements in the structure are very similar (Fig. 5). Again, the nitrogen atom is folded back towards the phosphate group. Hydrogen bonding of the type observed in phosphatidylethanolamine is not possible, and indeed the nitrogen atom of the bulky NMe_3^+ group comes no closer than 4 Å to the phosphate oxygens. Instead, the phosphate groups are separated and shielded by water molecules of hydration.

The head groups are not perfectly parallel to the bilayer surface; rather, the P–N vectors are inclined at angles of 17 and 27° to the surface in the two forms. The major difference between the two structures is that in molecule A

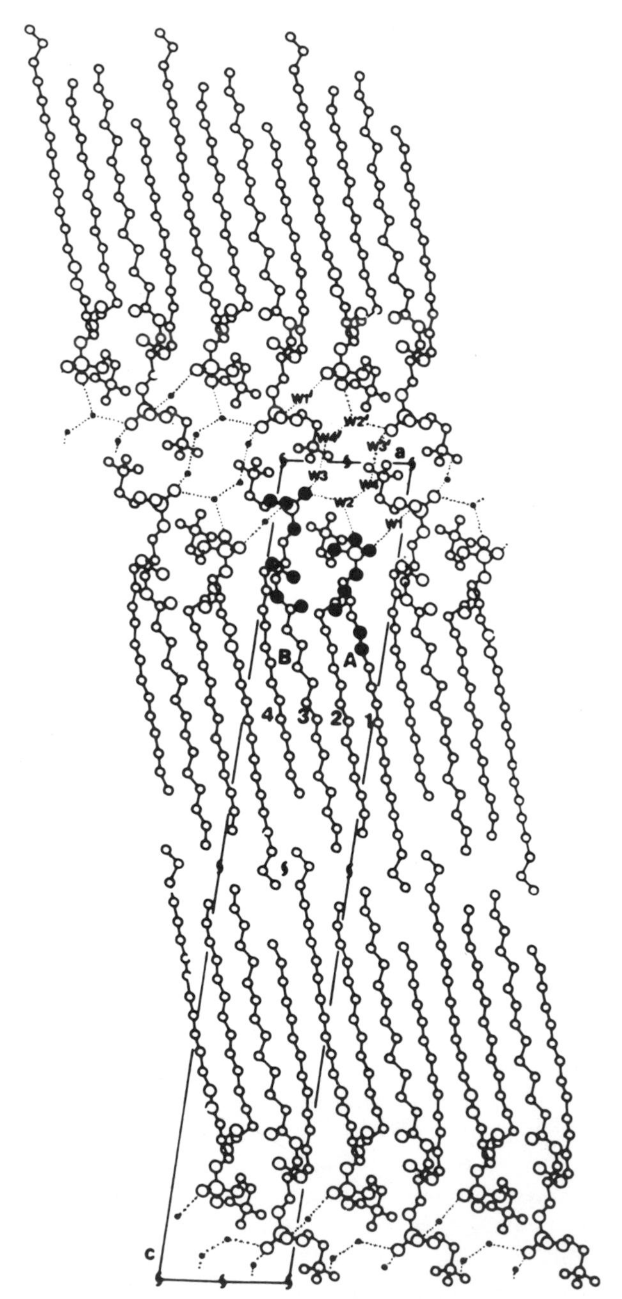

Fig. 4. Molecular arrangement of dimyristoyl phosphatidylcholine dihydrate. The two lipid arrangements are marked A and B. The hydrogen bonding system in the polar region is shown by dotted lines. From Pearson and Pascher (1979). Reprinted by permission from *Nature*, Vol. 281, pp. 499–501. Copyright © 1979 Macmillan Journals Limited.

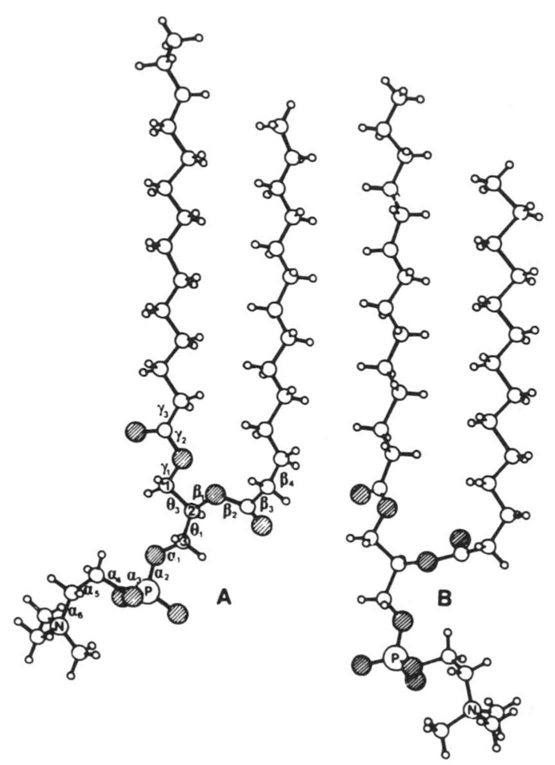

Fig. 5. The two molecular arrangements found in dimyristoyl phosphatidylcholine dihydrate. From Pearson and Pascher (1979). Reprinted by permission from Nature, Vol. 281, pp. 499–501. Copyright © 1979 Macmillan Journals Limited.

the phosphorylcholine group points away from its own glycerol group; in B, it is close to it (Fig. 5). These differences result from different conformations about the C-3—C-2 bond of the glycerol group. NMR studies suggest that this rotation occurs easily (Seelig, 1977).

The two fatty acyl chains are almost perpendicular to the bilayer surface, with a tilt angle of 12°. Interestingly, the initial part of the 2-fatty acyl chain extends parallel to the surface, then bends at the second carbon to become parallel to the first chain in a manner very similar to that observed for phosphatidylethanolamine. In these synthetic lipids, chain 1 consequently extends further into the bilayer than does chain 2. In natural lipids this difference is minimized, because longer fatty acyl chains are normally found at position 2 rather than at position 1 (Van Deenan, 1965).

As already described, if the lipid head group is large it will have a larger excluded area in the plane of the bilayer than the fatty acyl chains, and in dimyristoyl phosphatidylcholine dihydrate this is overcome by alternate stacking of the lipids to give a ripple structure. An alternative is to tilt the chains to "fill in" the potential space in the hydrocarbon region. Such a structure has been found in crystals of cerebroside (Pascher and Sundell, 1977; see Fig. 3). In cerebroside, the initial part of the sphingosine chain with the amide-bound fatty acyl carboxyl group has a conformation and orientation similar to the structurally corresponding glycerol dicarboxyl ester group. In cerebroside, a chain tilt of 41° is produced by a bend at carbons 6 and 2 of the sphingosine and fatty acyl chains, respectively. The bend in the sphingosine chain is facilitated by the presence of a trans double bond in the 4,5 position; this double bond is equivalent to the OCO ester group of the 1-chain in the glycerophospholipids. Interestingly, the chain tilt found in the cerebroside is the same as that in lysophosphatidylcholine (Fig. 3).

All the available evidence suggests that the structures of lipids in these crystals are good indications of their structures in bilayers. For dispersions of phosphatidylcholines in excess water, both in the gel and the liquid–crystalline state, studies using X ray (Franks, 1976), neutron diffraction (Worcester and Franks, 1976; Buldt *et al.*, 1978), and NMR techniques (Seelig, 1978) show that the preferred conformation of the P–N dipole is approximately parallel to the bilayer surface. Although the NMR studies are also consistent with a structure in which the rest of the head group is arranged much as that found in the crystal, the NMR data is not sufficient to prove that the structures actually are the same (Jacobs and Oldfield, 1981).

The organization of the head group in phosphatidylglycerols seems to be rather similar to that of phosphatidylcholine and phosphatidylethanolamine (Wohlgemuth *et al.*, 1980). For phosphatidylserine, however, the head group region apparently is significantly different; although details are not certain, the head group appears to be immobile relative to the other lipids, probably as a result of head group–head group interactions (Browning and Seelig, 1980). Sphingomyelin also forms a bilayer structure very similar to the other lipids (Khare and Worthington, 1978).

The common feature of the initial part of the 2-chain extending parallel to the surface of the bilayer has also been confirmed by NMR in hydrated bilayers of phosphatidylethanolamine, phosphatidylcholine (Seelig and Browning, 1978), and phosphatidylserine (Browning and Seelig, 1980). Neutron diffraction analysis of dipalmitoyl phosphatidylcholine in the gel state has shown that as a consequence the two chains are out of step by about 1.8 Å or 1.5 C—C bonds (Buldt *et al.*, 1978).

In fully hydrated dipalmitoyl phosphatidylethanolamine, the thickness of the bilayer in the gel phase, measured as the distance between phosphate groups, is 49 Å, corresponding to all trans chains oriented perpendicularly to the bilayer surface (McIntosh, 1980) as was observed for the crystal (Fig. 3).

In fully hydrated dipalmitoyl phosphatidylcholine in the gel phase at 20°C, below the pretransition, however, the distance between phosphate groups is only 42 Å. This suggests that the fatty acyl chains are tilted at an angle of 30° to the bilayer surface (McIntosh, 1980). In this way, the bulky phosphatidylcholine head groups can be accommodated within the plane of the bilayer. The structure is thus similar to that of lysophosphatidylcholine as shown in Fig. 3. For phosphatidylethanolamine, the tilt is not necessary because the head group is smaller.

The connection of the chain tilt with head group packing is also shown by experiments with phosphatidic acids (Jähnig *et al.*, 1979). The charge on the lipid head group can be varied by varying pH. With increasing charge, the lateral packing of the head groups was found to decrease at the same time that the tilt of the fatty acyl chains increased. Further confirmation comes from observations that decane and tetradecane remove the tilt from gel phase dipalmitoyl phosphatidylcholine. The idea here is that the fatty acyl chains normally tilt to fill the potential void created by the large head group. If the potential void is occupied by a hydrophobic molecule such as an alkane, there is no longer any need for the chains to tilt (Fig. 6) (McIntosh, 1980).

Lipids that show tilted chains show a pretransition at a temperature a few degrees below the main transition. This transition is associated with a change

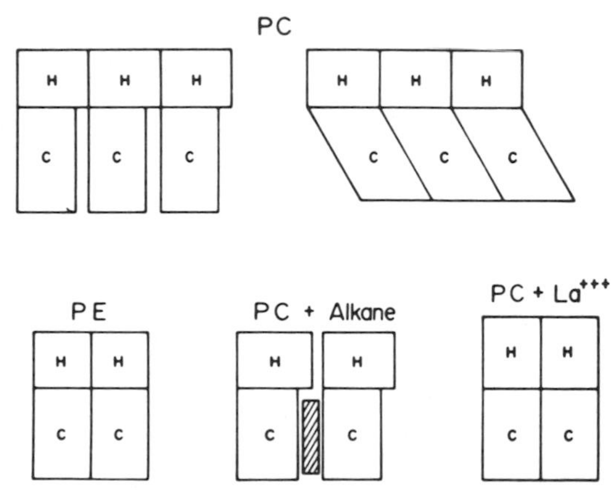

Fig. 6. A diagramatic scheme of lipid packing. For the phosphatidylcholines (PC), the large headgroup (H) prevents the chains (C) from coming into contact. The chains therefore tilt. For phosphatidylethanolamine (PE), the head group is smaller and allows contact of chains, even when untilted. For PC, the introduction of tetradecane (shaded rectangle, bottom center) fills in potential voids in the chain region and thus allows chains to straighten. Lanthanum also straightens the fatty acyl chains by changing the conformation of the PC head group. From McIntosh (1980).

to a lamellar structure distorted by a periodic undulation or ripple, as has been seen by freeze-fracture electron microscopy (Janiak *et al.*, 1976). Presumably the structure in this intermediate state is related to the structure observed for dimyristoyl phosphatidylcholine dihydrate (Fig. 4).

The pretransition will be described on page 61.

Nonbilayer Phases

A number of structures other than the familiar bilayer are possible for phospholipids: the most common are the micellar and hexagonal phases. In addition to the normal gel to liquid–crystalline bilayer phase transition, a bilayer to hexagonal (H_{11}) polymorphic phase transition has been observed for phosphatidylethanolamines at about 10° above the normal phase transition. This phase is characterized by a hexagonal array of cylinders which are formed with the phosphatidylethanolamine head groups located at the inner cylinder surface and the hydrocarbon side chains directed radially out from this surface. Water fills the central core of the cylinder. The transition has a very small enthalpy change (Cullis and de Kruijff, 1979).

Presumably the hexagonal phase occurs because the phosphatidylethanolamine head group is rather small compared to the area occupied by the fatty acyl chains in the liquid–crystalline phase, particularly when the chains are polyunsaturated. The data on glucosyl diglycerides from *Acholeplasma* are consistent with this interpretation (Wieslander *et al.*, 1978). Monoglucosyl diglyceride, with a small head group, adopts a hexagonal phase, whereas diglucosyl diglyceride, with a larger head group, adopts a lamellar phase.

Nonbilayer structures are also adopted by negatively charged phospholipids in the presence of divalent metal ions. Cardiolipin, in the absence of Ca^{2+}, forms a lamellar structure (Rand and Sengupta, 1972), but in the presence of Ca^{2+} the structure transforms to a hexagonal (H_{11}) phase (Cullis and de Kruijff, 1979; Vail and Stollery, 1979). Similarly, phosphatidylserine adopts a bilayer structure in the absence of calcium (Cullis and de Kruijff, 1979), but addition of calcium causes transformation to a more complex structure.

For mixtures of these lipids with lipids such as the phosphatidylcholines, which only form the lamellar phase, the situation is unclear. It had been suggested on the basis of ^{31}P-NMR studies that hexagonal phases were commonly formed in mixtures of phosphatidylcholines and phosphatidylethanolamines. It is now thought, however, that in mixtures containing more than about 15 mol % phosphatidylcholine, the structures are lamellar although containing regions of an inverted micellar nature (*lipidic intramembranous particles*) (de Kruijff *et al.*, 1979; Hui *et al.*, 1981). Earlier, based on ^{31} P-NMR studies, sphingomyelin was reported to adopt a hexagonal structure in the liquid–crystalline phase; this has been refuted by Hui *et al.* (1980).

Brain sulphatides, gangliosides, and triphosphoinositides form micelles in water (Howard and Burton, 1964; Hendrickson and Fullington, 1965; Abrahamsson *et al.*, 1972; Sugiwa, 1981). For the phosphatidylcholines and phosphatidylethanolamines, the lamellar phases show no chain interdigitation, and the two halves of the bilayer can be considered to be independent. With negatively charged phospholipids, however, this is not so. A number of gel phases have been observed for phosphatidylglycerol (for example) in which the chains interdigitate so that the methyl ends of the fatty acyl chains of one monolayer are near the head groups of the opposite layer (Ranck *et al.*, 1977). This structure apparently can occur because of a strong repulsion between the negatively charged head groups which increases the area per head group.

The Gel to Liquid Crystalline Phase Transition

Many studies of the thermodynamics of the phase transition have been published. A selection of data for the phosphatidylcholines and phosphatidylethanolamines is presented in Table I. Unfortunately, it is clear that there is a large scatter in the reported data; for example, reported enthalpies for the transition in dipalmitoyl phosphatidylcholine vary between 6.8 and 9.6 kcal/mol. However, a number of general features are clear. First, in a series of phosphatidylcholines or phosphatidylethanolamines transition temperatures and enthalpies increase with the length of the fatty acyl chains. Second, lipids with a trans double bond near the middle of the chain have lower transition temperatures and enthalpies than the corresponding lipids with saturated chains. Third, a cis double bond in the middle of the chain has a more marked effect on transition temperatures than does a trans double bond, but has no effect on transition enthalpies. The position of the cis double bond is important: the minimum transition temperature occurs when the double bond is in the middle of the chain (Fig. 7). Finally, for lipids with saturated fatty acyl chains transitions in phosphatidylethanolamines occur 20°C higher than in the corresponding phosphatidylcholine; each methyl group removed increases the transition temperature by about 4°C (Vaughan

TABLE I

Transition Data for Phosphatidylcholines (PCs) and Phosphatidylethanolamines (PEs)

Phospholipid[a]	T (°C)[b]	ΔH° (kcal/mol)
PC		
12:0	~0[1,2]	4.3[2], 1.7[3]
14:0	23[1], 23.8[4], 24.0[7]	6.7[1], 6.3[4], 6.8[2], 5.4[3], 5.0[5], 5.4[7]
16:0	41.1[8], 41.4[7], 41[1], 41.5[2], 41.8[4]	9.6[4], 8.7[1], 8.5[7], 6.8[8]
18:0	58[1], 54.2[4], 54.0[3]	10.7[1], 10.8[4]
22:0	75[1]	14.9[1]
16:1 (c,9)	−36[2]	9.1[2]
18:1 (c,9)	−14[2], −21[6], −22[9]	11.2[2], 7.6[9]
18:1 (t,9)	9.5[2]	7.3[2]
18:0/18:1 (c,6)	30[6]	
18:0/18:1 (c,12)	12[6]	
18:0/18:1 (c,16)	43[b]	
PE		
12:0	29[2], 30.5[7]	4.0[2], 3.5[7]
14:0	47.5[2], 49.1[7]	6.4[2], 5.7[7]
16:0	60[2], 63.1[7]	8.5[2], 8.8[7]
16:1 (c,9)	−33.5[2]	4.3[2]
18:1 (c,9)	−16[2]	4.5[2]
18:1 (t,9)	35[2]	7.0[2]

[a]Fatty acid chains are indicated by the number of carbon atoms, followed by the number of double bonds, the position of the double bond, cis or trans are in brackets.

[b]References: 1. Ladbrooke and Chapman (1969); 2. Van Dijck *et al.* (1976); 3. Mabrey and Sturtevant (1976); 4. Hinz and Sturtevant (1972); 5. Lentz *et al.* (1978); 6. Barton and Gunstone (1975); 7. Wilkinson and Nagle (1981); 8. Chen *et al.* (1980); 9. Phillips *et al.* (1969).

and Keough, 1974). It is interesting that this difference disappears if one of the fatty acyl chains contains a cis double bond.

The dependence of transition properties on chain length is not surprising. Increased chain length will result in increased van der Waals attractions, which tend to stabilize the ordered, high-density gel phase relative to the disordered, low-density liquid–crystalline phase. A cis double bond appears to prevent the close packing of the fatty acyl chains normally observed in the gel phase; the effect is greater when the double bond is in the middle of the chain. For bilayers of diphytanoyl phosphatidylcholine, no phase transition was observed at temperatures down to −120°C. This is consistent with a very disordered structure caused by the methyl groups of the phytol chains, which would prevent efficient lateral packing (Lindsey *et al.*, 1979).

Phase transition temperatures of many homologous series of hydrocarbon compounds show an alternation between compounds of odd and even carbon chain length. This occurs when at least one of the phases involved in the

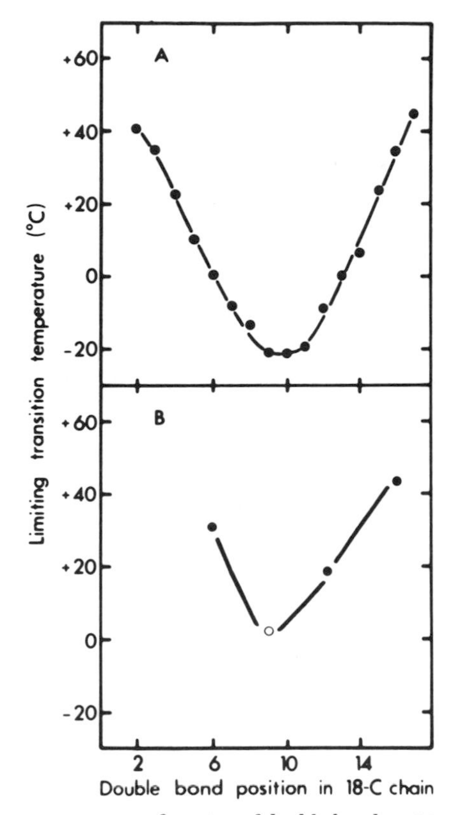

Fig. 7. Transition temperatures as a function of double bond position in the fatty acyl chains. A, 1,2-diatadec-*cis*-enoyl-phosphatidylcholines; B, 1,octadecanoyl-2-octadec-*cis*-enoyl phosphatidylcholine. From Barton and Gunstone (1975).

transition has the hydrocarbon chains so arranged that the major chain axis is not perfectly perpendicular to the plane of the end group of the hydrocarbon chain. The main transition in the phosphatidylcholines does not show an odd–even effect, which is consistent with a structure in which the fatty acyl chains are perpendicular to the bilayer surface immediately below and above the transition. Interestingly, however, the pre-transition does show an odd–even effect, suggesting that the lipid fatty acyl chains are tilted with respect to the bilayer normal below the pre-transition temperature (see page 61) (Silvius *et al.*, 1979).

The substitution of hydrogens by fluorines in the chain has effects analogous to the introduction of cis double bonds (Longmuir *et al.*, 1977; Sturtevant *et al.*, 1979). Because of the differences in conformation of the fatty acyl chains at position 1 and 2 of the glycerol backbone, transition properties of mixed acid phosphatidylcholines depend on the chain position. This is illustrated by the data in Table II (Keough and Davis, 1979). The longer chain at position 2 gives a better packing and thus a higher transition temperature.

The effect of the head group on the properties of the transition can be understood in terms of the structures described on page 47. Hydrogen bonding between head groups in phosphatidylethanolamines will inhibit the lateral expansion of the bilayer that occurs at the phase transition, and thus will raise the transition temperature. Nagle (1976) has suggested that the hydrogen bonding is only weak and transient, involving only a few percent of the molecules at any one time; therefore, there will be only a very small effect on lateral mobility.

The cis double bond in the phosphatidylethanolamines probably reduces the packing density of the lipids so that hydrogen bonding becomes impossible (see Berde *et al.*, 1980). Transition temperatures will then be the same for phosphatidylethanolamines and phosphatidylcholines.

Bach *et al.* (1978) have studied the effect of increasing the separation between the phosphate and quaternary ammonium groups in phosphatidylcholines:

```
R—(CO)—O—CH2
            |
R—(CO)—O—CH      O−
            |     |
          CH2O— P—O—(CH2)n—N+(CH3)3
                  |
                  O
```

Increasing n from the normal 2 to 7 leads to about a 3° increase in transition temperature, and further increase to 11 then leads to a decrease in transition temperature. Superimposed on this general trend is a small odd–even effect. These effects are relatively minor, and presumably reflect changes in the nature of the electrostatic interaction in the head group region of the bilayer.

TABLE II

Transition Data for *sn*-Glycero-3-phosphorylcholines[a]

Fatty acid			
1-position	2-position	T_c	ΔH (kcals/mol)
Palmitoyl	Myristoyl	27.3	6.5
Myristoyl	Palmitoyl	35.3	7.9

[a]Data from Keough and Davis (1979).

TABLE III

Transition Data for the Sphingomyelins[a]

Lipid	T (°C)	$\Delta H°$ (kcal/mol)
N-palmitoyl dihydrosphingosine phosphorylcholine	47.8	9.4
N-palmitoyl sphingosine phosphorylcholine	41.3	6.8
N-stearoyl sphingosine phosphorylcholine	52.8	17.9
N-lignoceryl sphingosine phosphorylcholine	48.6	15.3

[a]From Barenholz *et al.* (1976).

In normal pH ranges (3–12), both phosphatidylcholines and phosphatidylethanolamines are zwitterionic. At pH > 12.5, the ethanolamine group deprotonates, and the head group becomes $—PO_4^-(CH_2)_2NH_2$. This change cannot be studied in normal phospholipids because of ester hydrolysis, but in the corresponding ether analog it has been shown to result in a decrease in transition temperature, as is expected from charge–charge repulsion of the phosphate groups (Stumpel *et al.*, 1980). At pH values less than 3, the phosphate groups of both phosphatidylcholines and phosphatidylethanolamines become protonated to give $—PO_4H(CH_2)_2\ N^+(CH_3)_3$ and $—PO_4H(CH_2)_2NH_3^+$ respectively. Again, charge repulsion, now between the positively charged amine groups, would be expected to cause a decrease in transition temperature. This is not observed; instead, an increase in transition temperature is found. This has been attributed to the presence of protons in the membrane surface (Eibl and Woolley, 1979).

Data for the phase transitions of sphingomyelins are presented in Table III. In the sphingomyelins, the "interface" region of the molecule between the hydrophobic acyl chains and the zwitterionic head group is more complex than the glycerol diester group in the phosphatidylcholines. The corresponding group for the sphingomyelins is the 2-amido-3-hydroxy-trans-4-ene group. The presence of NH and OH groups could lead to hydrogen bond formation between the sphingomyelin molecules or to intramolecular hydrogen bond formation between these groups and the phosphate of the head group.

Little work has been done on the glycolipids. In the brain cerebrosides, the fatty acids are mainly tetracosanoic, 2-hydroxytetracosanoic, and 2-hydroxyoctadecanoic acids; phase transitions have been detected at about 60–70°C (Reiss-Husson, 1967; Oldfield and Chapman, 1972). In *N*-palmitoylgalactocerebroside, a gel to liquid–crystalline phase transition occurs at about 82°C, and the structure of the bilayer seems to be very similar to that of the phosphatidylcholines (Skarjune and Oldfield, 1979).

Marked environmental dependence is seen for phase transitions in charged lipids. Transition properties for the phosphatidylserines at pH 7.0 in the absence of divalent metal ions are not very different from those of the corresponding phosphatidylcholines (Table IV). With lipids such as phosphatidic acid, the head groups can be titrated to give single and doubly charged ions (Träuble and Eibl, 1974; Jacobson and Papahadjopoulos, 1975), resulting in approximately a 20° drop in transition temperature. This effect was first explained by Träuble and Eibl (1974) on the basis of the Gouy-Chapman double layer energy, but unfortunately these authors worked in terms of free energy per cm^2 rather than the correct free energy per mole; the latter does not change with area per molecule, so the proposed explanation is fallacious (Nagle, 1980). Rather, the explanation of the effect probably lies in the direct electrostatic charge–charge repulsions between the discrete charges on the lipid molecules in the membrane (Nagle, 1976; Forsyth *et al.*, 1977). Interestingly, the transition temperature even for doubly ionized dipalmitoyl phosphatidic acid is higher than that of the corresponding phosphatidylcholine and about equal to the phosphatidylethanolamine. Nagle (1976) suggested that hydrogen bonding interactions between the phosphate groups could be the explanation. Cevc *et al.* (1980) also conclude that similar interactions are important in explaining the effects of ionization on bilayers of phosphatidylglycerols. Gel to liquid–crystalline phase transitions have also been reported in phosphatidylserines (Träuble and Eibl, 1974; MacDonald *et al.*, 1976), phosphatidylinositol (Schnepel *et al.*, 1974), and cardiolipin (Hegner *et al.*, 1973).

Because of the marked dependences of the transition temperature of charged lipids on pH, it is possible to trigger the phase transition isothermally for these lipids by varying the pH. Thus, at 55°C the phase transition from gel to liquid crystalline can be triggered by increasing the pH to 7.8 (Träuble and Eibl, 1974). Binding of calcium to negatively charged lipids causes a large increase in transition temperature (Träuble, 1977), but it also causes large structural rearrangements, probably involving Ca^{2+} bridging

TABLE IV

Transition Properties for Phosphatidylserines in 0.01*M* NaCl, pH 7.0 plus EDTA[a]

Fatty acid	*T* (°C)	$\Delta H°$ (kcals/mol)
Palmitic	53	9.0
Myristic	36	7.0
Elaidic	22	—
Oleic	−11	8.8

[a]Data from Browning and Seelig (1980).

adjacent surfaces in the structure (Papahadjopoulos *et al.*, 1975). The relevance of such structures as models for regions of negatively charged lipid in biological membranes is unclear.

In summary, the (at first rather surprising) conclusion is that the lipid phase transition temperature is relatively insensitive to the nature of the lipid head group, and, therefore, to interactions in the head group region of the bilayer. The most significant portion of the lipid for determining its transition temperature is its fatty acyl chains, and the most important interaction is therefore the van der Waals interaction between the chains.

Thermodynamics of the Lamellar Phase

The Gel Phase

Although the gel phase is often referred to as solid-like, it is known that this is an oversimplification. NMR studies have shown that reorientation can occur about the long axes of the fatty acyl chains, the rate of the motion decreasing as the temperature is lowered; motion being finally frozen out at −7°C for dipalmitoyl phosphatidylcholine to give an essentially rigid system (Davis, 1979). NMR studies also suggest that the fatty acyl chains are not in a perfect all trans state in the gel phase, but that some gauche rotamers are present. This is consistent with Raman studies, which suggest that the number of gauche rotamers is one or two per chain just below the main transition (Yellin and Levin, 1977; Gaber *et al.*, 1978; Levin and Bush, 1981). At such temperatures, the chains are packed in a slightly distorted hexagonal form. Freeze-etch pictures of this state show a rippled structure. As described on page 48, it is probable that this rippled structure is adopted to provide a suitable match between the areas occupied by the chains and the large head groups. At a temperature about 8°C below the main transition there is a second transition, usually referred to by the unfortunate term *pretransition* (Table V). This is not a pretransition in the sense that the word is usually used in physics, but is in fact another first-order transition. In the low temperature form, the ripple structure reportedly disappears, and the chains tilt with respect to the bilayer surface to give a structure presumably much like that illustrated in Fig. 3. More recent papers have, however, suggested that a rippled structure is also present below the pretransition, although the amplitude of the rippling is greater (Copeland and McConnell, 1980). The transition therefore corresponds to a change from tilted chains

TABLE V

Pretransitions in Phosphatidylcholines[a]

Fatty acyl chains	T (°C)	$\Delta H°$ (kcal/mol)
Myristic	13.5	—
Palmitic	35.1	1.09
Stearic	49.1	—

[a] Data from Hinz and Sturtevant (1972); Chen *et al.* (1980).

below the transition to chains oriented perpendicular to the bilayer surface above the transition (Janiak *et al.*, 1976).

Those lipids that do not show a tilted phase will not, of course, exhibit this lower transition; thus, phosphatidylethanolamines do not exhibit a pretransition. Again, for charged lipids with increasing charge the head group area increases as a result of charge repulsion; the chain tilt increases, and a pretransition appears (Watts *et al.*, 1978; Harlos *et al.*, 1979). This probably also explains why the pretransition is so sensitive to the effect of additions. As illustrated in Fig. 6, addition of alkanes removes the chain tilt, and addition of long chain molecules also removes the pretransition (Lee, 1976a).

In the gel phase, the head group lies parallel to the plane of the bilayer and occupies an area of 57Å^2. Water penetrates into the structure as far as the level of the glycerol backbone, and this serves to reduce the electrostatic interactions between the choline and phosphate groups of adjacent head groups (Buldt *et al.*, 1979). The thickness of a bilayer of dipalmitoyl phosphatidylcholine in the gel state is 45.9 Å, measured between the ester bonds.

With decreasing temperature, the angle of tilt of the chains increases and the hexagonal packing becomes progressively more distorted (Harlos, 1978; Janiak *et al.*, 1979). Indeed, infrared studies suggest that in dipalmitoyl phosphatidylcholine at temperatures lower than about 10°C, the packing changes to orthorhombic or monoclinic (Fig. 8) (Cameron *et al.*, 1980). The transformation is gradual, and is not complete until −60°C. Chen *et al.* (1980) have reported another transition in the gel phase of dipalmitoylphosphatidylcholine, detected calorimetrically at 18°C. The relationship between this transition and that detected by infrared is at present unclear.

Transformation from the gel to liquid–crystalline phase can be associated with the presence of defects in the gel phase. These could be mosaic and grain boundaries where zones of differing orientation meet (see Lee, 1975) or lattice vacancies. These regions of poor packing can act as nuclei for the growth of the new, liquid–crystalline phase. Stresses around these imperfec-

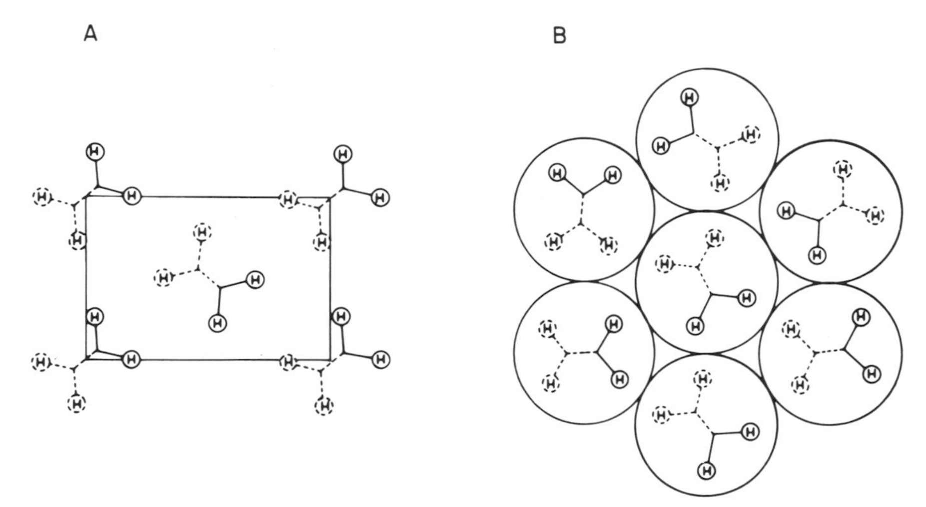

Fig. 8. Acyl chain packing. A, orthorhombic or monoclinic; B, hexagonal. The long axes of the acyl chains project from the page. In the hexagonal phase, motion about the long axes is such that the orientations of chain relative to each other are random at a given moment. From Cameron *et al.* (1980).

tions will also alter the effective pressure on the various domains and so will cause them to undergo the transition over a small temperature range. The transition will therefore have a finite width. These points have already been discussed at length and need not be repeated here (see Lee, 1975, 1977b).

The Liquid-Crystalline Phase

In the liquid–crystalline phase, the basic organization of the phospholipid molecules remains rather similar to that in gel phase. The head group in phosphatidylcholines remains parallel to the bilayer surface, although motion of the head group is faster in the liquid–crystalline phase than in the gel phase (Seelig, 1978). The two fatty acyl chains remain nonequivalent, with chain 2 adopting a conformation in which the first two carbons are nearly parallel to the plane of the bilayer (Seelig and Seelig, 1980). This is consistent with the suggestion, based on ^{13}C-NMR studies, that there are differences in the extent of hydration of the two ester carbonyl groups (Schmidt *et al.*, 1977).

The major difference between the two phases is that the lipid fatty acyl chains are disordered in the liquid–crystalline phase as a result of rapid trans–gauche isomerizations around carbon–carbon bonds. The most detailed information about the variation of order throughout the bilayer has come from ^{2}H-NMR studies (Seelig and Seelig, 1980). Order in the bilayer involves the alignment of the lipid molecules parallel to the bilayer normal. Figure 9A illustrates the principles involved for a simple rodlike molecule. At finite temperatures, thermal motion of the molecules will prevent perfect alignment, and the orientations of the molecules will be distributed over a range of angles; the bilayer normal is the most probable direction (see Fig. 9B).

We can characterize the position of a molecule by its angle, θ, with the bilayer normal. We are less interested in the instantaneous value of θ for a particular lipid than in its average for all the lipids in the sample, because this describes the extent of ordering in the bilayer. The angle θ itself, however, is not a particularly convenient way of describing order in the system. A more convenient description would be $\cos \theta$, but because there is an equal number of molecules on the two sides of the bilayer the sample average would always be zero. Thus $\cos^2\theta$ is used rather than $\cos \theta$, and the averaged quantities in which we are interested are written as $\overline{\cos^2\theta}$. If all the molecules are perfectly aligned, then $\theta = 0$ and $\overline{\cos^2\theta} = 1$. If on the other hand the molecules are randomly oriented, all values of θ are equally probable and $\overline{\cos^2\theta} = \frac{1}{3}$. The order in a system is usually described not by $\cos^2\theta$ per se, but by an order parameter S so defined that $S = 1$ in a perfectly ordered system and $S = 0$ in a totally disordered system. With the averaged values given above for $\cos^2\theta$, it can be seen that the appropriate order parameter is

$$S = \frac{1}{2}\,\overline{(3 \cos^2\theta - 1)}$$

Clearly $S = 1$ for the completely ordered system and $S = 0$ for the totally

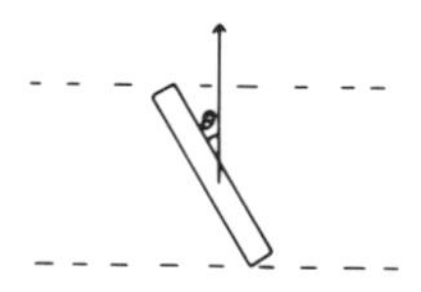

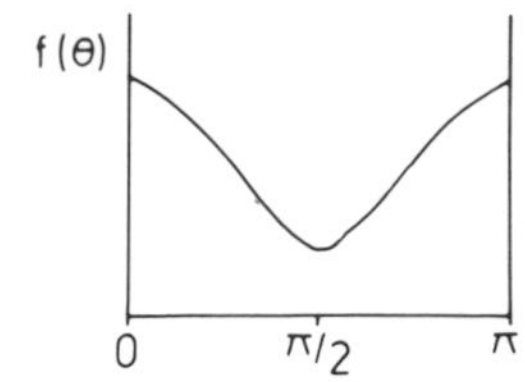

Fig. 9. A, the packing of rodlike molecules in a bilayer; B, the expected distribution of angles f(θ) made by the rods relative to the direction perpendicular to the bilayer surface.

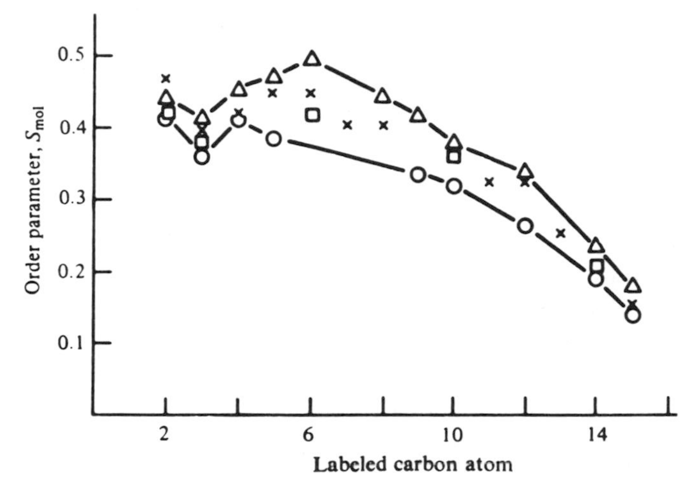

Fig. 10. Normalized order profiles of different bilayers. Variation of the molecular order parameter S_{mol} with the chain position. ○ is 1,2-dipalmitoylphosphatidylcholine; △ is 1-palmitoyl-2-oleoylphosphatidylcholine; □ is 1-2-dipalmitoylphosphatidylserine; x is *Acholeplasma laidlawi*. From Seelig and Browning (1978).

disordered system. This same formalism is used to describe the order of each C—H bond in the fatty acyl chain of a lipid in terms of the average angle between a particular C—H bond and the bilayer normal. The order parameter S_{CH} for a particular bond can be related to the molecular order parameter S_{mol} by (Seelig and Seelig, 1980):

$$S_{mol} = -2S_{CH}$$

If the chains are fixed in an all trans conformation and are rotating about the long molecular axis, then S_{mol} has the value 1. For movement through all angles, $S_{mol} = 0$.

The order parameter profiles for a number of bilayers in the liquid crystalline state, illustrated in Fig. 10, are remarkably similar. These parameters are compared at a fixed temperature of 18°C above the phase transition temperature of the respective lipid; this way of comparing lipids will be described later. In all of these lipids, order is fairly constant for the first half of the chain and then decreases rapidly toward the center of the bilayer.

The similarity of the order parameter profiles for the saturated and unsaturated lipids is particularly interesting. It shows that the order parameter is essentially determined by the distance from the lipid–water interface and not by the geometry of the chain. Seelig and Waespe-Šarčevic (1978) have described the hydrocarbon interior of the bilayer as a series of strata running parallel to the bilayer surface. Segments of the fatty acyl chains that are located in the same stratum are therefore characterized by the same segmental order parameter.

In addition to information about the time-averaged orientation of chain segments, we also need information about the rates of segmental motion. Unfortunately, order and mobility information are often combined in some general term such as fluidity. Order and rate are, in principle, distinct, and the correlation between chain order and chain mobility is not well understood. Interestingly, however, profiles of lipid mobility obtained using ^{13}C-NMR are rather similar to profiles of order parameter (Lee *et al.*, 1976). The least mobile part of the lipid molecule is the glycerol backbone. Motion increases in both directions from the backbone, towards the lipid head group and towards the methyl end of the fatty acyl chains. Further, motion is fairly constant for at least the first half of the fatty acyl chain but increases rapidly over the last few bonds.

As a result of the increased disorder of the chains, the thickness of a bilayer of phosphatidylcholine is about 20% smaller in the liquid–crystalline phase than in the gel phase; added to the 5% increase in the volume of the hydrocarbon region of the bilayer, this causes the area per molecule to increase approximately 25% at the transition.

The properties of different phospholipids in the liquid–crystalline state are difficult to compare because their phase transition temperatures are different. Is it appropriate to compare dipalmitoyl phosphatidylcholine with dimyristoyl phosphatidylcholine at 43°C, for example, when the former is only 1°C and the latter is 19°C, above their respective transition temperatures? Much data obtained for liquid crystals suggests that comparison of order parameters are best made by referring them to a reduced temperature. This is defined as

$$\theta = (T - T_c)/T_c$$

where θ is the reduced temperature, T is the temperature at which measurements are made, and T_c is the transition temperature. Thus at $\theta = 0$, the bilayer is at its phase transition temperature. The data of Seelig and Browning (see Fig. 10) show that order parameters for a variety of lipids are relatively independent of the lipid when calculated on a reduced temperature scale, which is usually referred to as the *principle of corresponding states*.

In contrast to order, it is not so clear how rates of motion should be compared. Watts *et al.* (1978) measured the fluidity of bilayers of phosphatidylglycerol by the amount of the small molecule Tempo that would dissolve in the bilayer. They observed that Tempo was more soluble in lipid in the charged than in the uncharged state when measured at some temperature above the phase transition temperature of both. When the solubility is referred to equal temperatures above the phase transition of that bilayer, then it seems the fluidities are equal.

Lipid Mixtures

The first question, when studying lipid mixtures, is, can the aqueous phase can be ignored? If the molecules are very hydrophobic, like lipids and membrane proteins, the amounts freely dissolved in the aqueous phase will be negligible; the aqueous phase need not be taken into account as a separate phase. However, if one or more components of the mixture are appreciably soluble in water, the aqueous phase must be explicitly considered, usually in terms of a partition between the aqueous and lipid phases.

For components within the membrane phase, the question is whether or not certain of the components will aggregate or remain dispersed within the membrane. This will be decided by the energy–entropy balance. In any mixture, the entropy of mixing always favors randomization of the components. However, steric hindrances and charge–charge interactions can cause association or segregation; thus, fluid lipid-like components should mix well with lipids. In contrast, relatively rigid molecules, steroids or proteins, would be expected to mix poorly with fluid lipids, and packing of sterols and proteins should be particularly poor. Packing of dissimilar species is also likely to be considerably different in the gel and liquid–crystalline phases. A rigid, "smooth" molecule could be expected to interact best with a rigid all trans chain and so favor the gel phase. Such a molecule will raise the transition temperature of the lipid, and a molecule that prefers the liquid–crystalline phase will correspondingly lower the transition temperature. For mixtures of lipid-like compounds, phase diagrams can be established that usefully describe many of the properties of the mixtures, as is described later. For components of very different properties, however, the concepts presented in the phase diagram are not particularly appropriate. For example, in a bilayer containing a large proportion of a relatively rigid molecule, the lipid molecules could become effectively isolated from each other. For such isolated lipids, any change in the rotational state of the chains with temperature will be gradual, and there will be no co-operative phase transition.

Mixtures with Water-Soluble Molecules

The thermodynamic properties of the bilayer described above can be used to derive information about mixtures of lipids with other components. Many small water-soluble compounds are soluble in lipid in the liquid–crystalline state but insoluble in lipid in the gel state. Such compounds will cause a decrease in the transition temperature of the lipid, described by the classical equation for the *depression of the freezing point*

$$\Delta T = (RT^2/\Delta H) \ln (1 - x), \tag{1}$$

where ΔT is the depression of the freezing point, ΔH is the enthalpy of the transition, and x is the mole fraction of the solute in the liquid–crystalline lipid. Knowing the aqueous concentration of the solute, a partition coefficient or binding constant can then be derived for the solute. For benzyl alcohol it has been shown that the above equation is accurate. From the decrease in transition temperature, a partition coefficient of 440 (mole fraction units) has been calculated. Assuming a volume of 550 cm^3/mol for the hydrocarbon region of the lipid, this is equivalent to 14.4 (molar units) (Ebihara *et al.*, 1979), and the directly measured value is 13.9 (molar units) (Katz and Diamond, 1974). For short chain *n*-alcohols agreement with theory is also good, but with increasing chain length up to *n*-octanol agreement becomes poor (Lee, 1977a). This has been attributed to an increasing solubility of the alcohol in the gel phase, which will result in a decreased effect on the transition temperature. In this case Eq. (1) can be replaced by

$$\Delta T = (RT^2/\Delta H) \ln\left[\frac{1 - x^l}{1 - x^g}\right], \tag{2}$$

where x^l and x^g are the mole fraction of solute in the liquid–crystalline and gel phases, respectively (Lee, 1977a). However, de Verteuil *et al.* (1981) have pointed out that since Eqs. (1) and (2) are valid only if the solute and the lipid form a random mixture, solute–solute interactions will also give behavior deviating from Eq. (1). It is likely that as chain length increases the interactions between the alcohols will strengthen, and their distribution within the lipid will become nonrandom. A consequence of Eq. 2 is that a species which partitions preferentially into the gel phase will increase the transition temperature.

For short-chain molecules, description of partitioning into the bilayer is relatively simple, because in general an insignificant amount of the small molecule will have partitioned into the membrane; the concentration of small molecule in the aqueous phase will be unchanged by the presence of lipid. For very hydrophobic molecules such as long-chain alcohols, the situation is also simple because all the molecules will be in the bilayer and the aqueous phase can be ignored (see page 76). The situation is more complex for molecules of intermediate hydrophobicity for which significant amounts will be present in both the lipid and water phases [(for examples, see Lee (1977a)].

Problems also arise for charged molecules, when it is no longer useful to describe binding by a partition coefficient. Binding of a charged molecule to the membrane will result in a buildup of charge on the membrane that will oppose further drug binding. These charge effects can be explained by the Gouy–Chapman theory (Lee, 1978b). Binding can be described best by a Langmuir adsorption isotherm of the type

$$\sigma^{S} = \left[\frac{1}{K}(\sigma^{\max} - \sigma^{S})\right][S]_{x=0}, \tag{3}$$

where σ^{S} is the number of molecules of solute adsorbed to the membrane per unit area, $\sigma^{\max}$ is the maximum number of solute molecules that can be adsorbed, K is the dissociation constant for binding, and $[S]_{x=0}$ is the concentration of solute at the surface of the membrane.

Many drug molecules have pK values close to physiological pH, and as a result they will exist as a mixture of charged and uncharged species. Both forms of the drug can bind to the membrane. In general, the uncharged form apparently binds more strongly than the charged form, presumably because the uncharged form can penetrate further into the hydrophobic interior of the membrane; this is equivalent to a change in pK for the drug on binding. The requisite equations (Lee, 1978b) utilize depression of transition temperatures to calculate binding constants for drugs to lipids. For those drugs studied by Lee (1978b), binding to phosphatidylcholines was found to be stronger than to phosphatidylethanolamines. It has been found that a number of small, general anesthetic molecules such as halothane, cyclopropane, nitrous oxide, etc., also decrease the transition temperatures of lipids by preferentially partitioning into lipid in the liquid–crystalline phase. The depression of transition temperature can be reversed by increasing pressure, as is predicted by the Clausius–Clapeyron equation. It is important to realize that this is a simple and classical effect of pressure; it does not mean that the increase in pressure "squeezes out" the anesthetic from the lipid–crystalline phase lipid (Mountcastle *et al.*, 1978; Kamaya *et al.*, 1979; MacNaughtan and MacDonald, 1980). Using the spin probe TEMPO, which is a local anaesthetic (Trudell *et al.*, 1973), it has been directly demonstrated that anesthetics are not displaced from the membrane by high pressure.

Most water-soluble drugs can be expected to bind to lipids in the liquid–crystalline phase with the hydrophilic groups of the drug in the head group region. Thus, for example, the aromatic ring of benzyl alcohol has been shown to affect the ^{1}H-NMR chemical shift of the NMe_3^+ protons of phosphatidylcholines (Colley and Metcalfe, 1972). Jain and Wu (1977) have shown that octan-1-ol has a greater effect on the transition properties of lipids than does octan-4-ol, which is also consistent with the alcohols anchored by the OH group in the lipid head group region. Despite earlier reports, Ebihara *et al.* (1979) demonstrated that benzyl alcohol has no effect on the thickness of lipid bilayers in the liquid–crystalline phase showing that insignificant amounts of benzyl alcohol could have penetrated to the middle of the bilayer. This contrasts to short chain alkanes, which do partition into the center of the bilayer (page 77).

Unfortunately, the concept of corresponding states cannot be applied in these systems. Many of the drugs that decrease transition temperature for the lipids, as has been described, also cause an increase in order and decrease in rates of motion (Pang and Miller, 1978; A. G. Lee, unpublished observations, 1981).

Mixtures of Lipids

Phase diagrams have been established for a variety of lipid mixtures. As shown in detail elsewhere, it is possible to compare the experimental phase diagrams with those calculated from the thermodynamic parameters of the component lipids (Lee, 1977b, 1978a). We will consider here only mixtures of two lipids in the presence of excess water. If the two component lipids are assumed to be completely immiscible in the gel phase, forming an ideal mixture in the liquid–crystalline phase, then the mole fraction x_A^{liq} of lipid A in the liquid–crystalline phase, in equilibrium with the gel phase at temperature T, is given by

$$\ln x_A^{liq} = \frac{H_A}{R}\left[\frac{1}{T_A} - \frac{1}{T}\right], \qquad (4)$$

where T_A and H_A are the transition temperatures and enthalpies of transition respectively for component A. This is the equation for the depression of the freezing point [Eq. (1)]. The corresponding mole fraction x_B^{liq}, of lipid B in the liquid–crystalline phase is

$$x_B^{liq} = 1 - x_A^{liq}. \qquad (5)$$

A phase diagram for a mixture of dilauroyl phosphatidylcholine and distearoyl phosphatidylcholine calculated in this manner is shown in Fig. 11.

If the components form ideal mixtures in both the liquid–crystalline and

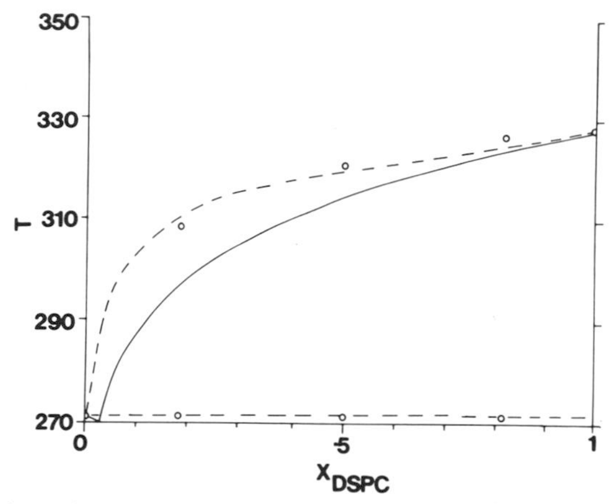

Fig. 11. The phase diagram for mixtures of dilauroyl phosphatidylcholine and distearoyl phosphatidylcholine. Circles are experimental points. Solid lines are phase diagrams calculated for ideal mixing in the liquid–crystalline phase. Broken lines are calculated for nonideal mixing in the liquid crystalline phase. From Lee (1978a).

the gel phases, then the equation for the mole fraction of A in the liquid–crystalline phase becomes

$$x_{\mathrm{A}}^{\mathrm{liq}} = \frac{1 - k_{\mathrm{B}}\exp(H_{\mathrm{B}}/RT)}{k_{\mathrm{A}}\exp(H_{\mathrm{A}}/RT) - k_{\mathrm{B}}\exp(H_{\mathrm{B}}/RT)}, \tag{6}$$

where

$$k_{\mathrm{A}} = 1/\exp(H_{\mathrm{A}}/RT_{\mathrm{A}}), \tag{7}$$

and

$$k_{\mathrm{B}} = 1/\exp(H_{\mathrm{B}}/RT_{\mathrm{B}}), \tag{8}$$

and T_{B} and H_{B} are, respectively, the temperature and enthalpy of the transition for lipid B. The mole fraction of lipid A in the gel phase is

$$x_{\mathrm{A}}^{\mathrm{gel}} = x_{\mathrm{A}}^{\mathrm{liq}}\, k_{\mathrm{A}}\, \exp(H_{\mathrm{A}}/RT). \tag{9}$$

A phase diagram calculated using these equations for mixtures of dimyristoyl phosphatidylcholine and dipalmitoyl phosphatidylcholine is shown in Fig. 12.

It is clear that two dissimilar lipids, although they might mix reasonably well in the liquid–crystalline phase, will be practically immiscible in the gel phase. Wilkinson and Nagle (1979) have demonstrated that dimyristoyl phosphatidylcholine and distearoyl phosphatidylcholine are almost immiscible in the gel phase and that dimyristoyl phosphatidylcholine and di-

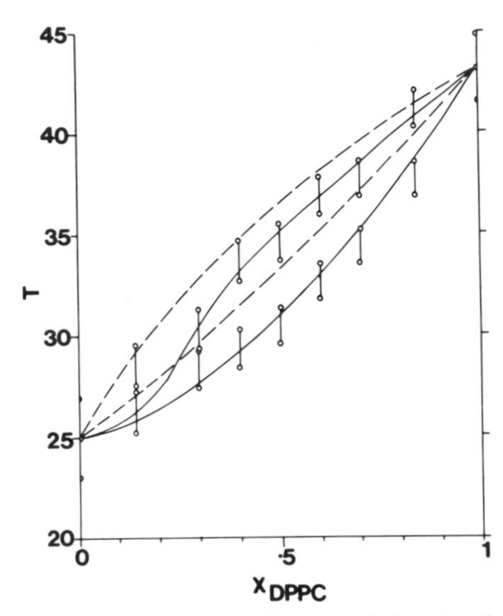

Fig. 12. The phase diagram for mixtures of dimyristoyl phosphatidylcholine and dipalmitoyl phosphatidylcholine. Circles are experimental points. Solid lines are phase diagrams calculated for ideal mixing in the liquid–crystalline and gel phases; broken lines are calculated for nonideal mixing in both phases. From Lee (1978a).

eicosanoyl phosphatidylcholine (C_{20} fatty acyl chains) are completely immiscible.

Figures 11 and 12 also demonstrate that mixing is not ideal, even in the liquid–crystalline phase. This non-ideality can be conveniently described by regular solution theory, which assumes that the entropy of mixing approaches ideal behavior and that the nonideality is caused by a nonzero enthalpy of mixing.

For two component lipids that are immiscible in the gel phase, Eq. (4) can be solved to give T^{ideal}, the temperature at which a mixture containing a mole fraction of x_A would have melted if behavior had been ideal. The temperature at which melting actually occurs for a nonideal mixture is given by (Lee, 1977b, 1978a)

$$P_0\left[\frac{(1 - x_A^{\text{liq}})^2}{H_A}\right] = \left[\frac{T}{T^{\text{ideal}}}\right] - 1 \tag{10}$$

where P_0 is a constant characterizing mixing behavior in the liquid–crystalline phase. Figure 11 illustrates that a good fit to the experimental phase diagram for mixtures of dilauroyl phosphatidylcholine and distearoyl phosphatidylcholine can be obtained in this way.

When mixing occurs nonideally in both liquid–crystalline and gel phase, then the necessary equations become

$$\ln\left[\frac{x_A^{liq}}{x_A^{gel}}\right] + \frac{P_0^{liq}(1 - x_A^{liq})^2 - P_0^{gel}(1 - x_A^{gel})^2}{RT} = \frac{H_A}{R}\left[\frac{1}{T_A} - \frac{1}{T}\right] \tag{11}$$

and

$$\ln\left[\frac{1 - x_A^{liq}}{1 - x_A^{gel}}\right] + \frac{P_0^{liq}(x_A^{liq})^2 - P_0^{gel}(x_A^{gel})^2}{RT} = \frac{H_B}{R}\left[\frac{1}{T_B} - \frac{1}{T}\right] \tag{12}$$

Because of their transcendental nature, Eqs. (11) and (12) have to be solved numerically. As shown in Fig. 12, it is possible to use these equations to obtain good fits to the experimental data.

The magnitudes of the P_0 values show that mixing is less ideal in the gel phase than in the liquid–crystalline phase, and it becomes less ideal the more the two lipids differ either in their fatty acyl chains or in their head groups. The nonideality is such that like lipids are more apt to be neighbors than are unlike lipids. The tendency of like lipids to cluster next to each other in the bilayer has been calculated in terms of athermal mixing theory, but this tends to overestimate the effect (Lee, 1978a). Von Dreele (1978) used an alternative approach and Freire and Snyder (1980) statistical modeling to interpret the nonideality of mixing; they came to very similar conclusions. Cheng (1980) has shown that it is possible to reduce the two nonideality parameters in Eqs. (11) and (12) to one when the change in area for the lipid at the phase transition is known.

The use of phase diagrams such as those shown in Figs. 11 and 12 to describe the compositions of the various lipid phases is described in Lee (1977b) (and others). It is perhaps worth emphasizing here, however, that although the phase transition for these mixtures occurs over a relatively broad temperature range, the transition is still first order. This can readily be seen, because the phase diagrams for these lipid mixtures have a geometric relation to the liquid–vapor phase diagrams for liquid mixtures. The liquid–vapor transition is undoubtedly first order, and the gel–liquid crystalline phase transition of a binary mixture is therefore also first order.

The phase diagrams presented above have been involved with the main transition only. It is also possible to derive phase diagrams for the pretransition region (Luna and McConnell, 1978).

The presence of Ca^{2+} causes a considerable enhancement of phase separations in mixtures containing negatively charged lipid, presumably as a consequence of the formation of rigid clusters of lipids with Ca^{2+} "bridging" between head groups. Phase diagrams have been drawn for mixtures of phosphatidylserine with phosphatidylcholine and phosphatidylethanolamine in the presence of Ca^{2+} (Tokutomi *et al.*, 1981).

Mixtures with Cholesterol

The studies of Franks (1976) and Worcester and Franks (1976) have established that cholesterol inserts into phosphatidylcholine bilayers with its OH group in the vicinity of the ester carbonyl groups of the lipid. The presence of the rigid sterol frame restricts flexing motions of neighboring fatty acyl chain regions in the liquid–crystalline phase. The ordering effect is largely restricted to the sterol backbone, because the hydrocarbon chain of cholesterol and the adjacent segments of neighboring fatty acyl chains undergo considerable flexing motions. In contrast, in the gel phase cholesterol increases chain flexibility by inhibiting good packing of the fatty acyl chains. For this reason it has been said to create a state of intermediate fluidity. Addition of cholesterol has little effect on the orientation of the head groups in phosphatidylcholines or phosphatidylethanolamines (Brown and Seelig, 1978).

Despite the good agreement about the effects of cholesterol on phospholipids, the phase diagram for the phospholipid–cholesterol system has not been determined. However, it is certain that mixtures of lipids and cholesterol will not be random, that is, they will not be ideal as has sometimes been assumed. Estep *et al.* (1978) and Mabrey *et al.* (1978) have studied the effects of cholesterol on the thermodynamics of the phase transition of dipalmitoyl phosphatidylcholine. With increasing cholesterol up to a mole fraction of 0.25, there is a decrease in the enthalpy of the main transition of dipalmitoyl phosphatidylcholine, which appears as a sharp component in the calorimetric scans. At a mole fraction of 0.25, this sharp component disappears. Simultaneously, a broad component appears in the scans, reaching a maximum intensity at the mole fraction of 0.25; after this point, it disappears. These experiments agree with those of Copeland and McConnell (1980) who by use of freeze-fracture microscopy observed that addition of cholesterol to a mole fraction of 0.2 causes an increase in repeat distance for the ripples seen when lipid is frozen from just below the main transition. Beyond the mole fraction of 0.2, the ripples disappear. Further, below a mole fraction of 0.2 cholesterol there is no sign of two phases (one with a normal ripple repeat distance and one with an abnormal one). Rather, a gradual increase in ripple repeat distance over the whole sample occurs. This suggests there is no simple microscopic phase separation into different phases of 0 and 0.2 mole fraction cholesterol. Instead, as was suggested by Copeland and McConnell, the adopted structure might be one in which thin strips of smooth 0.2 mole fraction cholesterol phase are separated by strips of pure phosphatidylcholine phase. Such a strip structure has been demonstrated in dark-field electron micrographs of lipid-cholesterol mixtures (Hui and Parsons, 1975). These structures would be in good agreement with the results (Rogers *et al.*, 1979) obtained using fluorescent cholesterol analogs, which suggested that sterols adopt a highly structured, nonrandom distribution in the lipid bilayer; the cholesterol molecules are essentially lined up in rows. Using molecular models, a large number of such structures are possible. Figure 13 shows one such structure, which is particularly appealing as it is possible to simply incorporate increasing amounts of lipid between the rows of cholesterol molecules to change the mole fraction of cholesterol from 0.5 to 0.33, to 0.25, and so on. It is clear that in the structure containing a mole fraction of 0.5 cholesterol the lipid molecules are effectively isolated in rows; thus, they will not be able to undergo a normal phase transition. Similarly, at 0.33 cholesterol, there are no lipid molecules in the structure solely surrounded by other lipids and thus no normal phase transitions will be present. Only with four "rows" of lipid present at a mole fraction of 0.2 could any lipid present be considered to be in a normal environment, and so it would be expected that at about this point a normal transition might appear. However, to what extent this model might be true is not clear.

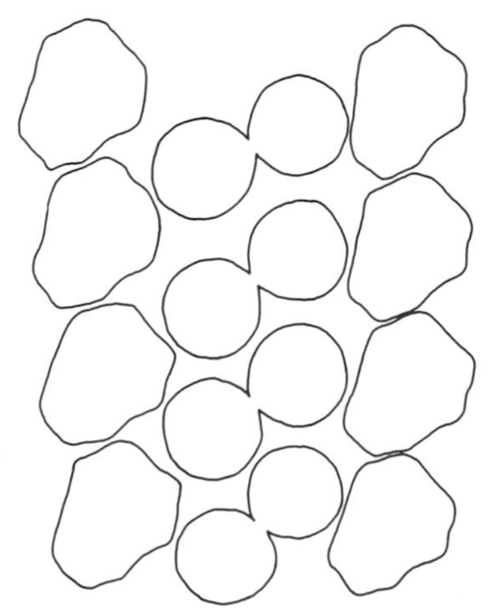

Fig. 13. Possible packing of cholesterol and lipid.

Mixtures of cholesterol with sphingomyelin (Estep *et al.*, 1979) and phosphatidylethanolamine (Blume, 1980) seem rather similar to those with phosphatidylcholines. Van Dijck (1979) has reported, however, that in mixtures of lipids cholesterol associates preferentially with the negatively charged lipid and prefers phosphatidylcholines to phosphatidylethanolamines. Blume (1980) could find no clear differences between the affinities of cholesterol for phosphatidylcholines and phosphatidylethanolamines, however.

Even less is known about the mixing of other sterols with lipids. Blume stressed the importance of a planar sterol α face for good mixing with lipids. He points out that lanosterol (a trimethylcholestane precursor of cholesterol) is synthesised in mammalian cells by the oxidative cyclization of squalene, which then undergoes three sequential demethylation steps to cholesterol; lanosterol itself does not normally accumulate as an end product in the cell (Dahl *et al.*, 1980). Certainly, the mixing properties of sterols and lipids seem to be dependent on the detailed three-dimensional structure of the sterol.

Mixtures with Other Hydrophobic Compounds

Although many hydrophobic molecules could be expected to be taken up by lipid bilayers, the effects of such compounds on the bilayer will be determined by the extent to which they mix with the lipids. Compounds very dissimilar in structure to lipids would be expected to mix poorly with the lipids and so to form essentially a separate phase; such compounds would have little effect on the properties of the lipid phase. They could, however, have very important biological effects. Thus it could be that a membrane protein would preferentially partition into such nonlipid regions, and consequently the hydrophobic species could have a considerable effect at a low molar ratio in the membrane.

The carotenoids are a good example. Below the lipid transition temperature, all the carotenoids are aggregated into a separate phase. In the liquid–crystalline phase, however, β-cryptoxanthin and β-zeaxanthin disperse and form a mixed phase with the lipid. These carotenoids carry OH groups which ensure some degree of mixing with the lipids. β-Carotene, on the other hand, has no polar groups and remains aggregated even in the liquid–crystalline phase (Yamamoto and Bangham, 1978; Mendelsohn and Van Holten, 1979). A similar example is provided by the cholesterol esters. Whereas cholesterol is miscible with lipids up to at least a molar ratio of 1:1, cholesterol esters are only miscible with lipid up to a few per cent beyond which a separate phase of solid ester separates out. The ester in the lipid phase is believed to adopt a "horseshoe" configuration with the ester near the aqueous surface, and the acyl chain and the cholesterol group extending into the bilayer (Gorrissen *et al.*, 1980).

Long chain alcohols and acids lead to increases in the phase transition temperatures for lipids (Lee, 1976a; Eliasz *et al.*, 1976; Mabrey and Sturtevant, 1978). Phase diagrams for these mixtures appear similar to those for mixtures of fairly miscible lipids (Lee, 1977b). However, it is not possible to calculate the expected form of such diagrams because there is no way to obtain data for the hypothetical gel-to-liquid–crystalline bilayer transition for the fatty acids or alcohols. Instead, if it is possible to assume that the mixed bilayer phase is nealy ideal, Eqs. (11) and (12) can be used "backwards" to obtain the enthalpy and temperature of the bilayer transition.

This interpretation is supported by experiments with hexadecanols (Pringle and Miller, 1979). Hexadecanol itself causes an increase in transition temperature for dipalmitoyl phosphatidylcholine. *Trans*-hexadecanol has relatively little effect, except at high concentration, and *cis*-hexadecanol causes a decrease in transition temperature. As observed for lipids, the cis double bond is more disruptive of packing than is the trans double bond, and van der Waals interactions are strongest for the saturated chain. It has been suggested that fatty acids are not dispersed in these lipid mixtures, but form some type of cluster. Apparently this is not so, however (Ptak *et al.*, 1980; E. Rooney and A. G. Lee, unpublished observations, 1981).

Short chain alkanes, although they partition into the bilayer, appear to form an essentially separate phase in the middle of the bilayer. This results in an increase in bilayer thickness (McIntosh *et al.*, 1980). Alkanes with C_{12} chains and longer, however, line up parallel to the lipid fatty acyl chains and have no effect on bilayer thickness. It was suggested that hexadecane at more than about a molar fraction of 0.14 in the gel phase formed a separate phase (McIntosh *et al.*, 1980).

LIPID–PROTEIN MIXTURES

It is certain, at least, that mixing of lipid and protein will not be ideal, (random). If the mixing of dimyristoyl phosphatidylcholine with dipalmitoyl phosphatidylcholine is nonideal, the mixing of dimyristoyl phosphatidylcholine with the (Ca^{2+}, Mg^{2+})-ATPase cannot be expected to be ideal. The questions that should concern us are the details of the nonidealities in lipid–protein mixtures. Are protein–protein contacts favored or are they avoided? As a general rule, it might be expected that they would be avoided as much as possible. Most membrane proteins are negatively charged, and electrostatic repulsion will tend to keep them apart. Models for some membrane proteins suggest that the hydrophilic regions of the proteins cover an area larger than that of the hydrophobic portion actually in the membrane. Even if the hydrophilic regions were in contact, therefore, the hydrophobic regions would not be; in terms of mixing within the lipid phase of the membrane, this would be equivalent to the suppression of protein–protein contact. Finally, one could argue that membrane proteins do aggregate when totally delipidated, but that this aggregation usually leads to irreversible denaturation. Would this be so common an occurrence if protein–protein contacts were generally favored in the membrane? Of course some protein–protein contacts will occur in the membrane, but if they are extensive, the interaction probably has a functional significance.

From this type of reasoning, a concept of a lipid shell or annulus surrounding the protein in a membrane was derived (see Lee, 1977b). The size of the annulus will depend on the size of the protein. The number of lipid molecules in instantaneous contact with a rhodopsin monomer in the membrane is roughly 25 (Jost and Griffith, 1980) and for the (Ca^{2+}, Mg^{2+})-ATPase it is about 30 (Warren *et al.*, 1974).

It is clear that lipid molecules adjacent to a protein will experience an environment different from that of lipids surrounded only by other lipids, and they will exhibit different motional properties. One question concerns the length of time that lipid molecules are adjacent to a membrane protein. If the lipid–protein interaction is long-lived, then it is logical to consider the protein and its surrounding shell or annulus of lipid as a separate component within the membrane. If on the other hand the lipid–protein interaction is short-lived and no particular lipid remains next to the protein for a long time, then the components in the membrane are simply lipid and protein. In fact, it appears that the lipid–protein interaction is usually quite short-lived. ESR experiments using spin-labeled lipids showed two kinds of environments in a variety of lipid–protein systems. One component of the spectra was typical of a normal lipid bilayer and the other suggested an immobilized environment as expected for the protein–lipid interface. This second immobilized component increased in intensity with increasing protein content (Jost and Griffith, 1980). A rather different picture has emerged from NMR studies. Spectra for deuterated lipids in a variety of lipid–protein systems yield evidence for only one component in the liquid–crystalline phase (Seelig and Seelig, 1980; Jacobs and Oldfield, 1981). The obvious explanation for these results lies in the different time scales of the ESR and NMR experiments. That of the ESR experiment is short, and if the rate of exchange of lipid between the bulk lipid phase and the lipid–protein interface were slower than approximately 10^7 Hz, a two-component spectrum would be observed. The NMR experiment has a much longer time scale, and two components could be observed only if the exchange rate were slower than approximately 10^4 Hz. A rate of exchange for the lipid between the two environments of 10^4–10^7 Hz would therefore be consistent with the observations. Experiments in which spin-labeled fatty acids were attached covalently to membrane proteins were originally considered to be in disagreement with this simple model, but this is no longer believed (Watts *et al.*, 1981).

It has been suggested, in the literature, that the concept of a lipid annulus is only meaningful for biochemical events if the lifetime of lipids in the annulus is long in comparison with the turnover number of the protein, which is about 10^{-1} sec for the (Ca^{2+}, Mg^{2+})-ATPase, but this argument is fallacious. It does not matter which particular lipid is in the annulus; it is only significant that the lipid that *is* in the annulus be immobilized. Oil flowing past the piston of a car is an effective lubricant, irrespective of the flow rate relative to the speed of movement of the piston.

ESR results suggest that the lipid next to the protein is *immobilized*. This term is used to indicate that the ESR spectrum is as would be expected for a powder, that is, a random array of spin labels moving only slowly. In other words, the annular lipid is not as well-oriented as is the bulk lipid, and it can occupy a broader distribution of orientations; it is disordered. At the same time, it is not totally immobile, although motion is highly restricted on the ESR time scale compared to the bulk lipid. Thus the annular lipid is both disordered spatially and restricted motionally. The studies of Stoffel *et al.* (1977) on the ($Ca^{2+}Mg^{2+}$)-ATPase using ^{13}C NMR are in agreement with a considerable immobilization of the fatty acyl chains by the protein. Deuterium NMR studies suggest that although annular lipid is moving more slowly than bulk lipid, the amplitude of the motion is the same or even larger (Kang *et al.*, 1979).

All these data can be interpreted fairly simply. The protein surface presented to the lipids is both rough and fluctuating. The fluorescence quenching experiments of Lakowicz and Weber (1973) show that proteins undergo considerable, large amplitude "breathing" motions. NMR studies show that residues on the outside surfaces of proteins have considerable mobility on a time scale of the order of 10^{-5} sec (Williams, 1978). X-ray data (Frauenfelder *et al.*, 1979; Artymiuk *et al.*, 1979) and theoretical calculations (McCammon *et al.*, 1977) also indicate considerable thermal motion for proteins.

Thus, the annular lipid "sees" a rough protein surface studded with projecting amino acid side chains, which is undergoing both localized density fluctuations and collective motions involving large parts of the protein. Lipid interacting with such a surface will therefore be disordered, but it will not be totally immobilized because fluctuations on the surface of the protein could be matched with fluctuations of the lipid fatty acyl chains.

In general, studies of lipid–protein interaction by both ^{2}H and ^{31}P NMR are complicated by the sensitivity of the spectra to any changes in the actual size of the lipid–protein vesicles or aggregates formed in the system (Burnell *et al.*, 1980). This problem has been overcome by McLaughlin *et al.* (1981), in their study of the (Ca^{2+},Mg^{2+})-ATPase, by using oriented multilayers. ^{31}P NMR spectra of this system show no evidence for immobilization of the head group of phosphatidylcholine by the protein, and the structure adopted appears to be identical to that in a normal lipid bilayer. It would be expected that because lipid adjacent to protein is in an unusual state of packing it would not be able to undergo the normal lipid phase transition. This is indeed so, as has been demonstrated by Gomez-Fernandez *et al.* (1979) for the dimyristoyl phosphatidylcholine–(Ca^{2+},Mg^{2+})-ATPase system.

It has been suggested that the disappearance of the lipid phase transition is caused by lipid which, at high protein molar ratios, is trapped in regions between the proteins and consequently is unable to undergo a normal transition (Chapman *et al.*, 1977; Cornell *et al.*, 1978). If these patches of trapped lipid occurred with a constant amount of trapped lipid per protein, then it would not be possible to distinguish it by calorimetric measurements from the annular lipid model. However, physically trapped lipid could be expected to vary in amount depending on the way in which protein molecules happened to come together during the heating or cooling process. It is therefore surprising that the decrease in enthalpy of the lipid transition caused by the protein lipophilin should be equal in heating and cooling scans (Boggs *et al.*, 1980).

The distribution of lipid and protein in the gel phase is clearly different from that in the liquid–crystalline phase. In freeze-fracture electron micrographs, patches of gel phase lipid and patches enriched with protein are observed (Kleeman and McConnell, 1976). However, the extent of any protein–protein interaction in these patches is unknown.

It is known that the activities of membrane proteins are sensitive to the physical state of the surrounding lipid. Thus, the (Ca^{2+},Mg^{2+})-ATPase reconstituted with dipalmitoyl phosphatidylcholine shows very little activity at low temperature when the surrounding lipid is in a nonfluid, gel-like state but shows a high activity in the fluid liquid–crystalline state (Warren *et al.*, 1974). This could be a simple consequence of the change in state of the surrounding lipid, or it could be a secondary consequence of, for example, protein aggregation following from the change in state of the lipid. Although no definitive evidence is available, the former seems most probable. Certainly, large-scale fluctuations in structure are probably significant to the activity of all proteins, including membrane proteins. If the surrounding lipid environment becomes highly viscous, such fluctuations will be severely damped and the activity will suffer. A second argument against large changes in aggregation at the transition is based on the studies of London and Feigenson (1981) on the lipid specificity of the (Ca^{2+},Mg^{2+})-ATPase, using a fluorescence quenching technique to study the relative binding affinities of the ATPase for different lipids. Interestingly, the preference for liquid–crystalline lipid over gel lipid was a factor of only two. Any major change in the state of aggregation of the protein should result in large changes in affinities for the corresponding lipid. However, London and Feigenson showed only relatively small differences in the affinity of the ATPase for different lipids. The interaction seemed relatively independent of chain length and lipid head group, although the interaction with phosphatidic acid in the presence of Ca^{2+} was markedly less favorable.

Detailed phase diagrams have not yet been delineated for lipid–protein systems and, indeed, it is not at all clear that phase diagrams would be the most suitable way of describing such systems. In particular, the transition as seen by the protein is very broad (Lee, 1977b). The (Ca^{2+},Mg^{2+})-ATPase reconstituted with dipalmitoyl phosphatidylcholine shows a broad transition from about 40 to 30°C (Warren *et al.*, 1974). If this change in activity corresponds solely to a change in the annular lipid, then it seems that the change in physical state for the annular lipid is continuous, rather than first-order and highly cooperative. Such a continuous change would be expected if the annular lipid were essentially isolated from other lipids.

Addendum

Since this chapter was written, we have completed a number of studies which provide new insight on the nature of the lipid–protein interaction in the ($Ca^{2+}$$Mg^{2+}$)-ATPase system. Using a fluorescence quenching technique similar to that of London and Feigenson (1981), it has been possible to measure relative binding constants for a variety of phospholipids to the ATPase (East and Lee, 1982). These studies have shown that binding constants of phospholipids to the ATPase are relatively independent of the lengths of the fatty acyl chains of the phospholipid. Thus, the binding constant for dimyristoylphosphatidylcholine relative to that for dioleolyphosphatidylcholine at 37°C is 0.8, despite the fact that the former lipid supports a fourfold-lower activity for the ATPase. Similarly, binding constants for phosphatidylcholines, phosphatidylethanolamines, and phosphatidylserines are all very similar despite markedly different effects on ATPase activity. The only large differences in binding seen are between phospholipids in the liquid–crystalline and gel phases, where the binding constant for lipid in the liquid–crystalline phase is 30 times greater than that for lipid in the gel phase (this preference is considerably larger than that reported by London and Feigenson, 1981).

The low lipid selectivity of the ATPase argues against any specific lipid "binding sites" on the ATPase; binding sites analogous to substrate binding sites on enzymes would be expected to show a marked structural specificity. It is probably more realistic, therefore, to think of the lipid–protein interface of the ATPase as consisting of a hydrophobic surface, probably located between two relatively polar surfaces, with which the lipid interacts. This surface is, however, not without some specificity, and it now appears that hydrophobic compounds that mix well with phospholipids also bind well at this interface. Thus, cholesterol binds very weakly to the ATPase at the lipid–protein interface, with a binding constant relative to that of dioleoylphosphatidylcholine of less than 0.1. The relatively low binding constant for the rigid cholesterol molecule is probably related to the similarly weak binding of phospholipid in the gel phase. Other hydrophobic molecules such as fatty acids, however, bind to the lipid–protein interface rather strongly (Simmonds *et al.*, 1982).

In binding to the lipid–protein interface of the ATPase, a fatty acid or other hydrophobic molecule will displace a phospholipid from the surface of the protein, and it is necessary to consider the stoichiometry of this displacement. In ESR studies, a 1:1 stoichiometry is commonly assumed, with one fatty acid displacing one phospholipid. A more likely stoichiometry would, however, be two fatty acids displacing one phospholipid, thus conserving the area of the lipid–protein interface. Binding to the surface of the protein can be described by a series of displacement reactions of the type

$$PL_n + L^* \rightarrow PL^* + nL$$

Such a "site model" will probably overemphasize order on the protein surface, but it may be considered as analogous to the lattice theory of liquids. An equilibrium constant for the displacement can then be written as

$$K = [PL^*]\,[L_f]^n/[PL_n]\,[L_f^*],$$

where $[L_f]$ and $[L_f^*]$ are the concentrations of unbound L and L*, respectively, and $[PL_n]$ and $[PL^*]$ are the corresponding bound concentrations. McGhee and von Hippel (1974) have shown that to describe the binding of a multivalent ligand, the free site concentration must be multiplied by a probability factor, P_n, giving the probability that a free site will be followed by at least $n-1$ other free sites, thus allowing binding of a ligand of valence n. In this case, it is possible to derive a binding expression analogous to the Scatchard equation (A. G. Lee, unpublished observations, 1982)

$$L_b^* = K\left[\frac{L_f^*(L_f + L_f^*)^{n-1}}{nL_f^n}\right](NP - nL_b^*)P_n$$

with

$$Pn = \left\{(NP - nL_b^*)/[NP - (n-1)\ L_b^*]\right\}^{n-1}.$$

Here L_b and L_b^* are the concentrations of L and L*, respectively, bound to the protein, and L_f and L_f^* are the corresponding unbound concentrations. P is the protein concentration and N is the number of binding sites per protein. These equations are readily solved to describe the binding of mixtures of monovalent (fatty acids, alcohols, amines, etc.) and divalent (phospholipids) ligands to the protein surface.

Because of the relatively low selectivity of the protein–lipid interface, the activity of the ATPase will be very sensitive to the presence of many extraneous hydrophobic compounds. Many of these compounds (including drugs) will be able to bind at the protein–lipid interface, displacing phospholipids; because minor changes in the chemical structure of the phospholipids in the annulus around the protein cause large changes in activity, so will replacement of phospholipids by other hydrophobic compounds. It is clear that such effects have to be considered in terms of the composition of the annulus around the protein. Effects on the bulk phospholipid component have only minor effects on the ATPase. Thus, although cholesterol considerably reduces the fluidity of the bulk phospholipid in the membrane, it has little or no effect on the ATPase in bilayers of dioleoylphosphatidylcholine because it is excluded from the annulus around the protein (Simmonds *et al.*, 1982).

To summarize, the observed low selectivity of binding suggests a relative weak binding of phospholipids to the ATPase and a lipid interacting nonspecifically with a hydrophobic surface. The question then arises as to whether it is possible to maintain a lipid annulus in such a situation. Most published theories of lipid–protein mixtures have concentrated on the nature of the lipid–protein interaction. Because the effect of protein (in these theories) is to order the annular lipid, there is an entropy-driven tendency for protein molecules to aggregate and reduce the amount of annular phospholipid (Marčelja, 1976; Schroeder, 1977). A very different picture emerges, however, if the dominant interaction for the mixing of lipids and proteins is the protein–protein interaction.

To maintain lipid annulae around membrane proteins, the protein– protein interaction needs to be repulsive. Mixing can be described in the quasi-chemical approach of Guggenheim (1935), where the probability of a protein–lipid interaction is related to the energy differences W between protein–lipid (W_{PL}), protein–protein (W_{PP}) and lipid–lipid (W_{LL}) pairs

$$W = 2W_{PL} - W_{PP} - W_{LL}$$

Clearly, protein–lipid interaction can be favored either by a strong protein–lipid interaction or by a very unfavorable protein–protein interaction. As we have argued against a strong protein–lipid interaction, we would suggest that mixing is dominated by a very unfavorable random protein–protein interaction in the membrane that serves to keep proteins apart and thus to maintain lipid annulae. Of course, some specific protein–protein interactions will be highly favorable and thus lead to specific aggregation of the proteins. A very clear example is bacteriorhodopsin, which exists in the purple membrane as a trimer but with little or no random contact between the trimers (Hayward and Shroud, 1981). A model of this type is consistent with a recent deuterium NMR study of the dimyristoylphosphatidylcholine–rhodopsin system (Bienvenue *et al.*, 1982). At low temperatures, in the presence of excess lipid the spectra clearly showed the presence of bulk lipid in the gel phase and a temperature-insensitive component corresponding to 30 lipids per rhodopsin attributable to annular lipid. Thus, although the bulk of the lipid is in the gel phase, it appears that protein–protein interactions are sufficiently unfavorable to prevent extensive protein aggregation and to maintain a lipid annulus around each rhodopsin molecule (this interpretation of the data is not exactly that proposed by Bienvenue *et al.*, 1982). The annular lipid is apparently unable to take part in a normal lipid phase transition, as expected for the model proposed above. When the bulk phospholipid transforms into the liquid–crystalline phase, exchange between the annular and the bulk lipid increases in rate, and separate components are no longer observed in the deuterium NMR spectrum.

Finally, it should be noted that protein–protein interfaces in protein oligomers provide a possible set of hydrophobic binding sites, distinct from sites at the lipid–protein interface. We have found evidence for a set of non-annular binding sites for fatty acids and sterols on the (Ca^{2+},Mg^{2+})-ATPase and have suggested that these could be at protein–protein interfaces in ATPase oligomers in the membrane (Lee *et al.*, 1982; Simmonds *et al.*, 1982).

References

Abrahamsson, S., Pascher, I., Larsson, K., and Karlsson, K. A. (1972). *Chem. Phys. Lipids* **8**, 152–179.

Albon, N., and Sturtevant, J. M. (1978). *Proc. Natl. Acad. Sci. U.S.A.* **75**, 2258–2260.

Artymiuk, P. J., Blake, C. C. F., Grace, D. E. P., Oatley, S. J., Phillips, D. C., and Sternberg, M. J. E. (1979). *Nature (London)* **280**, 563–568.

Bach, D., Bursuker, I., Eibl, H., and Miller, I. R. (1978). *Biochim. Biophys. Acta* **514,** 310–319.

Barenholz, Y., Suurkuusk, J., Mountcastle, D., Thompson, T. E., and Biltonen, R. L. (1976). *Biochemistry* **15,** 2441–2447.

Barton, P. G., and Gunstone, F. D. (1975). *J. Biol. Chem.* **250,** 4470–4476.

Berde, C. B., Andersen, H. C., and Hudson, B. S. (1980). *Biochemistry* **19,** 4279–4293.

Bienvenue, A., Bloom, M., Davis, J. H., and Devaux, P. F. (1982). *J. Biol. Chem.* **257,** 3032–3038.

Blaurock, A. E., and Gamble, R. C. (1979). *J. Membr. Biol.* **50,** 187–204.

Blume, A. (1980). *Biochemistry* **19,** 4908–4913.

Boggs, J. M., Clement, I. R., and Moscarello, M. A. (1980). *Biochim. Biophys. Acta* **601,** 134–151.

Brown, M. F., and Seelig, J. (1978). *Biochemistry* **17,** 381–384.

Browning, J., and Seelig, J. (1980). *Biochemistry* **19,** 1262–1270.

Buldit, G., Gally, H. U., Seelig, A., Seelig, J., and Zaccai, G. (1978). *Nature (London)* **271,** 182–184.

Buldit, G., Gally, H. U., Seelig, J., and Zaccai, G. (1979). *J. Mol. Biol.* **134,** 673–691.

Burnell, E. E., Cullis, P. R., and de Kruijff, B. (1980). *Biochim. Biophys. Acta* **603,** 63–69.

Cameron, D. G., Casal, H. L., Gudgin, E. F., and Mantsch, H. H. (1980). *Biochim. Biophys. Acta* **596,** 463–467.

Cevc, G., Watts, A., and Marsh, D. (1980). FEBS Lett. **120,** 267–270.

Chapman, D., Cornell, B. A., Eliasz, A. W., and Perry, A. (1977). *J. Mol. Biol.* **113,** 517–538.

Chen, S. C., Sturtevant, J. M., and Gaffney, B. J. (1980). *Proc. Natl. Acad. Sci. U.S.A.* **77,** 5060–5063.

Cheng, W. H. (1980). *Biochim. Biophys. Acta* **600,** 358–366.

Colley, M., and Metcalfe, J. C. (1972). FEBS *Lett.* **24,** 241–246.

Copeland, B. R., and McConnell, H. M. (1980). *Biochim. Biophys. Acta* **599,** 95–109.

Cornell, B. A., Sacré, M. M., Peel, W. E., and Chapman, D. (1978). FEBS *Lett.* **90,** 29–35.

Cullis, P. R., and de Kruijff, B. (1979). *Biochim. Biophys. Acta* **559,** 399–420.

Dahl, C. E., Dahl, J. S., and Bloch, K. (1980). *Biochemistry* **19,** 1462–1467.

Davis, J. H. (1979). *Biophys. J.* **27,** 339–358.

de Kruijff, B., Verkley, A. J., Van Echteld, C. J. A., Gerritsen, W. J., Mombers, C., Noordam, P. C., and de Gier, J. (1979). *Biochim. Biophys. Acta* **555,** 200–209.

de Verteuil, F., Pink, D. A., Vadas, E. B., and Zuckermann, M. J. (1981). *Biochim. Biophys. Acta* **640,** 207–222.

East, J. M., and Lee, A. G. (1982). *Biochemistry* **21,** 4144–4151.

Ebihara, L., Hall, J. E., MacDonald, R. C., McIntosh, T. J., and Simon, S. A. (1979). *Biophys. J.* **28,** 185–196.

Eibl, H., and Woolley, P. (1979). *Biophys. Chem.* **10,** 261–271.

Eliasz, A. W., Chapman, D., and Ewing, D. F. (1976). *Biochim. Biophys. Acta* **448,** 220–230.

Estep, T. N., Mountcastle, D. B., Biltonen, R. L., and Thompson, T. E. (1978). *Biochemistry* **17,** 1984–1989.

Estep, T. N., Mountcastle, D. B., Barenholz, Y., Biltonen, R. L., and Thompson, T. E. (1979). *Biochemistry* **18,** 2112–2117.

Forsyth, P. A., Marčelja, S., Mitchell, D. J., and Ninham, B. W. (1977). *Biochim. Biophys. Acta* **469,** 335–344.

Franks, N. P. (1976). *J. Mol. Biol.* **100,** 345–358.

Frauenfelder, H., Petsko, G. A., and Tsernoglou, D. (1979). *Nature (London)* **280,** 558–563.

Freire, E., and Snyder, B. (1980). *Biochemistry* **19,** 88–94.

Gaber, B. P., Yager, P., and Peticolas, W. L. (1978). *Biophys. J.* **21,** 161–176.

Gomez-Fernandez, J. C., Goni, F. M., Bach, D., Restall, C., and Chapman, D. (1979). FEBS *Lett.* **98,** 224–228.
Gorrisen, H., Talloch, A. P., and Cushley, R. J. (1980). *Biochemistry* **19,** 3422–3429.
Guggenheim, E. A. (1935). *Proc. R. Soc. London, Ser. A* **148,** 304–312.
Harlos, K. (1978). *Biochim. Biophys. Acta* **511,** 348–355.
Harlos, K., Stumpel, J., and Eibl, H. (1979). *Biochim. Biophys. Acta* **555,** 409–416.
Hauser, H., Pascher, I., and Sundell, S. (1980). *J. Mol. Biol.* **137,** 249–264.
Hayward, S. B., and Stroud, R. M. (1981). *J. Mol. Biol.* **151,** 491–517.
Hegner, D., Schummer, U., and Schnepel, G. H. (1973). *Biochim. Biophys. Acta* **307,** 452–458.
Hendrickson, H. S., and Fullington, J. G. (1965). *Biochemistry* **4,** 1599–1605.
Hinz, H. J., and Sturtevant, J. M. (1972). *J. Biol. Chem.* **247,** 6071–6075.
Hitchcock, P. B., Mason, R., Thomas, K. M., and Shipley, G. G. (1974). *Proc. Natl. Acad. Sci. U.S.A.* **71,** 3036–3040.
Howard, R. E., and Burton, R. M. (1964). *Biochim. Biophys. Acta* **84,** 435–440.
Hui, S. W., and Parsons, D. F. (1975). *Science* **190,** 383–384.
Hui, S. W., Stewart, T. P., and Yeagle, P. L. (1980). *Biochim. Biophys. Acta* **601,** 271–281.
Hui, S. W., Stewart, T. P., Yeagle, P. L., and Albert, A. D. (1981). *Arch. Biochem. Biophys.* **207,** 227–240.
Jacobs, R. E., and Oldfield, E. (1981). *Prog. NMR Spectrosc.* **14,** 113–136.
Jacobson, K., and Papahadjopoulos, D. (1975). *Biochemistry* **14,** 152–161.
Jähnig, F., Harlos, K., Vogel, H., and Eibl, H. (1979). *Biochemistry* **18,** 1459–1468.
Jain, M. K., and Wu, N. M. (1977). *J. Membr. Biol.* **34,** 157–201.
Janiak, M. J., Small, D. M., and Shipley, G. G. (1976). *Biochemistry* **15,** 4575–4580.
Janiak, M. J., Small, D. M., and Shipley, G. G. (1979). *J. Biol. Chem.* **254,** 6068–6078.
Jost, P. C., and Griffith, O. H. (1980). *Ann. N. Y. Acad. Sci.* **348,** 391–405.
Kamaya, H., Ueda, I., Moore, P. S., and Eyring, H. (1979). *Biochim. Biophys. Acta* **550,** 131–137.
Kang, S. Y., Gutowsky, H. S., Hsung, J. C., Jacobs, R., King, T. E., Rice, O., and Oldfield, E. (1979). *Biochemistry* **18,** 3257–3267.
Katz, Y., and Diamond, J. M. (1974). *J. Membr. Biol.* **17,** 101–120.
Keough, K. M. W., and Davis, P. J. (1979). *Biochemistry* **18,** 1453–1459.
Khare, R. S., and Worthington, C. R. (1978). *Biochim. Biophys. Acta* **514,** 239–254.
Kleemann, W., and McConnell, H. M. (1976). *Biochim. Biophys. Acta* **419,** 206–222.
Ladbrooke, B. D., and Chapman, D. (1969). *Chem. Phys. Lipids* **3,** 304–356.
Lakowicz, J. R., and Weber, G. (1973). *Biochemistry* **12,** 4171–4179.
Lee, A. G. (1975). *Prog. Biophys. Mol. Biol.* **29,** 3–56.
Lee, A. G. (1976a). *Biochemistry* **15,** 2448–2454.
Lee, A. G. (1977a). *Biochemistry* **16,** 835–840.
Lee, A. G. (1977b). *Biochim. Biophys. Acta* **472,** 237–344.
Lee, A. G. (1978a). *Biochim. Biophys. Acta* **507,** 433–444.
Lee, A. G. (1978b). *Biochim. Biophys. Acta* **514,** 95–104.
Lee, A. G., Birdsall, N. J. M., Metcalfe, J. C., Warren, G. B., and Roberts, G. C. K. (1976). *Proc. R. Soc. London, Ser. B* **193,** 253–274.
Lee, A. G., East. J. M., Jones, O. T., McWhirter, J., Rooney, E. K., and Simmonds, A. C. (1982). *Biochemistry* **21,** 6441–6446.
Lentz, B. R., Freire, E., and Biltonen, R. L. (1978). *Biochemistry* **17,** 4475–4480.
Levin, I. W., and Bush, S. F. (1981). *Biochim. Biophys. Acta* **640,** 760–766.
Lindsey, H., Peterson, N. O., and Chan, S. I. (1979). *Biochim. Biophys. Acta* **555,** 147–167.
Liu, N., and Kay, R. L. (1977). *Biochemistry* **16,** 3484–3486.
London, E., and Feigenson, G. W. (1981). *Biochemistry* **20,** 1939–1948.

Longmuir, K. J., Capaldi, R. A., and Dahlquist, F. W. (1977). *Biochemistry* **16,** 5746–5755.
Luna, E. J., and McConnell, H. M. (1978). *Biochim. Biophys. Acta* **509,** 462–473.
Mabrey, S., and Sturtevant, J. M. (1976). *Proc. Natl. Acad. Sci. U.S.A.* **73,** 3862–3866.
Mabrey, S., and Sturtevant, J. M. (1978). *Biochim. Biophys. Acta* **486,** 444–450.
Mabrey, S., Mateo, P. L., and Sturtevant, J. M. (1978). *Biochemistry* **17,** 2464–2468.
McCammon, J. A., Gelin, B. R., and Karplus, M. (1977). *Nature (London)* **267,** 585–590.
MacDonald, R. C., Simon, S. A., and Baer, E. (1976). *Biochemistry* **15,** 885–891.
McGhee, J. D., and von Hippel, P. H. (1974). *J. Mol. Biol.* **86,** 469–489.
McIntosh, T. J. (1980). *Biophys. J.* **29,** 237–245.
McIntosh, T. J., Simon, S. A., and MacDonald, R. C. (1980). *Biochim. Biophys. Acta* **597,** 445–463.
McLaughlin, A. C., Herbette, L., Blasie, J. K., Wang, C. T., Hymel, L., and Fleischer, S. (1981). *Biochim. Biophys. Acta* **643,** 1–16.
MacNaughtan, W., and MacDonald, A. G. (1980). *Biochim. Biophys. Acta* **597,** 193–198.
Marčelja, S. (1976). *Biochim. Biophys. Acta* **455,** 1–7.
Mendelsohn, R., and Van Holten, R. W. (1979). *Biophys. J.* **27,** 221–235.
Mountcastle, D. B., Biltonen, R. L., and Hakey, M. J. (1978). *Proc. Natl. Acad. Sci. U.S.A.* **75,** 4906–4910.
Nagle, J. F. (1976). *J. Membr. Biol.* **27,** 233–250.
Nagle, J. F. (1980). *Annu. Rev. Phys. Chem.* **31,** 157–195.
Oldfield, E., and Chapman, D. (1972). FEBS *Lett.* **21,** 303–306.
Pang, K. Y., and Miller, K. W. (1978). *Biochim. Biophys. Acta* **511,** 1–9.
Papahadjopoulos, D., Vail, W. J., Jacobson, K., and Poste, G. (1975). *Biochim. Biophys. Acta* **394,** 483–491.
Pascher, I., and Sundell, S. (1977). *Chem. Phys. Lipids* **20,** 175–191.
Pearson, R. H., and Pascher, I. (1979). *Nature (London)* **281,** 499–501.
Phillips, M. C., Williams, R. M., and Chapman, D. (1969). *Chem. Phys. Lipids* **3,** 234–244.
Pringle, M. J., and Miller, K. W. (1979). *Biochemistry* **18,** 3314–3320.
Ptak, M., Egret-Charlier, M., Janson, A., and Bouloussa, O. (1980). *Biochim. Biophys. Acta* **600,** 387–397.
Ranck, J. L., Keira, T., and Luzzati, V. (1977). *Biochim. Biophys. Acta* **488,** 432–441.
Rand, R. P., and Sengupta, S. (1972). *Biochim. Biophys. Acta* **255,** 484–492.
Reiss-Husson, F. (1967). *J. Mol. Biol.* **25,** 363–382.
Rogers, J., Lee, A. G., and Wilton, D. C. (1979). *Biochim. Biophys. Acta* **552,** 23–37.
Schmidt, C. F., Barenholz, Y., Huang, C., Thompson, T. E., and Martin, R. B. (1977). *Biophys. J.* **17,** 83a.
Schnepel, G. H., Hegner, D., and Schummer, U. (1974). *Biochim. Biophys. Acta* **367,** 67–74.
Schroeder, H. (1977). *J. Chem. Phys.* **67,** 1617–1619.
Seelig, J. (1977). *Q. Rev. Biophys.* **10,** 353–418.
Seelig, J. (1978). *Biochim. Biophys. Acta* **515,** 105–140.
Seelig, J., and Browning, J. L. (1978). FEBS *Lett.* **92,** 41–44.
Seelig, J., and Seelig, A. (1980). *Q. Rev. Biophys.* **13,** 19–61.
Seelig, J., and Waespe-Šarčevic, N. (1978). *Biochemistry* **17,** 3310–3315.
Silvius, J. R., Read, B. D., and McElhaney, R. N. (1979). *Biochim. Biophys. Acta* **555,** 175–178.
Simmonds, A. C., East, J. M., Jones, O. T., Rooney, E. K., McWhirter, J., and Lee, A. G. (1982). *Biochim. Biophys. Acta* **693,** 398–406.
Skarjune, R., and Oldfield, E. (1979). *Biochim. Biophys. Acta* **556,** 208–218.
Stoffel, W., Zierenberg, O., and Scheefers, H. (1977). *Hoppe-Seyler's Z. Physiol. Chem.* **358,** 865–882.
Stumpel, J., Harlos, K., and Eibl, H. (1980). *Biochim. Biophys. Acta* **599,** 464–472.

Sturtevant, J. M., Ho, C., and Reimann, A. (1979). *Proc. Natl. Acad. Sci. U.S.A.* **76,** 2239–2243.
Sugiwa, Y. (1981). *Biochim. Biophys. Acta* **641,** 148–159.
Tokutomi, S., Lew, R., and Ohnishi, S. I. (1981). *Biochim. Biophys. Acta* **643,** 276–282.
Träuble, H. (1977). *In* "Structure of Biological Membranes" (S. Abrahamsson and I. Pascher, eds.), pp. 509–550. Plenum, New York.
Träuble, H., and Eibl, H. (1974). *Proc. Natl. Acad. Sci. U.S.A.* **71,** 214–219.
Trudell, J. R., Hubbell, W. L., Cohen, E. N., and Kendig, J. J. (1973). *Anesthesiology* **38,** 207–211.
Vail, W. J., and Stollery, J. G. (1979). *Biochim. Biophys. Acta* **551,** 74–84.
Van Deenan, L. L. M. (1965). *Prog. Chem. Fats Other Lipids* **8,** 1–127.
Van Dijck, P. W. M. (1979). *Biochim. Biophys. Acta* **555,** 89–101.
Van Dijck, P. W. M., de Kruijff, B., Van Deenan, L. L. M., de Gier, J., and Demel, R. A. (1976). *Biochim. Biophys. Acta* **455,** 576–587.
Vaughan, D. J., and Keough, K. M. (1974). FEBS *Lett.* **47,** 158–161.
von Dreele, P. H. (1978). *Biochemistry* **17,** 3939–3943.
Warren, G. B., Toon, P. A., Birdsall, N. J. M., Lee, A. G., and Metcalfe, J. C. (1974). *Biochemistry* **13,** 5501–5507.
Watts, A., Harlos, K., Maschke, W., and Marsh, D. (1978). *Biochim. Biophys. Acta* **510,** 63–74.
Watts, A., Davoust, J., Marsh, D., and Devaux, P. F. (1981). *Biochim. Biophys. Acta* **643,** 673–676.
Wieslander, A., Ulmius, J., Lindblom, G., and Fontell, K. (1978). *Biochim. Biophys. Acta* **572,** 241–253.
Wilkinson, D. A., and Nagle, J. F. (1979). *Biochemistry* **18,** 4244–4249.
Wilkinson, D. A., and Nagle, J. F. (1981). *Biochemistry* **20,** 187–192.
Williams, R. J. P. (1978). *Proc. R. Soc. London, Ser. B* **200,** 353–389.
Wohlgemuth, R., Waespe-Šarčevic, N., and Seelig, J. (1980). *Biochemistry* **19,** 3315–3321.
Worcester, D. L., and Franks, N. P. (1976). *J. Mol. Biol.* **100,** 359–378.
Yamamoto, H. Y., and Bangham, A. D. (1978). *Biochim. Biophys. Acta* **507,** 119–127.
Yellin, N., and Levin, I. W. (1977). *Biochim. Biophys. Acta* **489,** 177–190.

Chapter **4**

The Hydrophobic and Electrostatic Effects of Proteins on Lipid Fluidity and Organization[1]

Joan M. Boggs

[1]This study was supported by awards from the Multiple Sclerosis Society and the Medical Research Council of Canada.

ISBN 0-12-053002-3

Introduction[2]

The effects of membrane proteins on lipid bilayer *fluidity* or *order* have been studied using many techniques including X-ray diffraction, measurement of the surface area and surface pressure of monolayers, measurement of the permeability of vesicles, differential scanning calorimetry (DSC), NMR and laser Raman spectroscopy, and use of spin-labeled and fluorescent probes. The terms fluidity and order may encompass such properties as the amplitude of motion of the lipid fatty acid chains, the rate of rotational motion of the lipid, the number of gauche–trans isomers of the fatty acid chains, and the degree of orientation of the lipid (or regions of it) with respect to the bilayer normal. Monolayer and permeability measurements, and measurements of the enthalpy and temperature of the phase transition, can suggest the occurrence of perturbing effects on the lipid without indicating the molecular mechanism. NMR, ESR, and laser Raman spectroscopy can provide more information about the mechanism. The term fluidity will be used in this review only in a qualitative sense to discuss the perturbing effect of proteins on the lipid. If the technique used provides more detailed information about the molecular structure of the lipid–protein complex, it will be described.

Indirect measurements of the contribution of proteins to the fluidity of the membrane have been made by comparing properties of the intact membrane with that of its lipid extract. For example, comparison of the intensities of the lipid resonances in the ^{13}C-NMR spectra from the sarcoplasmic reticulum membrane and from its lipid extract indicated that 25% of the intensity was lost in the membrane spectrum. The intact membrane also bound 27% less TEMPO (a small spin label that partitions between the aqueous phase and the bilayer) than the lipid extract (Robinson *et al.*, 1972). Similarly, the change in fluorescence of *N*-phenyl-1-naphthylamine during the phase transition of the intact *E. coli* membrane was only 80% of the change that occurred in the lipid extract (Träuble and Overath, 1973). Such studies indicate that either the fluidity of a major fraction of the lipid in the intact membrane is equal to that in the lipid extract (the remainder is either highly ordered or immobilized by interaction with protein) and/or that the average

[2] Abbreviations: ESR, electron spin resonance; DSC, differential scanning calorimetry; NMR, nuclear magnetic resonance; CD, circular dichroism; PC, phosphatidylcholine; PG, phosphatidylglycerol; PA, phosphatidic acid; PS, phosphatidylserine; PE, phosphatidylethanolamine; DMPX, dimyristoyl form of phospholipid; DPPX, dipalmitoyl form of phospholipid; DSPX, distearoyl form of phospholipid; CBS, cerebroside sulfate; SDS, sodium dodecyl sulfate; TEMPO, 2,2,6,6-tetramethyl piperidine-1-oxyl; 5-S-SL, 5-doxylstearic acid; 12-S-SL, 12-doxylstearic acid; 16-S-SL, 16-doxylstearic acid; CSL, 3-doxylcholestane (where doxyl stands for 4′,4′-dimethyl-oxazolidine-N-oxyl).

fluidity of all the lipid in the intact membrane is reduced by a fractional amount as a result of interaction with proteins.

The average fluidity of the lipid in intact membranes has been compared to that of the extracted lipid using spin labels. The amplitude of motion of fatty acid spin labels in the intact cell envelope membrane of *Halobacterium cutirubrum* (80 wt % protein) was much less than in the lipid extract. The fluidity gradient [increase in amplitude of motion as the spin label moiety is moved toward the interior of the bilayer (see Fig. 1A)] and phase transitions present in the lipid extract were abolished in the intact membrane, except in the interior of the bilayer (Esser and Lanyi, 1973). Fatty acids with the spin label bound to the eighth carbon down from the carboxyl group, but not those bound to the twelfth carbon, had a lower amplitude of motion in the membrane of Sindbis virus (66% protein) than in the lipid extract (Sefton and Gaffney, 1974). In chromaffin granule membranes (~38% protein), however, there was not much difference in spin-label motion between the lipid extract and the intact membrane, but the difference was greater in the interior of the bilayer than nearer to the polar head group, particularly at 37° (Fretten *et al.*, 1980).

Fatty acid spin labels also had a greater amplitude of motion in the lipid extract of the sarcoplasmic reticulum membrane than in the intact membrane. Return of the proteolipid of this membrane to the lipid extract at a concentration of 5 wt % caused a recovery of some of the order in the interior of the bilayer, although not near the polar head group (Laggner and Barratt, 1975). This approach was also used with the myelin membrane (20–30 wt% protein), for which fatty acid spin labels had a relatively greater amplitude of motion in the lipid extract than in the intact membrane near the polar head group region, although not in the interior of the bilayer (Boggs and Moscarello, 1978a). When the two major proteins of myelin were added to its lipid extract or to other lipids, the amplitude of motion of fatty acid spin labels in the bulk lipid decreased, near the polar head group region and in the interior of the bilayer in the case of the myelin proteolipid (lipophilin) (Boggs *et al.*, 1976), but predominantly near the polar head group region in the case of the myelin basic protein (Boggs and Moscarello, 1978b).

Reconstitution of purified membrane proteins into lipid vesicles of known lipids is a more direct, better defined way of determining the effects of proteins on lipids. Comparison of a lipid extract with the intact membrane requires the assumption that the lipids are organized in the same way in the extract as they are in the intact membrane. This may not be a valid assumption, and altered lipid organization could affect the fluidity regardless of the presence or absence of protein. Furthermore, some proteins may order the lipid and others disorder it, with effects differing in different regions of the bilayer.

The proteins of membranes have been classified as either extrinsic (peripheral) or intrinsic (integral), based on their solubility and ease of removal from the membrane. Intrinsic proteins have received wide attention because the lipid adjacent to the protein is in a state different from that of the bulk lipid (Jost *et al.*, 1973). This boundary lipid is restricted in motion (although its orientation relative to the bilayer normal is not necessarily increased), and exchanges with the bulk lipid at a rate greater than 10^4 sec^{-1} but less than 10^7 sec^{-1} (Jost and Griffith, 1980). It does not go through a cooperative phase transition (Boggs *et al.*, 1976). This lipid, which surrounds the protein, may regulate the activity or function of the protein independently of the properties of the bulk lipid. Intrinsic proteins also have a small ordering effect on the bulk lipid both near the polar head group and in the interior of the bilayer (Laggner and Barratt, 1975; Boggs *et al.*, 1976; Curatolo *et al.*, 1978; Knowles *et al.*, 1979; Utsumi *et al.*, 1980; Gomez-Fernandez *et al.*, 1980), which is probably due to steric hindrance of motion caused by the protein, an effect similar to that of cholesterol.

Extrinsic proteins which bind electrostatically to the lipid alter the fluidity by neutralizing the negatively charged lipid and decreasing the repulsion between neighboring lipids, or by a complexing effect on the acidic lipids similar to that of divalent cations. However, they also can have more complicated effects. It is now clear that there is no sharp demarcation between extrinsic and intrinsic proteins, and proteins that are considered extrinsic because they are water soluble and relatively easily removed also may interact hydrophobically with the bilayer (Kimelberg, 1976). These complex effects of extrinsic proteins are the major subject of this chapter.

Although the effects of membrane proteins on fluidity may be significant to membrane function, particularly in the localized domain of the protein, it is questionable that the mechanism of function of most proteins, many of which do alter fluidity or permeability, is accomplished by this effect. The main importance of understanding the effects of proteins on bilayer structure and fluidity is the indirect information it provides concerning the mode of interaction of the protein with the lipid and of the organization of the lipid–protein complex. These studies provide important clues to the conformation of the protein in the lipid environment and the manner by which the conformation responds to the lipid. It is presently difficult to obtain this information directly, although much progress is being made in this direction using a number of techniques.

Studies with central nervous system myelin basic protein, for example, suggest that its conformation depends on the particular type of lipid to which it is bound. Although this protein has no known enzymatic activity or dynamic function, immune response in myelin against it may occur in the demyelinating disease multiple sclerosis. Immune response would require recognition of the antigenic determinants of the protein on the myelin surface. If the conformation and mode of interaction of a membrane protein is controlled by the lipid environment, the antigenic determinants exposed may also depend on the lipid environment (Boggs and Moscarello, 1978a; Boggs *et al.*, 1981b,1982a). The activity of several enzymes also depends on the lipid environment, suggesting that the conformation of these proteins is also determined by the type of lipid. For example, a membrane glycosyltransferase was specifically activated by PE (Beadling and Rothfield, 1978); the activity of bovine milk galactosyltransferase was stimulated by some lipids and inhibited by others (Mitranic and Moscarello, 1980); β-Hydroxybutyrate dehydrogenase required PC for activation (Gazzotti *et al.*, 1975).

Because so few basic extrinsic membrane proteins have been purified and can be obtained in high yield, many studies have utilized other basic proteins such as lysozyme, ribonuclease, and polyamino acids. Proteins such as albumin have been used at a low pH, at which they have a net positive charge. These proteins are useful as model peptides with varying numbers and sequences of positively charged, hydrophobic, or other residues, and they help to elucidate the interaction of basic membrane proteins with lipid bilayers.

Evidence for Hydrophobic and Electrostatic Interactions of Extrinsic Membrane Proteins

Perturbing Effects of Proteins on Lipid Bilayer

Early X-ray diffraction studies on the interactions of lysozyme (Gulik-Krzywicki *et al.*, 1969) and albumin [at pH 3.3 (Rand, 1971)] with acidic lipids provided evidence that these proteins could interact hydrophobically as well as electrostatically with lipid, because the bilayer thickness did not increase sufficiently to accommodate the protein at the polar head group–aqueous interface. Proteins such as ferricytochrome *c*, lysozyme, albumin (at pH 4.5), spectrin, hemoglobin, and myelin basic protein were also found to

TABLE I

Summary of Effects of Proteins on Lipid Bilayers and Monolayers[a]

	Isoelectric point	Effect on lipid bilayer thickness[b]
Group 1: Electrostatic		
Polylysine		
Ribonuclease	9.6 (1)	
Albumin (pH 7.4)	4.9 (37)	
Insulin (pH 3)	5.6 (1)	No effect (12)
Group 2: Electrostatic and hydrophobic		
Ferricytochrome *C*	10.7 (2)	No effect (6,8,9)
Ferrocytochrome *C*	10.7 (2)	Decrease (9)
Albumin (pH 4.5)	4.9 (37)	Decrease (10)
Hemoglobin	7.1 (37)	
Myelin basic protein	>10.6 (5)	Decrease (11)
Polymixin		
Mellitin		
Spectrin	5.3 (3)	
Lysozyme	11.1 (1)	Decrease (pH 4) (6,7)
Group 3: intrinsic, hydrophobic		
Lipophilin (myelin proteolipid apoprotein)	9.2 (4)	No effect (13)
Gramicidin A		
Glycophorin		

[a]Numbers in parentheses are references. (1) Butler *et al.*, 1973; (2) Kimelberg and Papahadjopoulos, 1971b; (3) Sweet and Zull, 1970; (4) Chan and Lees, 1978; (5) Eylar and Thompson, 1969; (6) Gulik-Krzywicki *et al.*, 1969; (7) Mateu *et al.*, 1978; (8) Shipley *et al.*, 1969; (9) Letellier and Shechter, 1973; (10) Rand, 1971; (11) Brady *et al.*, 1981; (12) Rand and Sengupta, 1972; (13) Kimelberg and Papahadjopoulos, 1971a; (14) Papahadjopoulos *et al.*, 1973; (15) Papahadjopoulos *et al.*, 1975; (16) Quinn and Dawson, 1969a; (17) Quinn and Dawson, 1969b; (18) Demel *et al.*, 1973; (19) El Mashak and Tocanne, 1980; (20) Juliano *et al.*, 1971; (21) Mombers *et al.*, 1980; (22) Shafer, 1974; (23) Gould and London, 1972; (24) Chapman *et al.*, 1974; (25) Lippert *et al.*, 1980; (26) Boggs and Moscarello, 1978b; (27) Boggs *et al.*, 1980; (28) Boggs *et al.*, 1981a; (29) Boggs *et al.*, 1981c; (30) Hartmann *et al.*, 1978; (31) Mollay, 1976; (32) Mombers *et al.*, 1977; (33) Mombers *et al.*, 1979; (34) Galla and Sackmann, 1975; (35) Van Zoehlen *et al.*, 1978; (36) D. Chapman *et al.*, 1977; (37) Young, 1963.

[b]Thickness of hydrocarbon region of bilayer.

[c]Higher salt concentration and different lipids used than in (14).

[d]Number of (+) indicates relative degree of effect.

[e]Can cause lysis (23).

[f]Tested on PC only.

TABLE I *Continued*

Increase in surface pressure or area of monolayers[c]	Increase in permeability of liposomes[c]	Effect on phase transition	
		T_c	ΔH
No (18,22)	+ (2,13)[d]	Increase (15,28,34)	Increase (15)
+ (13,15)	No (13,2)	No effect (15)	Increase (15)
+ (13)	No (13,20)		
+++ (13–17)	++ (2,13), no (23)[b]	Decrease (7,15,24)	Decrease (15,24)
+++ (13)	++ (13,20)		
	+++ (14)	Decrease (15)	Decrease (15)
+++ (14,15,18)	+++ (14,23)	Decrease (15,26–29)	Decrease (15,29)
+++ (19)		Decrease (30)	
		No effect (31)[e]	Decrease (31)[e]
++ (20,21)	++ (3,20)	Small increase (32), decrease (33)	Decrease (32,33)
+++ (13,14)	+++ (2,13), no (23)[b]	Decrease (7,24,25)	—
+++ (14,15)	++ (14)	No effect (15)	Decrease (15)
	++ (14)	No effect (15,36)	Decrease (15,36)
		No effect (35)	Decrease (35)

increase the surface pressure or the surface area of acidic lipid monolayers and to increase the permeability of liposomes of acidic lipids in a nonlytic way; ribonuclease, albumin (at pH 7.4), and polylysine did not, suggesting that the first group interacted both hydrophobically and electrostatically with the lipid and the second group interacted only electrostatically (Quinn and Dawson, 1969a,b; Sweet and Zull, 1970; Kimelberg and Papahadjopoulos, 1971a,b; Juliano *et al.*, 1971; Papahadjopoulos *et al.*, 1973, 1975; Gould and London, 1972; Demel *et al.*, 1973). However, intrinsic proteins such as lipophilin (myelin proteolipid apoprotein) and gramicidin A also expanded monolayers and increased the permeability of liposomes (Kimelberg and Papahadjopoulos, 1971a; Papahadjopoulos *et al.*, 1973, 1975), and they had no effect on bilayer thickness (Rand *et al.*, 1976). Thus, intrinsic proteins and extrinsic proteins that also interact hydrophobically cannot be distinguished by these techniques. (See Table I for a summary of the effects of these proteins on lipids.)

In 1975, Papahadjopoulos *et al.* compared the effects of a number of proteins on the temperature T_c and enthalpy ΔH of the lipid phase transition, and showed that the proteins could be divided into three groups: (1) those which only interacted electrostatically, such as polylysine, increased T_c and ΔH; (2) those which interacted both hydrophobically and electrostatically, such as cytochrome *c*, hemoglobin, and myelin basic protein, decreased both T_c and ΔH; (3) intrinsic proteins decreased ΔH but had no effect on T_c (see Table I). However, it has recently been reported that an intrinsic protein that has a β sheet conformation in contrast to the α-helical structure of lipophilin and rhodopsin also decreased T_c (Dunker *et al.*, 1979). Therefore, it is also necessary to consider the solubility properties of the protein and other evidence for its location in the membrane, such as observation of intramembranous particles by freeze-fracture electron microscopy, to distinguish between groups 2 and 3.

Spectrin, an extrinsic protein, recently was shown to decrease the T_c and ΔH of the phase transition of dimyristoylphosphatidylserine (DMPS) (Mombers *et al.*, 1979). It also decreased ΔH of dimyristoyl-phosphatidylglycerol (DMPG) but increased T_c slightly (Mombers *et al.*, 1977). These effects were observed at pH 7.4, which is above the isoelectric point of spectrin (5.3). A decrease in T_c might have been observed at a pH lower than its isoelectric point and, indeed, this protein produced a greater increase in the surface pressure of lipid monolayers at pH 3.5 than at 7.4 (Juliano *et al.*, 1971). Thus, spectrin appears to interact both electrostatically and hydrophobically, belonging to group 2. Polymixin, which has five positive charges and a short hydrophobic tail, also belongs to this class; it is a useful model for the hydrophobic segments of extrinsic proteins which are thought to penetrate into the bilayer. Polymixin decreased the T_c of dipalmitoylphosphatidic acid (DPPA) (Hartmann *et al.*, 1978; Sixl and Galla, 1979), which is consistent with the increase in surface pressure it produced in phosphatidylglycerol (PG) monolayers (El Mashak and Tocanne, 1980). It is not yet clear to which class mellitin belongs, as it has been used primarily with phosphatidylcholine (PC); it might have greater effects on acidic lipids. It decreased ΔH of PC, but had no effect on T_c (Mollay, 1976).

Although polylysine is a classic example of a polypeptide which only interacts electrostatically when fully protonated, other polyamino acids that contain both basic and hydrophobic amino acids and may interact both electrostatically and hydrophobically are good models for this type of protein. For example, poly(Lys-Phe) and poly(Lys-Tyr) decreased ΔH of phosphatidylserine (PS), but increased T_c slightly (Bach and Miller, 1976). These authors also found that poly(Lys-Phe) increased the conductance of PS vesicles.

Electrostatic Binding

Most of the extrinsic type proteins bind to and interact with acidic lipids to a much greater extent than with PC, indicating that electrostatic interaction is required for significant binding even for those proteins which also interact hydrophobically. However, hemoglobin had as great an effect on the permeability of PC as on that of PS vesicles (Papahadjopoulos *et al.*, 1973). The polyamino acids containing hydrophobic amino acids such as poly(Ala_{16}-$Lys_{13.5}$) had a greater effect on PC than did those containing only basic amino acids (Vitello *et al.*, 1979), indicating that although binding to PC may be much less than to acidic lipids, hydrophobic interactions can follow if the protein contains hydrophobic amino acids.

A high affinity of myelin basic protein for acidic lipids was first demonstrated by Palmer and Dawson (1969), using a biphasic solvent system to detect complex formation. Acetylation of some of the lysines in the protein decreased by a proportionate amount the number of lipid molecules bound (Steck *et al.*, 1976). The interaction of this protein is stronger with phosphatidylethanolamine (PE) than with PC, but it is much weaker with these two lipids than with acidic lipids (Boggs and Moscarello, 1978b). The binding to PC is significant only at pH 8.9 (Smith, 1977a).

Addition of basic protein to lipid vesicles containing a random mixture of an acidic lipid and PC caused clustering of the two types of lipids because of preferential binding of the acidic lipids to basic protein and enrichment of the PC content of the unbound lipid. This was detected by DSC because the T_c of the mixture shifted in the direction of the PC component whether PC was the higher or lower melting lipid of the mixture (Boggs *et al.*, 1977b). From a phase diagram, it was estimated that the protein bound to 27–34 molecules of acidic lipid, consistent with the number of positively charged amino acids in the protein (31) at pH 7.4. Other extrinsic proteins, polylysine and cytochrome *c*, were demonstrated by use of a mixture of a spin-labeled lipid and an unlabeled lipid to bind acidic lipids preferentially; the amount of clustering of the spin-labeled component was measured by the dipolar broadening of its ESR spectrum (Galla and Sackmann, 1975; Birrell and Griffith, 1976). An intrinsic protein, lipophilin, was also shown by DSC to bind acidic lipids preferentially to its boundary layer through electrostatic interactions (Boggs *et al.*, 1977a). Some selectivity of rhodopsin for PS although not for phosphatidic acid (PA) or PG in a lipid mixture was found using spin-labeled phospholipids (Watts *et al.*, 1979). Selectivity of Na, K-ATPase for negatively charged spin-labeled hydrocarbon chains compared to neutral or positively charged ones has also been reported (Brotherus *et al.*, 1980). ^{31}P-NMR indicates preferential association of polyphosphoinositide with glycophorin A when isolated from the red cell membrane (Armitage *et al.*, 1977). Cardiolipin was copurified with cytochrome oxidase (Yu *et al.*, 1975).

Effect of Increase in Salt Concentration

Dissociation of proteins from lipids by increasing salt concentration is sometimes used as a criterion for distinguishing proteins that only interact electrostatically. However, binding of polylysine to PS–PC monolayers has been reported as increasing at higher salt concentrations because of shielding of the increased surface charge density contributed by the bound polylysine. However, binding to PC monolayers decreased at higher salt concentrations (Vitello *et al.*, 1979). Proteins that can also interact hydrophobically are not as greatly dissociated by increased salt; this was found for poly$(Ala)_{16}$-$(Lys)_{13.5}$ bound to PC (Vitello *et al.*, 1979).

Increased concentration of monovalent and even divalent cations did not cause dissociation of spectrin (Mombers *et al.*, 1977, 1979, 1980), polymixin (Sixl and Galla, 1979; El Mashak and Tocanne, 1980), or myelin basic protein (Papahadjopoulos *et al.*, 1973, 1975; Demel *et al.*, 1973; Boggs *et al.*, 1981a). Indeed, myelin basic protein displaced Ca^{2+} from lipid monolayers (Demel *et al.*, 1973) and Mn^{2+} from lipid vesicles if the Mn^{2+}/lipid mole ratio was less than 6:1 (Boggs *et al.*, 1981a). It also displaced protons from DPPG at pH 4, preventing increase in lipid T_c because of protonation (Boggs *et al.*, 1982b).

The effect of lysozyme on the permeability of PS liposomes was not reversed by increased Na^+ concentration, although the effect of cytochrome *c* was partly reversed (Kimelberg and Papahadjopoulos, 1971b). Increased Na^+ concentration caused dissociation of lysozyme from PA at pH 6, suggesting that it interacted electrostatically at that pH. However, at pH 4, it interacted hydrophobically, based on its effect on bilayer thickness determined by X-ray diffraction (Mateu *et al.*, 1978).

Adsorption of ferricytochrome c to lipid monolayers was prevented by an increase in Na^+ concentration; however, if the Na^+ was added after adsorption had occurred, the protein was not dissociated and the protein -induced increase in surface pressure occurred at a slower rate (Quinn and Dawson, 1969a). In other studies, increased Na^+ concentration partly reversed the effects of cytochrome *c* (Kimelberg and Papahadjopoulos, 1971b) and caused dissociation from the lipid (Gulik-Krzywicki *et al.*, 1969; Mateu *et al.*, 1978), suggesting that cytochrome *c* may not interact hydrophobically to as great a degree as do lysozyme and myelin basic protein. This is supported by the lack of effect of ferricytochrome *c* on the thickness of the hydrocarbon region of the bilayer, (as detected by X-ray diffraction (Gulik-Krzywicki *et al.*, 1969; Shipley *et al.*, 1969). Based on X-ray diffraction results in another study, it was concluded that ferricytochrome *c* interacted electrostatically and ferrocytochrome *c* interacted hydrophobically. The former could be dissociated by Na^+ whereas the latter was not (Letellier and Schechter, 1973). However, ferricytochrome *c* was found to have a greater effect than ferrocytochrome *c* on the optical and electrical properties of black lipid films, leading to the conclusion that ferricytochrome *c* penetrated further into the film than did ferrocytochrome *c* (Noll, 1976). Ferri- and ferrocytochrome *c* have not been compared by using other techniques.

Molecular Models Incorporating Hydrophobic Interaction

The detailed molecular structure of the lipid–protein complexes in which hydrophobic interaction occurs is not known, but it has been suggested that it involves either penetration of hydrophobic segments or side chains of the protein into the bilayer or deformation of the bilayer such that the acyl chains of the lipids associate with a hydrophobic surface or hydrophobic pocket of the protein; X-ray diffraction techniques cannot distinguish between these possibilities. Models have been presented by Gulik-Krzywicki *et al.* (1969), Kimelberg and Papahadjopoulos (1971a), Giannoni *et al.* (1971), London *et al.* (1973), Papahadjopoulos *et al.* (1975), Hartmann *et al.* (1978), Boggs and Moscarello (1978a), and Boggs *et al.* (1982a). The interactions with lipid of some of the extrinsic proteins which are included in Table I, particularly myelin basic protein, cytochrome *c*, lysozyme, spectrin, polymixin, mellitin, and polyamino acids, have been studied more extensively using techniques which can give more detailed information about both the lipid and protein structure and the protein environment in complexes with lipid. These studies will now be described in more detail.

Effect of Proteins on Fatty Acid Motion

Comparison of Effects of Basic Protein and Polylysine

The decrease in T_c caused by proteins of Group 2 suggested that they might fluidize the bilayer. Studies of the effect of myelin basic protein on fatty acid spin labels have shown that this is not entirely true, as will be

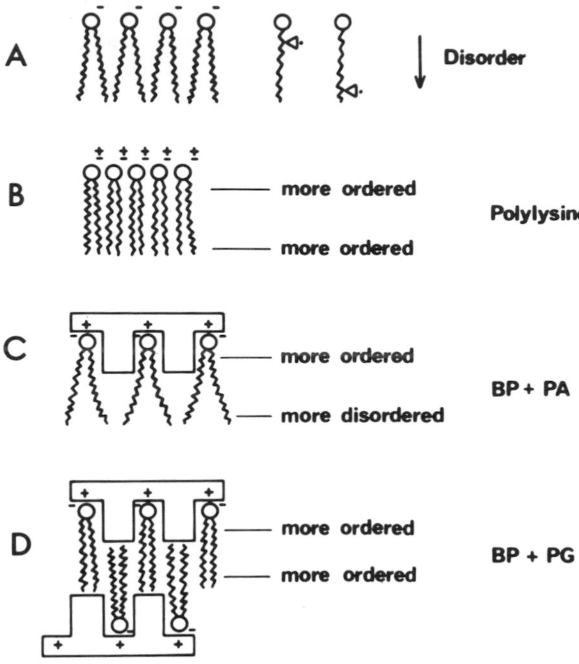

Fig. 1. Diagrammatic representation of predicted effects of different types of interactions with proteins on amplitude of motion of fatty acid spin labels. (A) In pure lipid bilayer a fatty acid spin labeled near the carboxyl group (indicated by △) has less motion (more ordered) than one labeled near the terminal methyl (more disordered). (B) Electrostatic binding of a positively charged polypeptide such as polylysine to a negatively charged lipid decreases the charge repulsion between the lipid polar head groups and causes closer packing of the lipids, resulting in a decrease in the amplitude of motion (increase in order) of fatty acids spin labeled both near the carboxyl and the terminal methyl. (C) Electrostatic binding, combined with penetration of hydrophobic segments of a protein such as basic protein (BP) into the bilayer of an acidic lipid such as PA, decreases the amplitude of motion near the carboxyl because of steric hindrance to motion by the protein segments; it also increases the amplitude of motion near the terminal methyl because of the increased volume available to the terminal ends of the fatty acid chains which is caused by the lipid expansion necessary to accommodate the protein. (D) Alternatively, interdigitation of the fatty acid chains may occur, occupying the increased volume in the center of the bilayer created when the protein penetrates partway into the bilayer. This would cause a decrease in the amplitude of motion both near the carboxyl end and at the terminal methyl. Because the terminal methyl now occupies the same position in the bilayer as a carbon partway down the chain, its motion should be restricted to a similar extent.

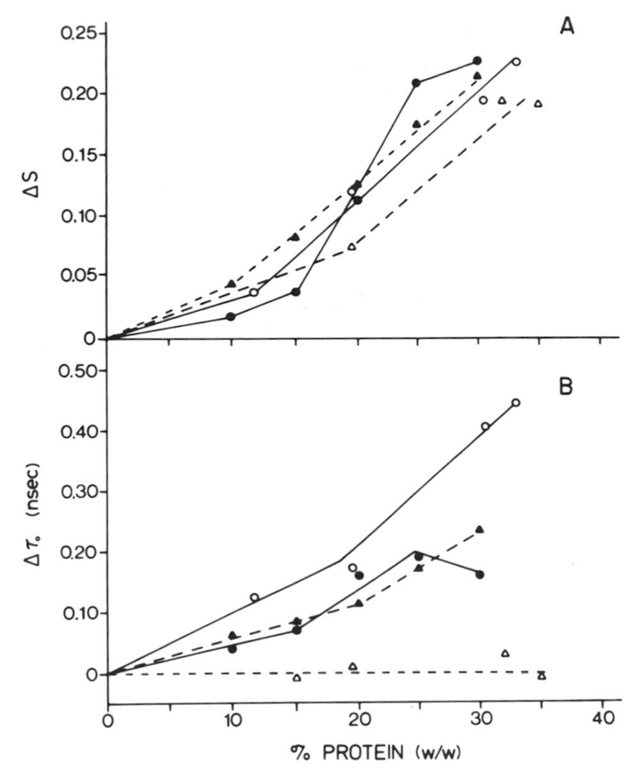

Fig. 2. Effect of increasing concentration of basic protein or polylysine on (A) change in order parameter, ΔS, of 5-S-SL and (B) change in motional parameter, $\Delta\tau_0$, of 16-S-SL relative to the pure lipid in phosphatidic acid (Δ) and phosphatidylglycerol (0) vesicles. Basic protein (open symbols); polylysine (closed symbols). Reprinted with permission from Boggs *et al.* (1981a). Copyright (1981) American Chemical Society.

discussed here, although the protein does significantly perturb the lipid organization. A protein that penetrates partway into the bilayer might be expected to decrease the amplitude of motion of the lipid fatty acids near the polar head group because of the increased packing density and steric hindrance to motion in this region but to increase the motion in the interior of the bilayer because of the lipid expansion caused by penetration of the protein (Fig. 1C). A protein that interacts only electrostatically, on the other hand, might be expected to decrease the amplitude of motion both near the polar head group and in the interior of the bilayer as a result of reduction of the surface charge density and possibly of a complexing effect that condenses the lipid packing (Fig. 1B). The effects of basic protein and polylysine on the amplitude of motion of fatty acids spin-labeled at the 5th carbon from the carboxyl (5-S-SL) and at the 16th carbon in the interior of the bilayer (16-S-SL) of PA and PG (both prepared from egg PC) are compared in Fig. 2. The T_c and ΔH of both lipids were decreased by basic protein and increased by polylysine (Papahadjopoulos *et al.*, 1975; Boggs *et al.*, 1977b, 1981a; Boggs and Moscarello, 1978b). Polylysine decreased the amplitude of motion of both 5-S-SL and 16-S-SL in both lipids as expected (indicated by an increase in the order parameter and motional parameter relative to the pure lipid, ΔS and $\Delta \tau_0$, respectively). In contrast, basic protein decreased the amplitude of motion of 5-S-SL in PA, but had no effect on the motion of 16-S-SL in this lipid. Basic protein has also been found to increase the motion of 16-S-SL in PA (Boggs and Moscarello, 1978b) or above the T_c of DPPA (Fig. 4C). This is more consistent with penetration partway into the bilayer (Fig. 1C). The effects of the proteins on 5-S-SL were greater than the effects of divalent cations or protonation of the lipid, indicating that their effect is more than simply charge neutralization of the lipid (Boggs *et al.*, 1981a). Basic protein added to lyso-PC micelles increased the mobility of the average phosphate group, as determined from the ^{31}P-NMR spectrum (Littlemore and Ledeen, 1979), which was attributed to disruption of the lipid organization by penetration of the protein.

In PG, however, basic protein decreased the amplitude of motion of both 5-S-SL and 16-S-SL and indeed had a greater effect on 16-S-SL than polylysine (Fig. 2). In view of the very different effects of these two proteins on the T_c and ΔH, it was considered unlikely that basic protein interacted only electrostatically with PG (Boggs *et al.*, 1981a). Further studies of the effect of basic protein on the liquid crystalline and gel phases of DPPG, discussed below, provided a possible explanation for the result shown in Fig. 2.

Dependence of Penetration of Basic Protein on Phase State of Lipid

Basic protein had a greater effect on the T_c and ΔH of dipalmitoylphosphatidylglycerol (DPPG) after the sample had been heated once, which suggested that greater penetration occurred in the liquid crystalline phase than in the gel phase (Papahadjopoulos *et al.*, 1975; Boggs and Moscarello, 1978b; Boggs *et al.*, 1980, 1981a). The first heating scan of the DPPG–basic protein complex (48 wt % protein) and a reheating scan are shown in Fig. 3 (d,e). The T_c on the first heating scan was similar to that of the pure lipid [Fig. 3 (c)]; the ΔH was slightly less, 7.4 kcal/mol compared to 8.3 kcal/mol for the pure lipid. On reheating from a low temperature (< 12°C), the main transition was reduced and a second lower temperature endothermic peak was present, followed by an exothermic transition [Fig. 3 (e)]. The heat released

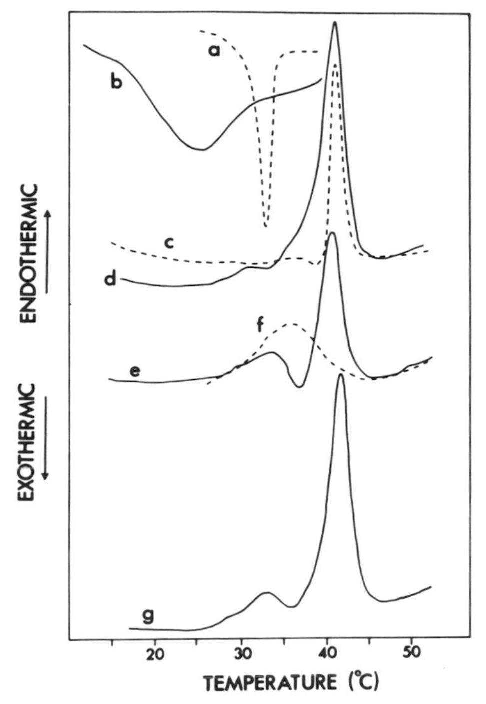

Fig. 3. DSC thermograms of DPPG alone (a,c) or complexed with 48% basic protein (b, d–g). Scans a and b are cooling and all others are heating scans; d is 1st heating scan of sample, heated from −3°C; e is a reheating scan, from −3°C; f (- - -), reheating scan, from 17°C; g, reheating scan from −3°C after cycling from 50°→3°→34°C and incubation at 34°C for 15 min to allow completion of the exothermic process. T_c on cooling scan is normally 7–10° below the T_c of the heating scan at the heating and cooling rates used (10°/min). Samples were prepared in HEPES buffer (10 mM) containing 10 mM NaCl, 1 mM EDTA at pH 7.4.

during this exothermic transition depended on the heating rate and the amount of protein present and in some samples went well below the baseline (Boggs *et al.*, 1981c; Boggs and Moscarello, 1982). The combined ΔH of both endothermic peaks in the reheating scan was 4.4 kcal/mol. The effect of the protein on the T_c of the freezing transition obtained from cooling scans was greater than the effect on the melting transition, as shown in Fig. 3 (a and b) for the pure lipid and lipid–protein complex respectively. The ΔH of the cooling scan was 5.5 kcal/mol.

If the sample was reheated from 12° or higher, the decrease in T_c and appearance of the heating scan [Fig. 3(f)] was similar to that of the cooling scan, and the ΔH was 4 kcal/mol. This suggests that supercooling to below 12° [as in Fig. 3(e)] altered the state of the complex, creating a metastable state which started to melt at 25°C [as in Fig. 3(f)], but refroze, giving off heat, into a state which melted at the same temperature as in the first heating scan. Incubation of the sample at a temperature near that of the exothermic transition, 34°C, for 15 min, allowed the exothermic process to go to completion, and the DSC scan obtained after this treatment resembled the 1st heating scan [as shown in Fig. 3(g)] with a T_c and ΔH only slightly reduced (7.7 kcal/mol) from that of the pure lipid. A lower temperature endothermic peak was still present [see Fig. 3(g)] and in most samples never disappeared. However, the appearance of the main transition suggests that the phase formed during the exothermic transition was identical to the gel phase present on the first heating scan. If the heating rate was fast [as in Fig. 3(e)], the transition did not have time to go to completion, and if the heating rate was very slow, the exothermic transition was not observed because it occurred over the temperature range of the lower temperature endothermic peak in Fig. 3(e). Incubation of the sample at a low temperature (<12°C) for a prolonged time (2–18 hr) also converted the DSC scan to one resembling the first heating scan. This behavior has not been observed for the complex of basic protein with dimyristoylphosphatidic acid (DMPA) or dipalmitoylphosphatidic acid (DPPA). In these lipids, the protein decreased the T_c and ΔH to a similar degree on heating and cooling and the effect was not reduced by incubation in the cold (Boggs and Moscarello, 1978b; Boggs *et al.*, 1980, 1981a).

The nature of the gel phase in the DPPG–basic protein complex was investigated using fatty acid spin labels. The protein was found to have a pronounced immobilizing effect on 16-S-SL in the gel phase of DPPG (Boggs and Moscarello, 1978b; Boggs *et al.*, 1981a,c). It also decreased the motion of 5-S-SL and 12-S-SL (spin-labeled at the 12th carbon) in the gel phase, but the motion of these probes was already relatively restricted in the gel phase of the pure lipid in contrast to 16-S-SL, which had anisotropic motion. The immobilizing effect of basic protein on 16-S-SL was much greater than that of Mg^{2+} or protonation of the lipid (Fig. 4B) as shown by the large increase in T_{11} at 10–35°C caused by the protein in Fig. 4A. Ca^{2+} also does not cause as great an increase in T_{11} as did basic protein (J. M. Boggs, unpublished data). Polylysine decreased T_{11} slightly, indicating a small disordering effect on the gel phase of the lipid (Fig. 4B). Basic protein had only a small ordering effect on 16-S-SL in the gel phase of DMPA (Boggs and Moscarello, 1978b) or DPPA at very low temperatures (Fig. 4C). On cooling, the decrease in T_c of DPPG caused by basic protein was evident from the fatty acid motion (Fig. 4A) and a significant increase in T_{11} did not occur until the sample was cooled to below 15°C. Even at 9°C, the value of T_{11} did not return to its original value until after several hours. Thus, 16-S-SL in the metastable state of the complex [which gives the reheating DSC scan in Fig.

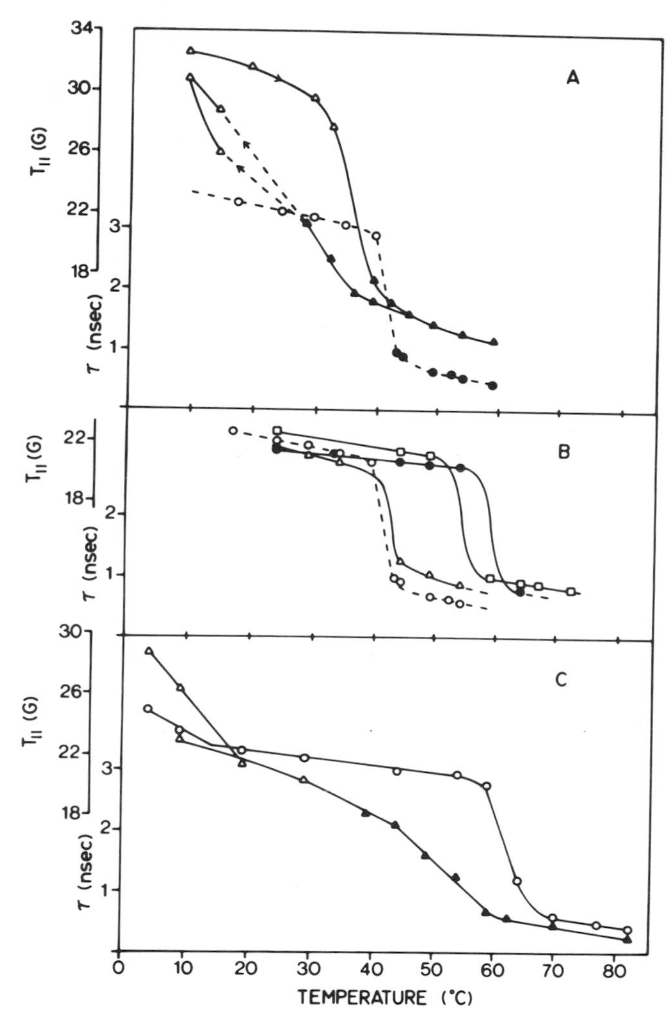

Fig. 4. Effect of temperature on ESR spectral parameters of 16-S-SL in lipid vesicles. The maximum hyperfine splitting, T_{11}, is plotted at lower temperatures, and τ_0 at higher temperatures where the spectrum becomes more isotropic. Both T_{11} and τ_0 increase as the amplitude of motion decreases. (A) DPPG only (0) and with 53.4% basic protein (Δ). Arrows indicate the direction of the temperature change. Spectra were multicomponent on cooling in the temperature range 10–30°C; this region is indicated by a dashed line. Filled symbols are τ_0 values and open symbols are T_{11}. (B) DPPG only (0) and with 30% polylysine (Δ), and with Mg^{2+} (□); DPPG at pH 2 (•). (C) DPPA only (0) and with 50% BP at pH 6. Basic protein causes 16-S-SL to become significantly more immobilized than in pure DPPA only at a temperature much below T_c; T_{11} never reaches value found in DPPG with basic protein. Buffer contained 10 mM NaCl, 1 mM EDTA; at pH 7.4 except where otherwise noted.

3 (e)] was not quite as immobilized as in the original state but was much more immobilized than in the gel state of the pure lipid (Fig. 4A).

Changes in the fatty acid motion during the exothermic transition observed by DSC were monitored at 31° at intervals after taking the sample through the cycle 59–9–31°C. The spectra of 16-S-SL at 31°C immediately after equilibration and after 45 min are shown in Fig. 5A and 5B. The spin label was initially relatively mobile; indeed, it was more mobile than in pure DPPG at 31°C (Fig. 5A, dashed line). However, it became more immobilized than in pure DPPG with time (Fig. 5B). The rate of this change depended on the temperature used within the range 29–34°C, but was nearly complete in 3 min at 32 and 33°C.

After this treatment, the T_{11} value of the spin label at 9°C regained its original value, as shown by comparison of the spectrum at 9°C obtained after this treatment (Fig. 5C, dashed line) with the spectrum obtained on cooling directly to 9°C from above the phase transition temperature (Fig. 5C, solid line). This is consistent with the DSC results indicating that this treatment

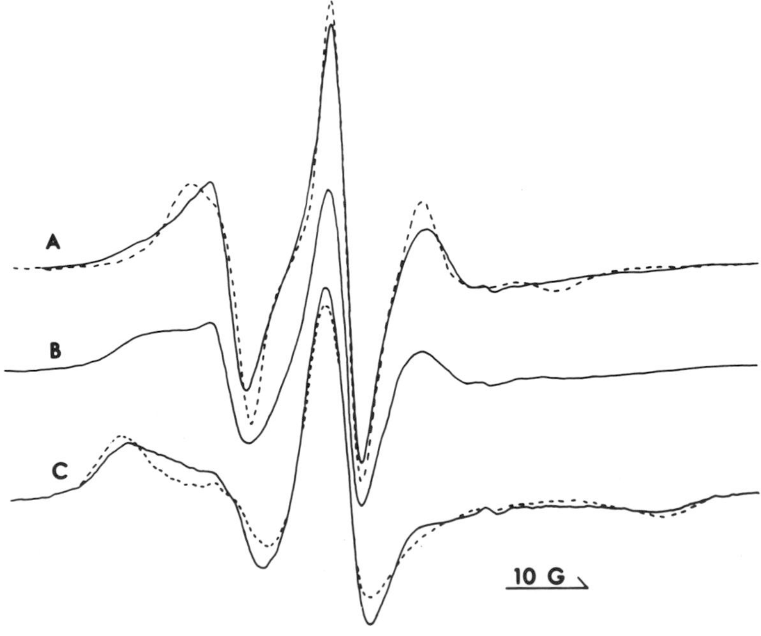

Fig. 5. ESR spectra of 16-S-SL in (A) DPPG only (- - -) and in complex with 53.4% basic protein (——) at 31°C immediately after equilibrium at 31°C; (B) in complex with 53.4% basic protein at 31°C after 45 min at this temperature; (C) in complex with 53.4% basic protein at 9°C on cooling directly from 59°C (——) and after incubation at 31°C for 45 min (- - -).

converted the sample back to the gel state in which it existed on the first heating scan.

A possible explanation for the DSC results is that supercooling creates a distorted and defective gel phase in which the fatty acid chains try to recrystallize around the hydrophobic segments of the protein that penetrated into the bilayer when it was in the liquid crystalline phase. On reheating to 30–34°C, the lipid starts to melt, allowing molecular rearrangement into a more stable structure. This may involve simply squeezing the protein hydrophobic segments out of the bilayer as the liquid refreezes, releasing heat and resulting in immobilization of 16-S-SL. The complex, with the protein interacting primarily electrostatically, then melts at a temperature and with an enthalpy almost as great as the pure lipid.

However, the high degree of immobilization of 16-S-SL caused by basic protein in the gel phase of DPPG indicates a restriction in motion of the terminal methyl groups of the lipid fatty acids that is comparable to that near the polar head group. It is difficult to reconcile this high degree of immobilization with electrostatic interaction of the protein on the surface of the bilayer, because polylysine and divalent cations do not have this effect. It is of course possible that the arrangement of positively charged amino acids on the surface is such that the protein can greatly increase the packing density of the lipid. An alternative mechanism may also occur. The hydrophobic segments may not be completely squeezed out during the exothermic process; instead the fatty acid chains may refreeze by interdigitating as represented in Fig. 1(D). The hydrophobic segments of the protein are apparently not squeezed out of DMPA and DPPA, because the effect on the T_c is similar on heating and on cooling; it is difficult to see why they should be squeezed out of DPPG. On the other hand, lack of immobilization of 16-S-SL suggesting the absence of interdigitation in DPPA could be rationalized by considering the hydrogen bonding properties of PA. Any PA not involved in direct electrostatic binding to basic protein would be involved in intermolecular hydrogen bonding with an adjacent lipid, which would prevent interdigitation. If interdigitation occurs it must involve all of the lipid in the bilayer, not just the lipid bound to the protein. Therefore it would not be expected to occur for PA at pH values at which hydrogen bonding occurs.

Interdigitation of DPPG has been shown to occur by X-ray diffraction in the presence of choline and polymixin (Ranck *et al.*, 1977; Ranck and Tocanne, 1982a,b). There is no firm evidence that interdigitation occurs for the basic protein–DPPG complex. However, it is a reasonable model because it would compensate for the lipid expansion caused by penetration of the protein partway into the bilayer and also would stabilize the complex by allowing greater van der Waals interactions between the fatty acid chains below T_c. Above T_c, greater penetration of the protein and correspondingly less interdigitation probably would occur. Even a small degree of interdigitation would be enough to decrease the amplitude of motion of 16-S-SL in the liquid crystalline phase to the extent observed.

Effect of pH and Lipid Fatty Acid Chain Length in PG. The rate of the exothermic process and changes in the motion of 16-S-SL during the exothermic process depended on the pH and on the fatty acid chain length of PG (Boggs *et al.*, 1982b). No exothermic transition was observed for the DPPG–basic protein complex at pH 6; all heating scans indicated only slight perturbation of the lipid. The T_c of the cooling scan, however, was well below that of the pure lipid, indicating that penetration occurred in the liquid crystalline phase as it did at higher pH, decreasing the temperature of the freezing transition. Immobilization of 16-S-SL occurred immediately upon completion of the freezing transition. These results indicate that whichever mechanism is responsible for the immobilization (i.e., squeezing out of the protein or interdigitation of the fatty acid chains), it occurs very rapidly at pH6; supercooling and partial remelting were not necessary.

In contrast, the process was apparently very slow at pH 8. The T_c on heating was decreased as much as on cooling, and no higher temperature peak near that of DPPG was observed. Incubation for many hours at the T_c of the phase transition of the complex was necessary to obtain a small peak at the higher temperature. 16-S-SL was not as immobilized in the gel phase of the complex at pH 8 as it was at pH 7.4 or lower.

These results are probably related to protonation of some of the 10 histidines in the protein, which would decrease the number and size of the hydrophobic segments which penetrate into the bilayer. This in turn would allow the lipid to refreeze and the hydrophobic segments to be squeezed out more readily. Alternatively, if the degree of penetration is less, the lipid packing will be less distorted, and interdigitation might occur more easily. When all the histidines and possibly some of the other basic residues are deprotonated at pH 8, greater penetration occurs, resulting in more distortion of the lipid packing. As the lipid chains could not refreeze as easily, it is less likely that the protein would be frozen out; interdigitation would also not occur as readily.

In dimyristoylphosphatidylglycerol (DMPG), basic protein decreased T_c as much on heating as on cooling, even at pH 5. No exothermic transition or higher temperature peak were observed. 16-S-SL was not significantly immobilized in the gel state at $pH > 5$; its behavior resembled that of DPPG at pH 8. This suggests that even when the histidines are protonated, the relationship between the fatty acid chain length of DMPG (14 carbons) and the degree of penetration which occurs is similar to that between the fatty acid chain length in DPPG and the degree of penetration at pH 8. The protein is not frozen out and/or significant interdigitation does not occur below T_c.

In distearoylphosphatidylglycerol (DSPG), on the other hand, the greater effect of basic protein on cooling scans compared to heating scans was observed at all pH values, and an exothermic transition was observed even at pH 8. The exothermic transition was even more pronounced in DSPG than in DPPG at pH 7. 16-S-SL was immobilized below the T_c as in DPPG. The rate of the exothermic process was faster than in DPPG at similar pH but still decreased with increase in pH. Consistent with the earlier reasoning, this suggests that the penetration which occurs even at pH 8 is less significant with respect to the fatty acid chain length (18 carbons) than in DPPG at pH 7 or higher. Fatty acid packing is less distorted, the protein can be frozen out, and/or interdigitation can occur more readily. These results indicate that for the lipid to refreeze into a stable state, with or without interdigitation, the fatty acid chain extending beyond the hydrophobic protein segments must be of a certain length in order to have sufficient van der Waals interactions between the chains. Thus, hydrophobic segments can extend further into DSPG than DMPG without preventing refreezing.

Other Proteins: Effect on Fatty Acid Motion and Dependence on Lipid Phase State

A dependence of penetration on the lipid packing has been found with a number of proteins bound to lipid monolayers; that is, no effect on the surface pressure occurred above a certain initial value of surface pressure (Quinn and Dawson, 1969a,b; Mombers *et al.*, 1980; Juliano *et al.*, 1971; Demel *et al.*, 1973; Kimelberg and Papahadjopoulos, 1971a). The interaction of lysozyme with vesicles has also been found to depend on the phase state. Below T_c, the protein interacted electrostatically and had little effect on the laser Raman spectrum of the lipid fatty acid chains. The laser Raman spectrum of the protein itself was characteristic of an α-helical conformation, in contrast to its random solution conformation (Lippert *et al.*, 1980). Above T_c, the laser Raman spectrum of the protein indicated that an irreversible conformation change to β sheet had occurred. Following this the protein lowered the T_c, suggesting that the protein interacted hydrophobically. This resulted in an increase in the number of trans segments (as judged from an increase in the ratio of intensities at the wavelengths 1129 and 1098 cm, I_{1129}/I_{1098}) in the fatty acid chains above T_c, compared to the pure lipid, however.

Smaller effects on fatty acid motion have been observed for other proteins. Polyglutamate and polytyrosine increased T_{11} of 5-S-SL in PC (Yu *et al.*, 1974). A laser Raman spectroscopic study showed that polylysine increased the number of gauche conformers of dimyristoylphosphatidylcholine (DMPC) below T_c, indicating a disordering effect, although it had no effect above T_c (Susi *et al.*, 1979). The effect on the Raman spectrum of acidic lipids has not been studied. Polymixin decreased the amplitude of motion of 5-S-SL in DPPA at pH 9 above the T_c; the effect on 16-S-SL was not studied (Hartmann *et al.*, 1978). Mellitin has been reported to decrease the mobility of spin-labeled cholestane (CSL) and 5-S-SL in PE and PC multilayers, but it also decreased the orientation dependence of their spectra which indicates a disordering effect. Raman spectra also indicated a decrease in fatty acid chain mobility (Verma *et al.*, 1974). In a more recent study of the effect of mellitin on the Raman spectra of DMPC, it was found that mellitin disordered the fatty acid chains in the interior of the bilayer (I_{2839}/I_{2850} increased) both above and below T_c. The methylene segments were also more disordered (I_{2885}/I_{2850}) above T_c, but were more ordered below T_c (Verma and Wallach, 1976). Lavialle *et al.* (1980) also found an increase in I_{2940}/I_{2880} (disorder) of DMPC–mellitin above and below T_c, but found an increase in gauche conformers below T_c (increase in I_{1090}/I_{1130}) and a decrease above T_c, indicating more disorder below and more order above T_c. Ferricytochrome *c* (at pH 9.8) did not have much effect on fatty acid spin labels in cardiolipin–PC vesicles, even at a 1:11 protein/lipid mole ratio (Van and Griffith, 1975). However, in a ^{1}H-NMR and ^{13}C-NMR study, ferricytochrome *c* caused restriction of lipid motion near the polar head group but increased the mobility slightly in the interior of the bilayer at a protein/lipid mole ratio of 1:100 (Brown and Wüthrich, 1977).

Effect of Proteins on Lipid Orientation

It must be remembered that proteins may affect the orientation of the lipids or the width of the distribution of orientations, as well as the rate and amplitude of motion. This is difficult to determine from the ESR spectra of spin labels in isotropic vesicles. A decrease in the orientation of the lipid would also be consistent with penetration of proteins into the bilayer, and this may occur. The latter effect can be determined from the orientation dependence of the ESR spectra in planar-oriented films of the lipid–protein complex, although there are technical difficulties in preparing such films and in assessing whether the protein has been incorporated into the bilayer without distorting the entire film. Butler *et al.* (1973) and Smith *et al.* (1974) examined the effect of a number of proteins on the orientation dependence of CSL in planar lipid films, and found that proteins such as cytochrome *c*, albumin, hemoglobin, poly-Lys-Phe, poly-Lys-Ala, and poly-Lys-Leu decreased the orientation dependence of the probe. Lysozyme (pH 2–9), ribonuclease, insulin (pH 3–3.5), and polylysine produced highly orientation-dependent spectra, indicating a high degree of order of the lipid. The disordering effect of the copolymers of Lys with Ala or Phe was found to increase with the amount of hydrophobic amino acid in the polymer.

It is difficult to predict what the effect on fatty acid motion would be if the lipid fatty acid chains extended upward to interact with a hydrophobic surface or a hydrophobic pocket of the protein. It is improbable that the lipid would go through a detectable co-operative phase transition at high protein concentrations, however, and it is also difficult to reconcile such a model with the effects of basic protein on DPPG described earlier. It could also be argued that proteins such as basic protein remain on the surface of the bilayer but disorder the lipid by forcing it to accommodate to the specific arrangement of positively charged groups on the protein. However, basic protein at least is a relatively flexible molecule and should be able to accommodate to the lipid bilayer. This mechanism would also not account for the large decrease in motion of 16-S-SL in the gel phase of the basic protein–DPPG complex.

Evidence for Location of Some Portions of Proteins in Hydrophobic Region of Bilayer

Evidence for Sequestration

There is evidence for sequestration of some regions of basic protein and other proteins when bound to lipids, which supports the model of penetration of the protein into the bilayer. Five specific sites on basic protein, normally hydrolyzed by trypsin when the protein is in solution, were protected from tryptic hydrolysis when the protein was bound to lipid vesicles or monolayers; this suggests that these regions may be sequestered from the aqueous phase by penetration into the bilayer (London and Vossenberg, 1973; London *et al.*, 1973). Three of the five basic residues at the protected sites (Arg-25, Arg-33, and Lys-58) are within 1–3 residues from an acidic amino acid and could be neutralized via intramolecular salt formation so that these regions might penetrate into the bilayer. In an α-helical coil, a positively charged amino acid can be neutralized by a negatively charged amino acid if it is not more than three residues away and can be embedded in the bilayer, as occurs in rhodopsin (Engelman *et al.*, 1980). A fourth protected arginine in basic protein is the methylated arginine at position 107, which may be sufficiently hydrophobic to penetrate into the bilayer. Binding of antibody (prepared against the intact protein) to basic protein was decreased when the protein was bound to the outer monolayer of lipid vesicles, suggesting that some antigenic determinants were either sequestered or altered in conformation (Boggs *et al.*, 1981b).

In order to determine whether hydrophobic residues in each half of basic protein could penetrate into the bilayer, the protein has been cleaved at tryptophan 116 with BNPS-skatole, and the ability of both fragments to increase the surface pressure of monolayers (London *et al.*, 1973), decrease the phase transition temperature of lipid vesicles, and decrease the motion of fatty acid spin labels in lipid vesicles (Boggs *et al.*, 1981a) has been studied. Both fragments were found to have a qualitatively similar effect although the effect of the smaller C-terminal fragment was quantitatively less. Thus, sites in both halves of the molecule can penetrate into the bilayer.

Effect of Lipid Environment on Different Residues of Protein

There is also evidence that hydrophobic residues of proteins bound to lipid are in a hydrophobic environment or are affected more than basic and polar residues by the lipid. Changes in tryptophan and tyrosine fluorescence of basic protein in the presence and absence of sodium dodecyl sulphate (SDS) indicated that the tryptophan moved into a less polar environment when the protein was bound to SDS micelles and tyrosine was in a more neutral environment (Jones and Rumsby, 1975). Similar results were found for the protein bound to lipid vesicles (Stollery and Epand, unpublished data). Interaction of the protein with SDS (Liebes *et al.*, 1976), GM_1 ganglioside, and lysophosphatidylcholine micelles (Littlemore and Ledeen, 1979) broadened aliphatic and aromatic residues, the methyl resonance from methionine-20, and the only *N*-methylarginine at position 107. Peaks from leucine β-CH_2, lysine δ-CH_2, and ϵ-CH_2 and glutamic γ-CH_2 groups were only slightly affected by interaction with lipid. Tyrosine was less affected than phenylalanine.

Exposure of Residues of Proteins to Aqueous Phase

The exposure of certain residues of proteins to the aqueous phase when bound to lipids has been probed by measuring the exchange of labile protons with deuterium. In mellitin bound to micelles (at pH 2.8), the labile protons of Lys, Arg, Gln, and Trp were found to exchange rapidly, indicating accessibility to the solvent; 15 of the backbone amide protons exchanged very slowly, indicating that they were buried (Brown, 1979). Tryptophan fluorescence indicated that this residue was in a less polar environment on binding to lipid, however (Dufourcq and Faucon, 1977). Binding of mellitin to lyso PC caused a shift of both the Trp-19 ^{1}H-NMR lines of the protein and the lipid resonances of the first and second methylene groups below the ester linkage, suggesting that the tryptophan of the protein penetrated into the micelle of lyso PC as far as the second methylene group of the chain (De Bony *et al.*, 1979). The ^{1}H-NMR resonances of Ala-4 and -15, Thr-10 and -11, Val-5 and -8, and Ile-17 or -20, Trp 19, and one of the lysines were greatly shifted downfield when mellitin was bound to detergent micelles; resonances from Gln-25 and -26, Arg-22 and -24, and two of the lysines were shifted very little which suggested they were on the surface of the micelle (Lauterwein *et al.*, 1979).

It has been predicted that residues 2–11 and 15–21 of mellitin would form amphipathic helices, and a model for its location in the bilayer has been proposed in which the polar sides of the helices are exposed to water and the hydrophobic side penetrates into the bilayer. This would greatly perturb the lipid, explaining its lytic properties (Dawson *et al.*, 1978). Mellitin has also been considered to pass through the bilayer as an intrinsic protein and to possess boundary lipid (Lavialle *et al.*, 1980).

Environment of Spin-Labeled Sites on Protein

The location of regions of cytochrome *c* and basic protein has been probed by covalently binding a spin label to them, at methionine 65 in the case of cytochrome *c* and at methionines 21 and 167 for basic protein. These are in hydrophobic regions for basic protein but not for cytochrome *c*. The hyperfine splitting of the ESR spectra of spin-labeled cytochrome *c* bound either to lipid vesicles (Brown and Wüthrich, 1977) or to mitochondria (Azzi *et al.*, 1972) indicated that in both cases the spin label was located in a polar environment. However, its mobility was not affected by an increase in viscosity of the aqueous phase (resulting from addition of sucrose), suggesting that it was not exposed. The mobility of the probe on the protein in solution, however, was sensitive to the addition of sucrose. The effect of the spin label on the spin lattice relaxation rate of nuclei of the lipid molecules was measured from the ^{1}H-NMR spectrum of the lipid, which provided information on the depth of the spin label in the bilayer. The results suggested that the spin label lay near the plane of the second methylene group of the lipid, although the hyperfine splitting suggested a polar environment (Brown and Wüthrich, 1977). This group also performed the converse of this experiment in glucagon-lyso-PC micelles containing fatty acids spin-labeled at different positions (Brown *et al.*, 1981). The effect of the fatty acid spin labels on the ^{1}H- and ^{13}C-NMR spectrum of different residues of glucagon provided information about the depth of these residues in the bilayer. It was concluded that the apolar residues were oriented toward the interior of the micelle and the polar residues toward the aqueous phase.

The spin labels bound to the methionines of basic protein were also in a relatively polar environment, although not quite as polar as water [hyperfine splitting A_0 was 16.51 G for the spin-labeled protein bound to DPPG, compared to 16.64 G for the spin-labeled protein in aqueous solution and 16.85 G for the free spin label in solution (Boggs *et al.*, 1980)]. The spin-label motion, monitored from the peak height, was sensitive to the lipid phase transition (see Fig. 6A). This could occur if the spin-labeled groups are on the surface of the bilayer, however, because the motion of the phosphate of the lipid is also sensitive to the phase transition (Cullis and de Kruijff, 1976). The spin-labeled regions were most sensitive to the phase transition of DPPG on the 1st heating scan (Fig. 6A, curve 1), for which DSC evidence described earlier suggested that little penetration occurred until the sample had been heated once. The spin label motion was much less sensitive to the phase transition on cooling (curve 2) and reheating (curve 3) scans, suggesting that penetration of some regions of the protein (not necessarily the spin-labeled methionine regions) above the phase transition had perturbed the lipid in the microenvironment of the spin-labeled methionines, so that they no longer sensed a sharp phase transition. In this type of experiment, the protein was used only as a probe, and it was at such a low concentration (4 wt %) that no effect on fatty acid spin label motion or on the T_c would have been observed by DSC. However, the lipid in the environment of the protein appeared to be perturbed by the protein in the same way as was deduced from results with high concentrations of protein. The spin-labeled methionines were probably buried no further than the ester linkage or possibly the 2nd methylene segment, considering the high value for the hyperfine splitting and the results described above for cytochrome *c*; however, they were sensitive to the perturbing effects of other regions of the protein which may penetrate more deeply into the bilayer.

The motional parameter τ_0 was much less sensitive to the phase transition, although it did change in a complex way throughout the phase transition, especially on the first heating scan (Fig. 6B, curve 1). Comparison with the smooth curve obtained for the protein in solution indicates that the abrupt changes in the presence of lipid are caused by changes in the interaction of the protein with the lipid. These changes were attributed to the opposing effects of immobilization resulting from increasing penetration as the lipid melted, plus the mobilizing effect caused by the increased fluidity of the lipid at higher temperatures (Boggs *et al.*, 1980). The protein was also en-

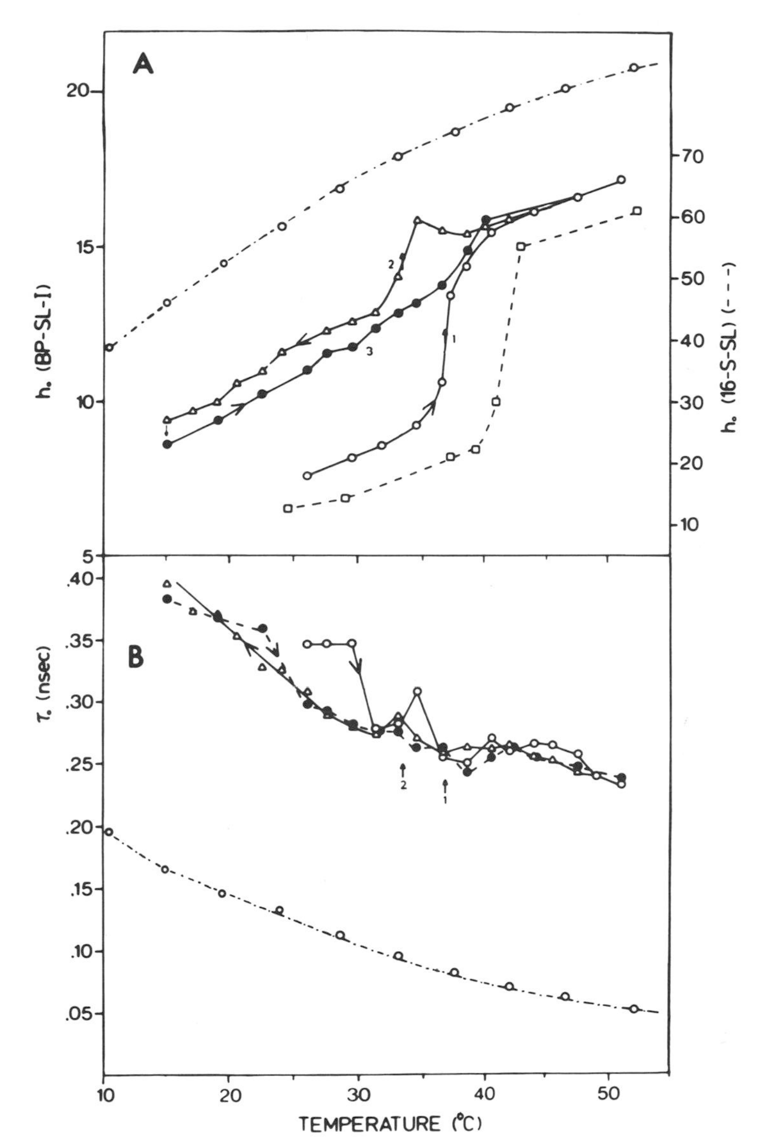

Fig. 6. Temperature dependence of (A) the height of the center line, h_0, of the ESR spectrum of an iodoacetamide spin label bound covalently to methionines 21 and 167 of basic protein, and (B) the motional parameter, τ_0, for the spin-labeled protein in aqueous solution (-·-·-) or in DPPG vesicles (—). Curve 1, first heating scan (0); curve 2, cooling scan (Δ); curve 3, second heating scan (•). In (A), vertical arrows indicate the midpoints of the steepest portions of curves 1 and 2 and have been placed at the same temperatures in (B). In (A), h_0 for 16-S-SL in pure DPPG is also plotted against temperature (- - -). Reprinted with permission from Boggs *et al.* (1980). Copyright (1980) American Chemical Society.

riched with ^{13}C by methylation of the two methionyl residues (Deber *et al.*, 1978) and the motion of the methionine regions was shown to increase during the phase transition by ^{13}C-NMR spectroscopy (Stollery *et al.*, 1980b).

Dependence of Hydrophobic Interaction of Protein on Type of Lipid

Evidence for Variable Hydrophobic Interaction with Different Lipids

Basic protein was first reported to have variable effects on different lipids by London *et al.*, who suggested that the protein might penetrate to a different extent in different lipids. They found that the effect of the protein on the surface pressure of lipid monolayers (at pH 5) decreased in the order cardiolipin > acidic lipid extract from myelin > cerebroside sulfate >> total myelin lipid extract = PS > PE >> neutral lipids (sphingolipids, PC, cerebroside) (Demel *et al.*, 1973). However, the effect on the surface pressure depends on the compressibility of the lipid, and the surface pressure of a lipid that is already tightly packed will be affected more by a small degree of penetration than one which is not tightly packed and thus is more compressible.

This group also found that vesicles of cerebroside sulfate (CBS) protected the protein from tryptic hydrolysis to a greater extent than did PS, suggesting greater penetration into CBS (London and Vossenberg, 1973). However, using monolayers they found that all of the lipids studied (CBS, acidic lipid extract, PS, cerebroside, and PC) partially protected the protein to a similar extent from enzymatic hydrolysis (London *et al.*, 1973).

We have also found variable interaction of basic protein with different lipids; the results were remarkably consistent regardless of the technique used. The effect on the lipid T_c was greatest for DPPG, for which the maximum decrease found was 13°C, followed by DMPA or DPPA (decrease, 10°C) and then DMPS (decrease, 8.5°C). The effect on DMPA or DPPA and DMPS was similar on heating and cooling in contrast to the results with DPPG. This suggested that the protein was not frozen out or that interdigitation did not occur in DPPA and DMPS. The protein had little or no effect on the T_c of CBS, PE or dimyristoylphosphatidylethanolamine (DMPE). The effect of basic protein on the T_c suggested that the degree of penetration into these lipids decreased in the order PG > PA > PS >> CBS ≃ PE (Boggs and Moscarello, 1978b; Boggs *et al.*, 1980, 1981a). However, it must be kept in mind that, as with surface pressure measurements, the T_c of a tightly packed lipid such as PA could be more affected by a small perturbation than would other lipids.

The protein also had variable effects on the amplitude of motion of 5-S-SL in the liquid crystalline phase of different lipids. This probe should be sensitive to the perturbing effects of penetration, whether or not interdigitation occurs. The effect of high protein concentrations (lipid to protein mole ratio 25–35:1) on this probe decreased in the order PG = PA > PS > CBS > PE (Boggs and Moscarello, 1978b; Stollery *et al.*, 1980a) in approximate agreement with the order obtained from measurement of the T_c above.

The spin-labeled protein was sensitive to the phase transition of DMPA, DMPE and DMPS as it was to that of DPPG (Fig. 7). However, the degree of sensitivity to the transition (width of the transition) was similar on the first heating, cooling, and second heating scans for all three lipids, in contrast to DPPG and in agreement with DSC results (Boggs *et al.*, 1980, and unpublished data). The protein spin label motion was highly sensitive to the

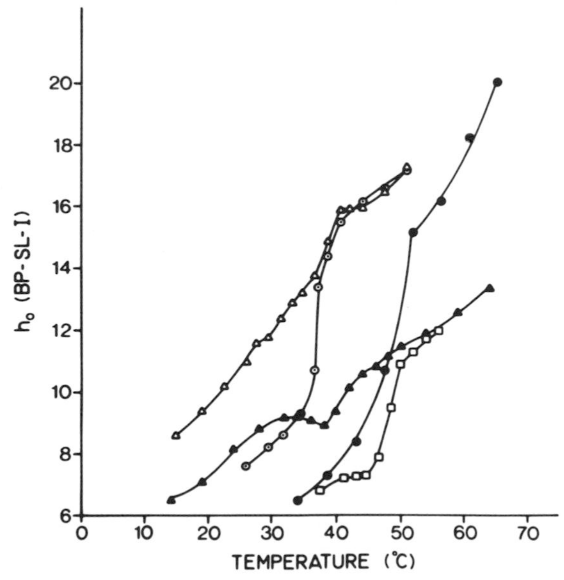

Fig. 7. Comparison of temperature dependence of the height of the center line, h_0, of the ESR spectrum of the spin-labeled basic protein in DPPG, 1st heating scan (0) and DPPG, 2nd heating scan (Δ); and the 2nd heating scans of DMPA (•), DMPE (□), and DMPS (▲) at pH 7.4. h_0 has been normalized so that it is nearly identical for all curves at the low temperature end.

phase transition of DMPE, as on the first heating scan of DPPG, suggesting the absence of a significant perturbing effect on DMPE in agreement with the lack of effect of higher protein concentrations on the T_c of PE and DMPE. The protein was less sensitive to the phase transition of DMPA and DMPS than of DMPE (Fig. 7). In DMPS, the peak height began to increase at a temperature well below the T_c, then leveled off and decreased slightly before increasing near the T_c of DMPS. The increase in mobility (increase in height) at the low temperature may be caused by melting of some of the lipid well below its T_c, indicating a significant perturbing effect on the lipid resulting from penetration of the protein. The decrease in motion just before the T_c may be due to increased penetration as the rest of the lipid melts, and the increase in motion at the T_c may result from the increased fluidity of the lipid as the remainder melts. This behavior was seen on all scans of DMPS, and, to a small degree, on heating scans of DMPE (Boggs *et al.*, 1980).

τ_0 changed in a complex way during the phase transition for the spin-labeled protein bound to all the lipids (as was discussed for DPPG, Fig. 6B), suggesting that it was sensitive to both the lipid fluidity and the degree of penetration of the protein. Deeper penetration of the probe itself, even if it was still within the ester region of the lipids, and/or deeper or increased penetration of other regions of the protein, should increase τ_0 of the spin label bound to the protein, just as the amplitude of motion of 5-S-SL is decreased by these events. Thus, τ_0 was used as a measure of the degree of penetration of the protein into the different lipids in the liquid crystalline phase. It was found to decrease in the order PG $\simeq$ PA $>$ PS $\simeq$ CBS $>$ PE $>$ BP in solution (Stollery *et al.*, 1980a). The hyperfine splitting of the spectra indicated a polar environment in all cases. An immobilized component was also seen for the spin-labeled protein bound to PG and PA at low temperatures, although not for the other lipids. This immobilization could be caused by either interaction at the polar head group region or penetration of the spin-labeled region itself; which has not yet been determined.

These results suggested that the tertiary structure of the protein should vary with different lipids, which would mean that the exposure of the antigenic determinants of the protein on the bilayer surface might depend on the type of lipid. This was tested using antibody raised against the intact basic protein in aqueous solution. The antibody-induced precipitation of the protein bound to the outer layer of lipid vesicles increased in the order PA $<$ PS $<$ PG $<$ CBS $<$ PE $\simeq$ basic protein in solution (Boggs *et al.*, 1981b), which was in fair agreement with the DSC and ESR results suggesting variable degree of penetration into different lipids; the results with PG, which caused greater exposure of the antigenic determinants than might be expected from the apparently high degree of penetration of basic protein into it, were an exception.

Dependence on Intermolecular Hydrogen Bonding Properties of Lipids

Differences in the degree of penetration of the protein into different lipids must be caused primarily by differences in the polar head groups. At certain pH values, PE, PS, and PA are capable of intermolecular hydrogen bonding through the polar head groups (Papahadjopoulos and Weiss, 1969; Jacobson and Papahadjopoulos, 1975; Eibl, 1977; MacDonald *et al.*, 1976; Eibl and Blume, 1979; Eibl and Woolley, 1979; Blume and Eibl, 1979; Boggs, 1980). CBS also can form intermolecular hydrogen bonds through the sphingosine amide and fatty acid hydroxyls (Pascher, 1976) (Fig. 8). These interactions increase the phase transition temperature and increase the packing density of the lipids. The interactions are weaker for PS and CBS than for PA and PE at pH 7.4, because of the net negative charge on these lipids at this pH. However, electrostatic interaction with the protein will abolish the net negative charge and allow hydrogen bonding to occur. This prevents penetration of the protein in the case of CBS.

The situation with PS is complicated by the possibility that binding of high concentrations of protein will lower the surface charge density of the bilayer and may lower the pK of the amine of PS. If the amine becomes unprotonated at pH 7.4, hydrogen bonding would be abolished and penetration of the protein could then occur. This may also happen to some extent with PE. However, because there is only one negatively charged group in PE, a group which is primarily involved in hydrogen bonding with the amine of an adjacent lipid, much less basic protein binds to PE than to the other lipids (Boggs and Moscarello, 1978b). The lipid to which the protein is bound, however, can no longer hydrogen bond, which may allow some penetration in localized areas. This is evidently not sufficient to affect the T_c observed by DSC or the fatty acid mobility. However, in the study with the spin-labeled basic protein (Boggs *et al.*, 1980), the T_c sensed by the spin label bound to the protein in DMPE was less than the T_c of the pure lipid and was several degrees lower on cooling than on heating, suggesting that some penetration did occur which perturbed the lipid in the microenvironment of the protein. Nevertheless, intermolecular interactions for PE appear to be more favorable than binding to the protein. Binding of the protein to PA, on the other hand, abolishes its ability to hydrogen bond; the protein is able to penetrate to a similar degree as in PG which probably cannot hydrogen bond in either the presence or the absence of the protein.

The penetration of spectrin into monolayers of different lipids at various pH values has been monitored from the increase in surface area produced at constant pressure (Mombers *et al.*, 1980). At pH 7.4, the degree of penetration was judged to decrease in the order PG > PA >> PS > PE, PC, and

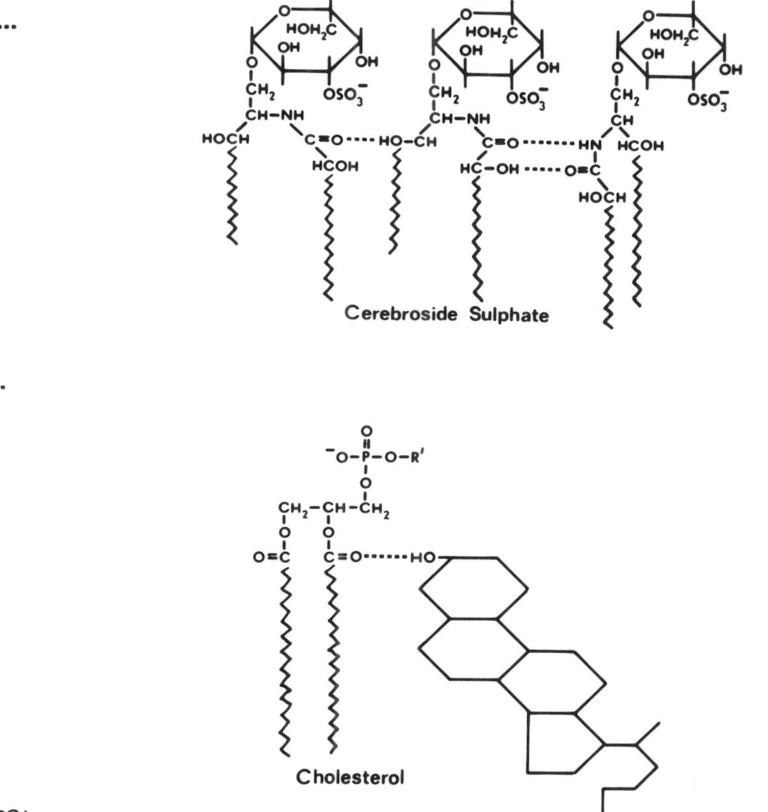

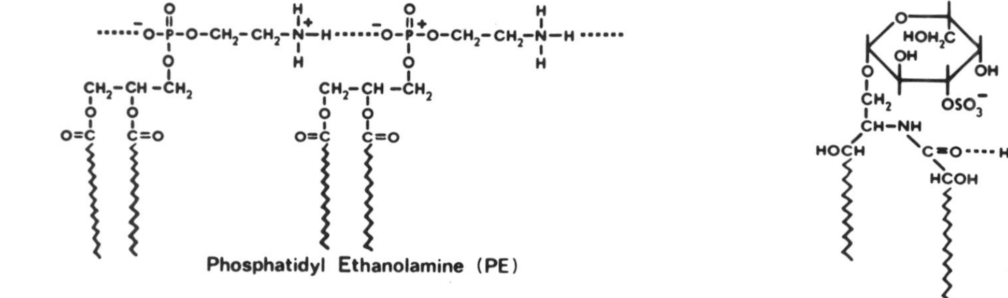

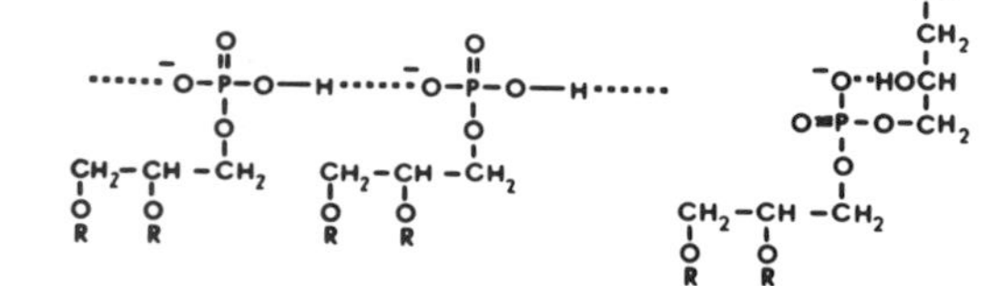

Fig. 8. Molecular structures of PE, PS, PA, PG, and CBS depicting groups that participate in intermolecular hydrogen bonding at neutral pH. Possible hydrogen bonding between cholesterol and a phospholipid is also shown. PG may hydrogen bond intramolecularly as shown. R = fatty acid chain esterified to glycerol.

cholesterol, which is consistent with the results found for basic protein. However, Mombers *et al.* (1980) attributed their results to the closer packing density of PS relative to PG which prevented penetration into PS. The role of intermolecular hydrogen bonding in causing the closer packing of PS was not considered nor was the reason for the high degree of penetration into PA, which also forms closely packed monolayers by itself (Hartmann *et al.*, 1978). According to our interpretation, binding of the protein would abolish hydrogen bonding in PA, allowing penetration and increase in surface area.

The effect of cytochrome *c* on the surface pressure of monolayers of several lipids (PE, PC, PA, and cardiolipin) has also been studied. It increased the surface pressure of PE only at a low initial surface pressure, 24 dynes/cm (Quinn and Dawson, 1969a), which is lower than the surface pressure of lipids in sonicated liposomes; penetration into PA and cardiolipin monolayers occurred up to initial surface pressures of 40 dynes/cm (Quinn and Dawson, 1969b). At 24 dynes/cm, PE may be too expanded to hydrogen bond significantly, thus allowing penetration to occur. Just as the strength of hydrogen bonding interactions may be decreased at lower surface pressures in monolayers, they are probably much less strong in the liquid crystalline phase than in the gel phase and less strong in lipids having unsaturated fatty acids than in synthetic lipids with saturated fatty acids. We found that less basic protein binds to DMPE than to egg PE indicating the greater ability of the basic residues of the protein to compete with the positively charged amine of the lipid for the negatively charged phosphate in the more expanded egg PE than in DMPE (Boggs *et al.*, 1980). Cytochrome *c* caused a much greater increase in the surface pressure of PE at pH 9 (where hydrogen bonding no longer occurs) even though less protein was bound.

The penetration of cytochrome *c* into PA decreased on raising the pH to 4–5 (Quinn and Dawson, 1969b). This was attributed to ionization of the carboxyl groups of the protein. However, there was a second maximum in the surface pressure at pH 6–7 that preceded the ionization of the second phosphate group of the pure lipid monolayer but not that of the lipid–protein monolayer. The authors concluded that binding of the protein lowered the p*K* of this group in PA. Ionization of this group would abolish any hydrogen bonding for this lipid and allow the increased penetration observed.

Dependence of p*K* Values of Ionizable Groups on Environment

It is known that the p*K* values of the ionizable groups of lipids depend strongly on the concentration of univalent and divalent cations and also on the presence of other lipids in the bilayer, because of their effects on the surface charge density of the bilayer (Träuble and Eibl, 1974; Träuble, 1976; Abramson *et al.*, 1964; Papahadjopoulos, 1968; Hauser and Dawson, 1967; Kolber and Haynes, 1979). Thus, it would not be surprising if electrostatic binding of proteins could also affect the p*K* of the ionizable groups of lipids (Boggs, 1980). Similarly, binding of protein to lipid may also affect the p*K*s of the ionizable groups on the protein, because of the contribution of charged groups on the protein to the surface charge density of the bilayer. This has not been specifically studied, but there is evidence suggesting that it may occur at least for the highly charged polylysine, whose positively charged groups are not all neutralized by binding to lipid.

At pH 9, the increase in T_c produced by polylysine in DPPA (Galla and Sackman, 1975) was not as great as that produced in DMPA at pH 7.4 (Boggs *et al.*, 1981a), which suggested that some penetration may occur at pH 9. This was supported by the hyperfine splitting of the ESR spectrum of a maleimide spin label bound to polylysine, indicating location in a hydrophobic environment when bound to DPPA at pH 9 (Hartmann and Galla, 1978). The motion of the spin label was sensitive to the phase transition. These results suggested that the pK_a of the ϵ amino group was lowered as a result of the high surface charge density acquired by the bilayer upon binding polylysine. Thus, it is apparently uncharged lysine groups that penetrate into the bilayer. A similar suggestion was made by Campbell and Pawagi (1980) who found that at pH 7.8 polylysine, but not polyglutamate, decreased the intensity of the UV absorption of the double bonds in the fatty acid chains of PC suggesting that penetration into the bilayer to the vicinity of the double bonds had occurred. However, polylysine did not affect the UV absorption of charged lipids. Binding of polylysine to PC would increase the surface charge density more than when bound to acidic lipids because fewer lysines are neutralized by PC. Thus it is conceivable, although not proven, that the p*K* might be lowered sufficiently that at least some of the lysines would be neutral at pH 7.8 and could penetrate into the bilayer.

Direct Evidence for Conformational Changes of Protein upon Binding to Lipid

The evidence obtained from studies of the effect of proteins on different lipids above and below the phase transition suggests that the conformation of proteins may vary with the lipid environment and with the phase state of the lipid. Antibody binding studies provide more evidence for this. More direct methods for conformational analysis such as circular dichroism spectroscopy, NMR and laser Raman spectroscopy, and other techniques are also being applied to determine protein conformation in lipid–protein vesicles.

Basic protein in solution has an open or random conformation with a bend probably occurring at a triproline segment (residues 99–101) (Brostoff and Eylar, 1971; Epand *et al.*, 1974; Krigbaum and Hsu, 1975; Martenson, 1978). Evidence for structured regions near residues 85–116 has been found by NMR spectroscopy; phenylalanine and tryptophan resonances were broadened in the aqueous solution NMR spectrum of the protein (Littlemore, 1978; B. E. Chapman *et al.*, 1977; Chapman and Moore, 1976, 1978). The tryptophan of the protein in aqueous solution was partly protected from quenching of its fluorescence, although the wavelength was indicative of a polar environment (Jones and Rumsby, 1975; Feinstein and Felsenfeld, 1975). There is a high probability of β bends at many sites throughout the molecule, based on comparison of the sequence with the known structure and sequence of other proteins (Epand *et al.*, 1974). Interaction with detergents and lipids has been found to induce some α helical and β structure in basic protein, which has a random conformation in solution (Anthony and Moscarello, 1971; Liebes *et al.*, 1976; Smith, 1977b; Keniry and Smith, 1979). It has recently been reported that this conformational transition is greatest with PG, PA, and PS (20–27% α and 10–12% β structure) and is much less with PE (1% α, 5% β) and even less with PC (Keniry and Smith, 1981). The effect of CBS could not be tested because it has an intrinsic CD spectrum. The correlation between the structure-inducing ability of the lipids and degree of penetration of basic protein into these lipids found by other methods suggests that the helical regions may penetrate into the bilayer.

Many other proteins and polypeptides have been shown, by CD spectroscopy, to undergo conformational changes upon binding to lipids. Polylysine and poly(Lys-Phe) were converted from random coil to α helix by PS, poly(Lys-Tyr) was converted from random coil to β structure by PS or PC, and PS induced a small increase in α helical content in poly-Orn (pH 10) and poly(Glu-Lys-Ala) (pH 5) (Hammes and Schullery, 1970; Bach and Miller, 1976; Bach *et al.*, 1975). Poly-Glu, which is α-helical in solution, was converted to a random coil by PC (Yu *et al.*, 1974); poly Lys-Ser remained random coil upon binding to PS or PC (Bach *et al.*, 1975). Mellitin was converted from extended chain to α helix by binding to PC or to inorganic phosphate (Dawson *et al.*, 1978; Drake and Hider, 1979; Lauterwein *et al.*, 1979). Anionic detergents increased the α helical content of cytochrome *c*, lysozyme, and other proteins; the effect was greater at pH 2 where the acidic amino acids are protonated (Wu *et al.*, 1981). However, infrared spectroscopy was used to detect an increase in β structure in cytochrome *c* in the presence of acidic lipids (Moshkov *et al.*, 1975). The CD spectrum of ferricytochrome *c* bound to cardiolipin resembled a state intermediate between the native and denatured states. The intermediate state could be induced in solution by pH extremes or ligand binding. Ferrocytochrome *c* bound to cardiolipin had a CD spectrum characteristic of the native structure in solution. Many of these changes in CD spectrum are probably caused by the electrostatic interaction and charge neutralization that occurs upon binding to lipid, because the presence of charged amino acids is helix-disrupting.

Evidence for changes in tertiary structure above and below the phase transition, or due to a different lipid environment, is more scarce. Small differences in the degree of penetration into the bilayer might not affect the amount of α helix if all the charged groups remain neutralized. The laser Raman spectrum of lysozyme has recently been measured above and below the T_c of DMPC–DPPA. In solution, the protein had a random conformation. The laser Raman spectrum of the protein bound to the lipid was characteristic of α helix below the T_c and of a β sheet above the T_c. This conformation change was irreversible on cooling below the T_c (Lippert *et al.*, 1980). The absorption caused by tryptophan indicated location in a more polar environment above T_c, although the protein was considered to interact electrostatically below T_c and hydrophobically above T_c.

Another approach to conformational analysis of proteins is measuring the nearness of two groups in the protein. The nonradiative energy transfer between two fluorescent groups in a membrane glycosyl-transferase was used to detect changes in the tertiary conformation when bound to different lipids. The two fluorescent groups were tryptophan and pyridoxal 5′-phosphate, which was bound to a free amino group of the protein. The results suggested that PE induced a different conformation from PA and PC, and that the conformation of the protein in all three lipids was different from that in solution (Beadling and Rothfield, 1978). PE is required for activation of this enzyme, possibly by a modification of its conformation.

The location of proteins within the bilayer, rather than their tertiary conformation, has also been monitored. Vertical displacement of membrane proteins by decreased lipid fluidity has been detected, using impermeable reagents to label the protein. In a study of red cell proteins, both intrinsic and extrinsic proteins were affected, although some were not (Borochov *et al.*, 1979). In a study of the membrane proteins of *Acholeplasma laidlawii*, the degree of labeling was altered by changes in membrane composition and fluidity; however, the results could not be interpreted in terms of increased exposure with decreased fluidity (Amar *et al.*, 1979). Particle density measurements of freeze-fracture micrographs of the alga *Anacystis nidulans* suggested that vertical displacement of intrinsic membrane proteins occurred below the T_c of the lipids, because many particles disappeared from the fracture face and reappeared when the temperature was raised above the T_c (Armond and Staehlin, 1979).

Conclusion

Studies of the interaction of purified proteins and polypeptides with chemically defined lipid bilayers of varying composition have revealed mechanisms by which the proteins can affect the fluidity and organization of lipids both directly and indirectly. Intrinsic proteins cause a pronounced decrease in the amplitude of motion of the lipid fatty acid chains adjacent to the protein (but do not order this lipid) and have a smaller motional restricting effect on the bulk lipid. They can also bind acidic lipids (and possibly specific acidic lipids) preferentially, thereby altering the lipid composition and the fluidity of the bulk lipid relative to that of a total lipid extract.

Many extrinsic membrane proteins may bind to other proteins rather than to lipids. The ways in which this might affect lipid organization are not known. Other extrinsic membrane proteins do bind directly to the lipids, however. Furthermore, some intracellular proteins may bind temporarily to membrane lipids in order to carry out a specific function. The effects of only a few extrinsic membrane or nonmembrane proteins have been studied to date. The results of these studies suggest that proteins which bind electrostatically without penetration into the bilayer should decrease lipid fluidity in the way that divalent cations do, by decreasing the charge repulsion between the lipid head groups and by tightening up the lipid packing through a bridging effect.

If the proteins also penetrate partway into the bilayer, they may decrease the amplitude of motion near the polar head group because of increased steric hindrance to motion, but may increase the amplitude of motion in the interior of the bilayer. Because this is a destabilizing effect, particularly in the gel phase, the protein may be frozen out below T_c, allowing the lipid to refreeze, or the lipid may refreeze by partial interdigitation of the fatty acid chains. This would occupy part of the space made available by expansion of the lipids caused by penetration of segments of the protein into the bilayer. The occurrence of interdigitation would result in a net decrease in the amplitude of motion of the fatty acid chains even in the interior of the bilayer, as has been observed for the complex of DPPG and basic protein. Decrease in the amplitude of motion of fatty acid spin labels does not necessarily imply an ordering effect, however. Indeed, the lipid packing may be more disordered than in the absence of the protein, and the bilayer may be more permeable.

Extrinsic proteins can also bind acidic lipids preferentially, which will alter the composition and fluidity of the bulk lipid. Proteins might also affect the fluidity indirectly by altering the pK_a of ionizable groups of the lipid, as do divalent cations, by decreasing the surface charge density of the lipid domain to which they are bound. This could induce a lipid phase transition and affect the hydrogen bonding properties of the lipid, which control the degree to which the protein can penetrate.

The effects of extrinsic proteins on lipid fluidity are probably not directly related to their mechanism of action, except in such cases as the lytic peptide mellitin and antibiotic peptides (e.g., polymixin). However, studies of the effect of proteins on lipid fluidity do provide indirect information about their mode of interaction with the bilayer, and their conformation or degree of exposure in the bilayer. Such studies indicate that the conformation or degree of exposure of the protein may depend on the phase state and type of lipid. Often, the protein may only penetrate significantly into the bilayer when the lipid is in the liquid crystalline phase and it is bound to a lipid that cannot participate in intermolecular hydrogen bonding. Because both extrinsic and intrinsic proteins can bind certain lipids preferentially, the fluidity and composition of the lipid in localized domains is more important to the membrane function than the average properties of the bulk lipid or of total lipid extracts.

The lipids and proteins in a membrane are in dynamic equilibrium with each other as well as with monovalent and divalent cations, hydrogen ions, and the other ligands in the aqueous environment. Changes in lipid composition or phase state can modulate the conformation, activity, and antigenicity of membrane proteins; concomitantly, proteins can control lipid composition by causing phase separation as do divalent cations, and may control the phase state of the lipid by altering its state of ionization. This allows numerous opportunities for control of membrane functions through alterations in both lipid fluidity and organization and protein conformation.

Acknowledgments

I would like to thank Dr. M. A. Moscarello and Dr. D. Papahadjopoulos for introducing me to the study of myelin basic protein and to acknowledge their collaboration on many of the projects discussed in this review. I also thank the Multiple Sclerosis Society of Canada for supporting this work with a Career Development Award.

References

Abramson, M. B., Katzman, R., Wilson, C. E., and Gregor, H. P. (1964). *J. Biol. Chem.* **249,** 4066–4072.

Amar, A., Rottem, S., and Razin, S. (1979). *Biochim. Biophys. Acta* **552,** 457–467.

Anthony, J. S., and Moscarello, M. A. (1971). *Biochim. Biophys. Acta* **243,** 429–433.

Armitage, I. M., Shapiro, D. L., Furthmayr, H., and Marchesi, V. T. (1977). *Biochemistry* **16,** 1317–1320.

Armond, P. A., and Staehlin, L. A. (1979). *Proc. Natl. Acad. Sci. U.S.A.* **76,** 1901–1905.

Azzi, A., Tamburro, A. M., Farnia, G., and Gobbi, E. (1972). *Biochim. Biophys. Acta* **256,** 619–624.

Bach, D., and Miller, I. R. (1976). *Biochim. Biophys. Acta* **433,** 13–19.

Bach, D., Rosenheck, K., and Miller, I. R. (1975). *Eur. J. Biochem.* **53,** 265–269.

Beadling, L., and Rothfield, L. I. (1978). *Proc. Natl. Acad. Sci.* **75,** 3669–3672.

Birrell, G. B., and Griffith, O. H. (1976). *Biochemistry* **15,** 2925–2929.

Blume, A., and Eibl, H. (1979). *Biochim. Biophys. Acta* **558,** 13–21.

Boggs, J. M. (1980). *Can. J. Biochem.* **58,** 755–771.

Boggs, J. M., and Moscarello, M. A. (1978a). *Biochim. Biophys. Acta* **515,** 1–21.

Boggs, J. M., and Moscarello, M. A. (1978b). *J. Memb. Biol.* **39,** 75–96.

Boggs, J. M., and Moscarello, M. A. (1982). *Biophys. J.* **37,** 57–59.

Boggs, J. M., Vail, W. J., and Moscarello, M. A. (1976). *Biochim. Biophys. Acta* **448,** 517–530.

Boggs, J. M., Wood, D. D., Moscarello, M. A., and Papahadjopoulos, D. (1977a). *Biochemistry* **16,** 2325–2329.

Boggs, J. M., Moscarello, M. A., and Papahadjopoulos, D. (1977b). *Biochemistry* **16,** 5420–5426.

Boggs, J. M., Stollery, J. G., and Moscarello, J. A. (1980). *Biochemistry* **19,** 1226–1233.

Boggs, J. M., Wood, D. D., and Moscarello, M. A. (1981a). *Biochemistry* **20,** 1065–1073.

Boggs, J. M., Clement, I. R., Moscarello, M. A., Eylar, E. H., and Hashim, G. (1981b). *J. Immunol.* **126,** 1207–1211.

Boggs, J. M., Stamp, D., and Moscarello, M. A. (1981c). *Biochemistry* **20,** 6066–6072.

Boggs, J. M., Moscarello, M. A., and Papahadjopoulos, D. (1982a). *In* "Lipid-Protein Interactions," (O. H. Griffith and P. Jost, eds.), Vol. 2, pp. 1–50. Wiley (Interscience), New York.

Boggs, J. M., Stamp, D., and Moscarello, M. A. (1982b). *Biochemistry* (submitted for publication).

Borochov, H., Abbott, R. E., Schachter, D., and Shinitzky, M. (1979). *Biochemistry* **18,** 251–255.

Brady, G. W., Murthy, N. S., Fein, D. B., Wood, D. D., and Moscarello, M. A. (1981). *Biophys. J.* **34,** 345–350.

Brostoff, S., and Eylar, E. H. (1971). *Proc. Natl. Acad. Sci. U.S.A.* **68,** 765–769.

Brotherus, J. R., Jost, P. C., Griffith, O. H., Keana, J. F. W., and Hokin, L. E. (1980). *Proc. Natl. Acad. Sci. U.S.A.* **77,** 272–276.

Brown, L. R. (1979). *Biochim. Biophys. Acta* **557,** 135–148.

Brown, L. R., and Wüthrich, K. (1977). *Biochim. Biophys. Acta* **468,** 389–410.

Brown, L. R., Bösch, C., and Wüthrich, K. (1981). *Biochim. Biophys. Acta* **642,** 296–312.

Butler, K. W., Hanson, A. W., Smith, I. C. P., and Schneider, H. (1973). *Can. J. Biochem.* **51,** 980–989.

Campbell, I. M., and Pawagi, A. B. (1980). *Can. J. Biochem.* **58,** 345–351.

Chan, D. S., and Lees, M. B. (1978). *J. Neurochem.* **30,** 983–990.

Chapman, B. E., and Moore, W. J. (1976). *Biochem. Biophys. Res. Commun.* **73,** 758–765.
Chapman, B. E., and Moore, W. J. (1978). *Aust. J. Chem.* **31,** 2367–2385.
Chapman, B. E., Littlemore, L. T., and Moore, W. J. (1977). *In* "Myelination and Demyelination" (J. Palo, ed.), pp. 207–220. Plenum, New York.
Chapman, D., Urbina, J., and Keough, K. M. (1974). *J. Biol. Chem.* **249,** 2512–2521.
Chapman, D., Cornell, B. A., Eliasz, A. W., and Perry, A. (1977). *J. Mol. Biol.* **113,** 517–538.
Cullis, P. R., and de Kruijff, B. (1976). *Biochim. Biophys. Acta* **436,** 523–540.
Curatolo, W., Verma, S. P., Sakura, J. D., Small, D. M., Shipley, G. G., and Wallach, D. F. H. (1978). *Biochemistry* **17,** 1802–1807.
Dawson, C. R., Drake, A. F., Helliwell, J., and Hider, R. C. (1978). *Biochim. Biophys. Acta* **510,** 75–86.
Deber, C. M., Moscarello, M. A., and Wood, D. D. (1978). *Biochemistry* **17,** 898–903.
De Bony, J., Dufourcq, J., and Clin, B. (1979). *Biochim. Biophys. Acta* **552,** 531–534.
Demel, R. A., London, Y., Geurts van Kessel, W. S. M., Vossenberg, F. G. A., and Van Deenen, L. L. M. (1973). *Biochim. Biophys. Acta* **311,** 507–519.
Drake, A. F., and Hider, R. C. (1979). *Biochim. Biophys. Acta* **555,** 371–373.
Dufourcq, J., and Faucon, J.-F. (1977). *Biochim. Biophys. Acta* **467,** 1–11.
Dunker, A. K., Williams, R. W., Gaber, B. P., and Peticolas, W. L. (1979). *Biochim. Biophys. Acta* **53,** 351–357.
Eibl, H. (1977). In "Polyunsaturated Fatty Acids" (W. H. Kunau and R. T. Holman, eds.), pp. 229–244. Am. Oil Chem. Soc., Champaign, Illinois.
Eibl, H., and Blume, A. (1979). *Biochim. Biophys. Acta* **553,** 476–488.
Eibl, H., and Woolley, P. (1979). *Biophys. Chem.* **10,** 261–271.
El Mashak, E. M., and Tocanne, J. F. (1980). *Biochim. Biophys. Acta* **596,** 165–179.
Engelman, D. M., Henderson, R., McLachlan, A. D., and Wallace, B. A. (1980). *Proc. Natl. Acad. Sci. U.S.A.* **77,** 2023–2027.
Epand, R. M., Moscarello, M. A., Zirenberg, B., and Vail, W. J. (1974). *Biochemistry* **13,** 1264–1267.
Esser, A. F., and Lanyi, J. K. (1973). *Biochemistry* **12,** 1933–1939.
Eylar, E. H., and Thompson, M. (1969). *Arch. Biochem. Biophys.* **129,** 468–479.
Feinstein, M. B., and Felsenfeld, H. (1975). *Biochemistry* **14,** 3049–3056.
Fretten, P., Morris, S. J., Watts, A., and Marsh, D. (1980). *Biochim. Biophys. Acta* **598,** 247–259.
Galla, H. J., and Sackmann, E. (1975). *J. Am. Chem. Soc.* **97,** 4114–4120.
Gazzotti, P., Bock, H. G., and Fleischer, S. (1975). *J. Biol. Chem.* **250,** 5782–5790.
Giannoni, G., Padden, F. J., and Roe, R. J. (1971). *Biophys. J.* **11,** 1018–1029.
Gomez-Fernandez, J. C., Goni, F. M., Bach, D., Restall, C. J., and Chapman, D. (1980). *Biochim. Biophys. Acta* **598,** 502–516.
Gould, R. M., and London, Y. (1972). *Biochim. Biophys. Acta* **290,** 200–218.
Gulik-Krzywicki, T., Shechter, E., Luzzati, V., and Faure, M. (1969). *Nature (London)* **223,** 1116–1121.
Hammes, G. G., and Schullery, S. E. (1970). *Biochemistry* **9,** 2555–2563.
Hartmann, W., and Galla, H.-J. (1978). *Biochim. Biophys. Acta* **509,** 474–490.
Hartmann, W., Galla, H.-J., and Sackmann, E. (1978). *Biochim. Biophys. Acta* **510,** 124–139.
Hauser, H., and Dawson, R. M. C. (1967). *Eur. J. Biochem.* **1,** 61–69.
Jacobson, K., and Papahadjopoulos, D. (1975). *Biochemistry* **14,** 152–161.
Jones, A. J. S., and Rumsby, M. G. (1975). *J. Neurochem.* **25,** 565–572.
Jost, P. C., and Griffith, O. H. (1980). *Ann. N.Y. Acad. Sci.* **348,** 391–405.
Jost, P. C., Griffith, O. H., Capaldi, R. A., and Vanderkooi, G. (1973). *Proc. Natl. Acad. Sci. U.S.A.* **70,** 4756–4763.

Juliano, R. L., Kimelberg, H. K., and Papahadjopoulos, D. (1971). *Biochim. Biophys. Acta* **241,** 894–905.
Keniry, M. A., and Smith, R. (1979). *Biochim. Biophys. Acta* **578,** 381–391.
Keniry, M. A., and Smith, R. (1981). *Biochim. Biophys. Acta* **668,** 107–118.
Kimelberg, H. K. (1976). *Mol. Cell. Biochem.* **10,** 171–190.
Kimelberg, H. K., and Papahadjopoulos, D. (1971a). *Biochim. Biophys. Acta* **233,** 805–809.
Kimelberg, H. K., and Papahadjopoulos, D. (1971b). *J. Biol. Chem.* **246,** 1142–1148.
Knowles, P. F., Watts, A., and Marsh, D. (1979). *Biochemistry* **18,** 4480–4487.
Kolber, M. A., and Haynes, D. H. (1979). *J. Membr. Biol.* **48,** 95–114.
Krigbaum, W. R., and Hsu, T. S. (1975). *Biochemistry* **14,** 2542–2546.
Laggner, P., and Barratt, M. D. (1975). *Arch. Biochem. Biophys.* **170,** 92–101.
Lauterwein, J., Bosch, C., Brown, L. R., and Wüthrich, K. (1979). *Biochim. Biophys. Acta* **556,** 244–264.
Lavialle, F., Levin, I. W., and Mollay, C. (1980). *Biochim. Biophys. Acta* **600,** 62–71.
Letellier, L., and Shechter, E. (1973). *Eur. J. Biochem.* **40,** 507–512.
Liebes, L. F., Zand, R., and Phillips, W. D. (1976). *Biochim. Biophys. Acta* **427,** 392–409.
Lippert, J. L., Lindsay, R. M., and Schultz, R. (1980). *Biochim. Biophys. Acta* **500,** 32–41.
Littlemore, L. A. T. (1978). *Aust. J. Chem.* **31,** 2387–2398.
Littlemore, L. A. T., and Ledeen, R. W. (1979). *Aust. J. Chem.* **32,** 2631–2636.
London, Y., and Vossenberg, F. G. A. (1973). *Biochim. Biophys. Acta* **307,** 478–490.
London, Y., Demel, R. A., Guerts van Kessel, W. S. M., Vossenberg, F. G. A., and Van Deenen, L. L. M. (1973). *Biochim. Biophys. Acta* **311,** 520–530.
MacDonald, R. C., Simon, S. A., and Baer, E. (1976). *Biochemistry* **15,** 885–891.
Martenson, R. E. (1978). *J. Biol. Chem.* **253,** 8887–8893.
Mateu, L., Caron, F., Luzzati, V., and Billecocq, A. (1978). *Biochim. Biophys. Acta* **508,** 109–121.
Mitranic, M. M., and Moscarello, M. A. (1980). *Can. J. Biochem.* **58,** 809–814.
Mollay, C. (1976). *FEBS Lett.* **64,** 65–68.
Mombers, C., van Dijck, P. W. M., Van Deenen, L. L. M., de Gier, J., and Verkleij, A. J. (1977). *Biochim. Biophys. Acta* **470,** 152–160.
Mombers, C., Verkleij, A. J., de Gier, J., and Van Deenen, L. L. M. (1979). *Biochim. Biophys. Acta* **551,** 271–281.
Mombers, C., de Gier, J., Demel, R. A., and Van Deenen, L. L. M. (1980). *Biochim. Biophys. Acta* **604,** 52–62.
Moshkov, D. A., Severina, Ye. P., Lazarev, Ya. D., and Borovyagin, V. L. (1975). *Biofizika* **20,** 233–237.
Noll, G. G. (1976). *J. Membr. Biol.* **27,** 335–346.
Palmer, F. B., and Dawson, R. M. C. (1969). *Biochem. J.* **111,** 629–636.
Papahadjopoulos, D. (1968). *Biochim. Biophys. Acta* **163,** 240–254.
Papahadjopoulos, D., and Weiss, L. (1969). *Biochim. Biophys. Acta* **183,** 417–426.
Papahadjopoulos, D., Cowden, M., and Kimelberg, H. (1973). *Biochim. Biophys. Acta* **330,** 8–26.
Papahadjopoulos, D., Moscarello, M. A., Eylar, E. H., and Isac, T. (1975). *Biochim. Biophys. Acta* **401,** 317–335.
Pascher, I. (1976). *Biochim. Biophys. Acta* **455,** 433–451.
Quinn, P. J., and Dawson, R. M. C. (1969a). *Biochem. J.* **113,** 791–803.
Quinn, P. J., and Dawson, R. M. C. (1969b). *Biochem. J.* **115,** 65–75.
Ranck, J. L., Keira, T., and Luzzati, V. (1977). *Biochim. Biophys. Acta* **488,** 432–441.
Rand, R. P. (1971). *Biochim. Biophys. Acta* **241,** 823–834.
Rand, R. P., and Sengupta, S. (1972). *Biochemistry* **11,** 945–949.

Rand, R. P., Papahadjopoulos, D., and Moscarello, M. A. (1976). *20th Annu. Meet.—Biophys. Soc. Abstra.*

Rank, J. L. and Tocanne, J. S. (1982a). *FEBS Lett.* **143,** 171–174.

Rank, J. L. and Tocanne, J. S. (1982b). *FEBS Lett.* **143,** 175–178.

Robinson, J. D., Birdsall, N. J. M., Lee, A. G., and Metcalfe, J. C. (1972). *Biochemistry* **11,** 2903–2909.

Sefton, B. M., and Gaffney, B. J. (1974). *J. Mol. Biol.* **90,** 343–358.

Shafer, P. T. (1974). *Biochim. Biophys. Acta* **373,** 425–435.

Shipley, G. G., Leslie, R. B., and Chapman, D. (1969). *Nature* **222,** 561–562.

Sixl, F., and Galla, H.-J. (1979). *Biochim. Biophys. Acta* **557,** 320–330.

Smith, I. C. P., Williams, R. E., and Butler, K. W. (1974). *Protides Biol. Fluids* **21,** 215–222.

Smith, R. (1977a). *Biochim. Biophys. Acta* **470,** 170–184.

Smith, R. (1977b). *Biochim. Biophys. Acta* **491,** 581–590.

Steck, A. J., Siegrist, H. P., Zahler, P., and Herschkowitz, N. N. (1976). *Biochim. Biophys. Acta* **455,** 343–352.

Stollery, J. G., Boggs, J. M., and Moscarello, M. A. (1980a). *Biochemistry* **19,** 1219–1225.

Stollery, J. G., Boggs, J. M., Moscarello, M. A., and Deber, C. M. (1980b). *Biochemistry* **19,** 2391–2396.

Susi, H., Sampugna, J., Hampson, J. W., and Ard, J. S. (1979). *Biochemistry* **18,** 297–301.

Sweet, C., and Zull, J. E. (1970). *Biochem. Biophys. Res. Commun.* **41,** 135–141.

Träuble, H. (1976). *In* "Structure of Biological Membranes" (S. Abrahamsson and I. Pascher, eds.), pp. 509–550. Plenum, New York.

Träuble, H., and Eibl, H. (1974). *Proc. Natl. Acad. Sci. U.S.A.* **71,** 214–219.

Träuble, H., and Overath, P. (1973). *Biochim. Biophys. Acta* **307,** 491–512.

Utsumi, H., Tunggal, B. D., and Stoffel, W. (1980). *Biochemistry* **19,** 2385–2390.

Van, S. P., and Griffith, O. H. (1975). *J. Membr. Biol.* **20,** 155–170.

Van Zoelen, E. J. J., van Dijck, P. W. M., de Kruijff, B., Verkleij, A. J., and Van Deenen, L. L. M. (1978). *Biochim. Biophys. Acta* **514,** 9–24.

Verma, S. P., and Wallach, D. F. H. (1976). *Biochim. Biophys. Acta* **426,** 616–623.

Verma, S. P., Wallach, D. F. H., and Smith, I. C. P. (1974). *Biochim. Biophys. Acta* **345,** 129–140.

Vitello, L., Kresheck, G. C., Albers, R. J., Erman, J. E., and Vanderkooi, G. (1979). *Biochim. Biophys. Acta* **557,** 331–339.

Watts, A., Volotovski, I. D., and Marsh, D. (1979). *Biochemistry* **18,** 5006–5013.

Wu, C.-S. C., Ikeda, K., and Yang, J. T. (1981). *Biochemistry* **20,** 566–570.

Young, E. G. (1963). *Compr. Biochem.* **7,** 25.

Yu, C., Yu, L., and King, T. E. (1975). *J. Biol. Chem.* **250,** 1383–1392.

Yu, K.-Y., Baldassare, J. J., and Ho, C. (1974). *Biochemistry* **13,** 4375–4381.

Chapter 5

Lateral Phase Separations and the Cell Membrane

Chris W.M. Grant

Introduction

The concept of *phase separation* will be discussed relative to its significance to the architecture and fluidity of the eukaryotic plasma membrane. The term phase separation will be defined as spontaneous coexistence of membrane domains with different composition. Excellent review articles now exist covering many of the problems that must be addressed; it remains only to apply such data to conditions known or suspected in real membranes. Almost certainly the same logic may then be applied to eukaryotic subcellular membranes and to prokaryotes by deemphasizing the role of cholesterol and cytoskeletal elements.

The problem of fluidity becomes a corollary of the problem of lipid distribution for a given membrane at a particular temperature. Integral pro-

ISBN 0-12-053002-3

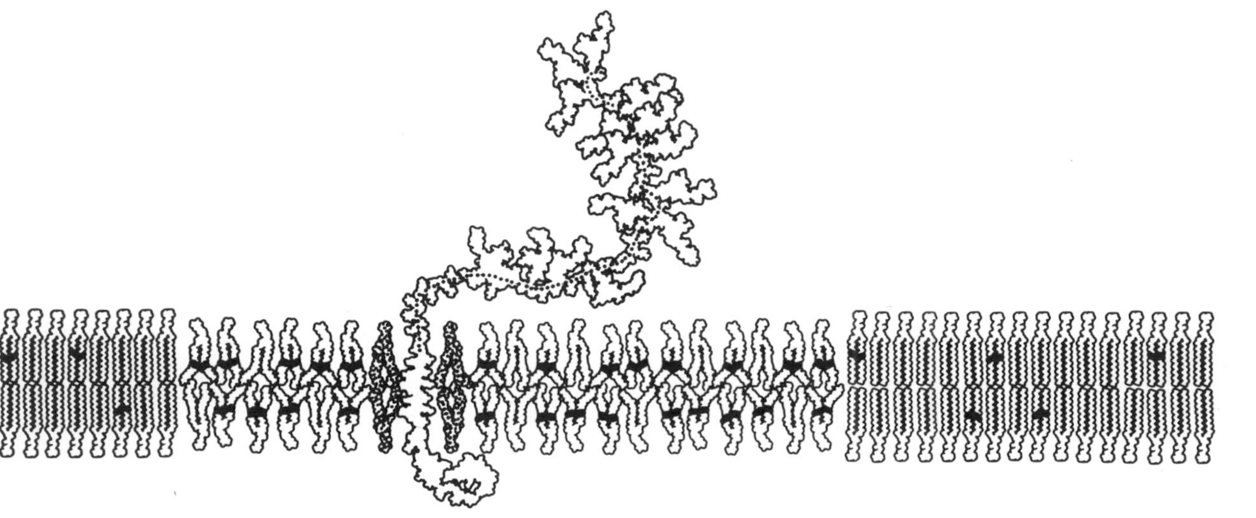

Fig. 1. Cross section through a bilayer membrane showing a fluid/rigid phase separation in a hydrated mixture of two phosphatidylcholines [such as dioleoyl (with unsaturated positions of acyl chains in black) and dipalmitoyl phosphatidylcholine]. Note that coexisting phases are not usually composed of pure species, but simply show selective enrichment. Scale: fluid phase is 58.3 Å thick. An integral protein has been included, drawn to accurately satisfy the specifications of the 30,000 MW glycoprotein, glycophorin (Tomita *et al.*, 1978). Its headgroup carries 16 oligosaccharide chains that extend from a single unstructured length of polypeptide backbone (dotted line). The protein has been placed in the fluid phase (see Section IV). Protein annular lipid (fine dotted shading) will often have composition and properties detectably different from surrounding lipid.

teins and cholesterol will be seen to have the potential for rigidifying fluid regions of bilayer, and for disordering crystalline regions. It must be emphasized that the physiological temperature range for most human cells is not a narrow band centered at 37°C, but extends several degrees higher and many degrees lower.

Illustrations have been based on Corey–Pauling–Koltum (CPK) models where possible and are rigorously to scale. One may begin with the classic lipid bilayer in Fig. 1, considering the fact that cell membranes contain mixtures of lipids with different headgroup, backbone, or acyl chain characteristics. As will be described later, based on such differences hydrated lipid mixtures usually will show a measurable tendency to separate into coexisting, compositionally distinct patches (phases). At this early stage, the bilayer has been left symmetric with regard to the two monolayer leaflets that comprise it, and headgroup interactions have not been specifically shown. An integral membrane protein has been added to the phase separated bilayer: it may be expected to (a) preferentially occupy the more fluid phase, and (b) take with it a certain number of lipids with thermodynamically preferred (but not absolutely required) characteristics. Although a solid–fluid phase separation has been depicted in Fig. 1, coexisting phases may also be both rigid or both fluid.

One must eventually progress to models that include lipid bilayer asymmetry and a divalent cation role in structure (Fig. 2). Unequal distribution of lipid types between the monolayers making up the bilayer, with relative independence of the two monolayers, could well be a universal phenomenon. The cation effect is by far most striking in the case of "negatively

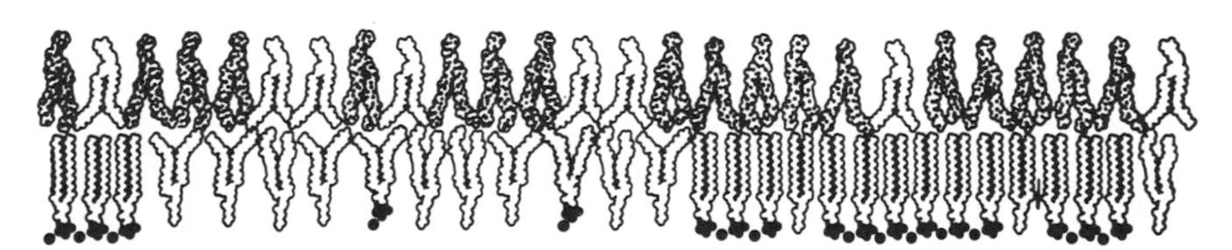

Fig. 2. Cross section through an asymmetric lipid bilayer whose composition reflects qualitatively certain features of the human erythrocyte: sphingomyelin (shaded) and phosphatidylcholine (white) predominance at the outer surface (upper monolayer), phosphatidylserine (black headgroups) and phosphatidylethanolamine (white) at the cytoplasmic surface (lower monolayer) (Chap *et al.*, 1977). A phase separation based on Ca^{2+}-induced headgroup cross-linking of phosphatidylserine (Ca^{2+}, black dots) is shown at the cytoplasmic surface. Ca^{2+}-cross-linking of the acidic phosphatidylserine headgroups has led to patchy rigidification. Each phosphatidylserine has been shown with a bound Ca^{2+}, although at cytoplasmic Ca^{2+} concentrations this will not be the case. The outer leaflet has been drawn as fluid with a tendency toward regional enrichment in one lipid or another. Domains will be impure and transient in such a system.

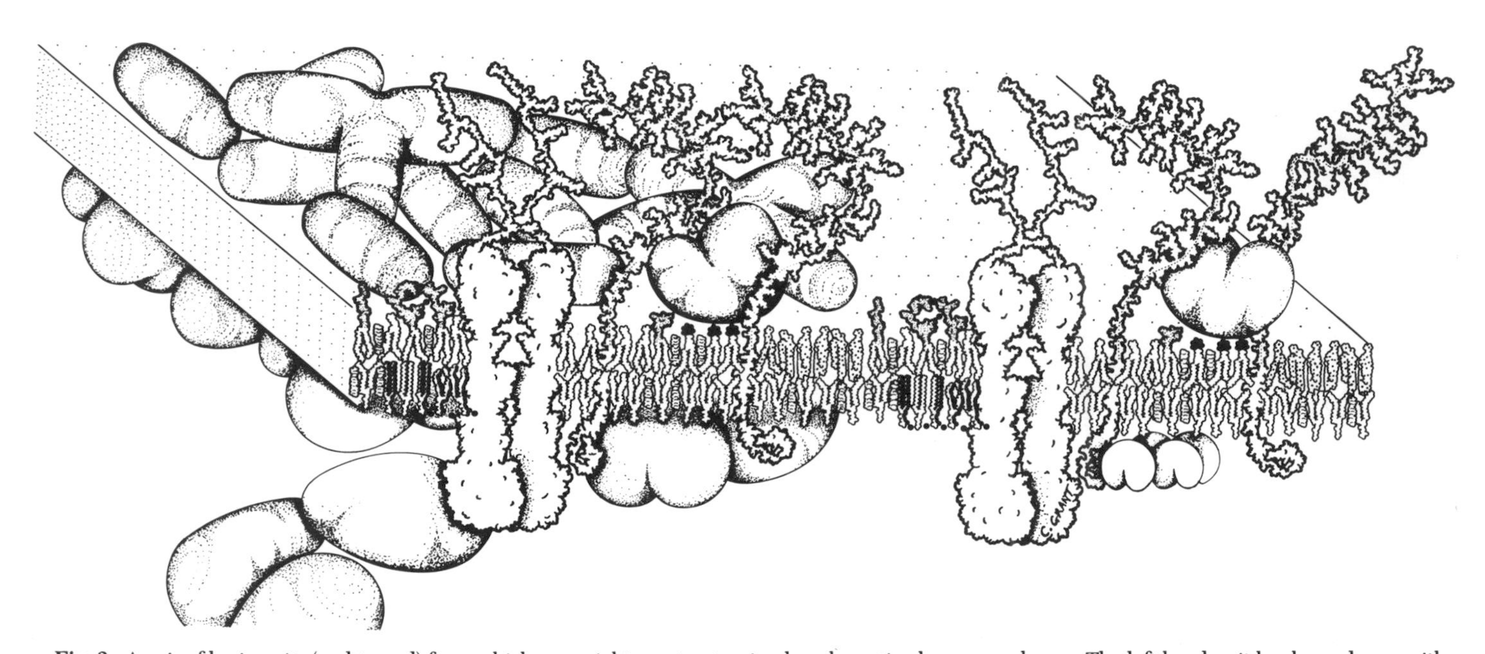

Fig. 3. A pair of basic units (end to end) from which one might construct a simple eukaryotic plasma membrane. The left hand unit has been drawn with adsorbed albumin and IgG on the upper (external) surface, and cytoplasmic proteins at the lower surface (see Section VII).

charged" phospholipids; a phase-separated patch of phosphatidylserine maintained by Ca^{2+} is shown in Fig. 2.

Figures 1 and 2 suggest a major role of acyl chain "melting" in the lipid phase separation. Certainly this phenomenon can be expected to have important consequences for membrane component distribution, as will be discussed. However, these consequences will be significantly altered by the presence of cholesterol and integral proteins and because real membranes possess complex mixtures of lipids. Phase separation of various lipids may in many cases be more realistically based on headgroup interactions as in Fig. 3; or a Ca^{2+} induced separation like that in Fig. 2, for example, might well exist without the dramatic acyl chain "freezing" shown.

Other factors to be considered with regard to membrane phase separations are subsequently developed and incorporated into Figs. 3 and 4, including the presence of membrane proteins at cytoplasmic and external surfaces. Both integral proteins and peripheral proteins may be expected to make extensive ionic, hydrogen bonding, and hydrophobic interactions with (often specific) membrane lipids, although in this chapter hydrophobic contacts have not been shown for the latter class of proteins. Oligosaccharide headgroups may extend hundreds of angstroms from the outer surface. Cytoskeletal elements such as microfilaments and microtubules attach directly or indirectly to the inner surface. And lastly, but importantly, all membrane surfaces must be expected to be liberally covered with nonspecific protein: species reversibly adsorbed from the fluid in contact with the membrane surface. In this last category, serum species such as albumin and the globulins commonly occur at the eukaryotic external surface, and all cytoplasmic species should be represented at the inner surface. Wherever proteins contact lipids there will be effects on lipid lateral distribution, packing, and mobility.

Various techniques have been employed to monitor phase separation phenomena in membranes; each approach has its devotees and its detractors. Many arguments that have arisen are succinctly summarized in one scientist's rueful comment that "the limit of zero perturbation is often the limit of zero information." Fortunately, such considerations are not within the scope of this chapter.

Basic Lipid Arrangement

It is interesting that the cell membrane with its intricate compartmentalized functions is held together almost entirely by noncovalent forces. Many individual membrane features and functions have been successfully regenerated

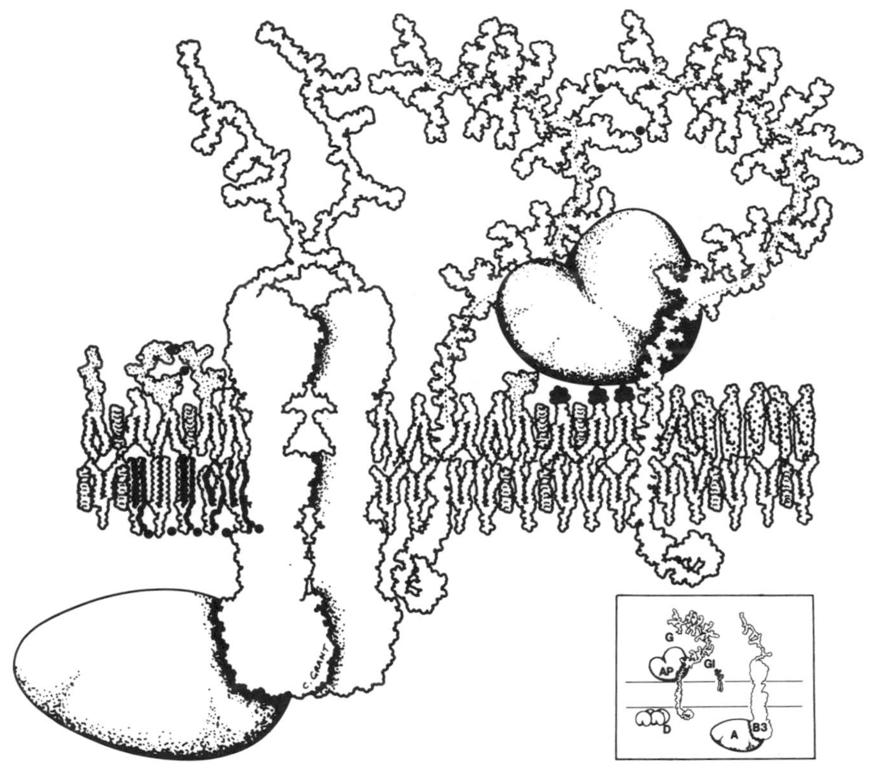

Fig. 4. Leading edge of a single basic unit enlarged to show details of component structure/arrangement. Each unit contains about 900 lipid molecules in a bilayer segment 315 Å square. Each unit also contains four integral proteins (two copies of B3 plus one G in a stable trimeric cluster, and one G not involved in the cluster), and two peripheral proteins (AP at the outer surface, and A or D at the inner surface). The protein components have been chosen to correspond to species of the human erythrocyte, but the basic features undoubtedly are fairly ubiquitous. The basic erythrocyte membrane could be roughly approximated by a sealed sheet of 5×10^5 pairs of basic units illustrated in Fig. 3; a more complete picture would require a few additional membrane proteins, a cytoskeletal network, and more nonspecific adsorbed protein at each surface. A more complex eukaryotic cell would combine various proportions of as many as six types of basic unit although each type would have features similar to those shown. Because each separate 450-lipid monolayer of the basic unit will be composed of several major lipid families, a monolayer microdomain of 100 lipid molecules of a particular family is about the maximum expected in a given repeat unit (a patch of 10×10 lipids) even if the driving force for phase separation were strong enough to pull them together. In many subunits, only vestigial lipid–lipid phase separations may exist, with a slight local enrichment of one lipid type over another. One reason is that in many subunits, because of contact with cholesterol and contact with integral proteins, the cooperativity of acyl chain melting will be reduced almost to zero. Cooperative chain melting must be expected to be confined to leftover lipids (ones not adjacent to cholesterol or integral proteins). The fraction of lipid shown in strikingly phase-separated domains may have been over-estimated by attempting to show all the various possibilities. In general, a given lipid patch will be a transient structure. [The following notes refer to boxed

from isolated components by simply recombining them under conditions such that they can spontaneously seek their arrangement of minimal Gibbs free energy (e.g., dialytic removal of detergent from aqueous solutions). In fact, probably the only cell membrane feature that cannot be accurately reproduced in this way is the overall topographic array of the various subsystems. Directed by this kind of logic, extensive physical studies have been made of purified lipids in an effort to determine their role and arrangement in cell membranes. A bewildering body of data has resulted. However, it is this writer's opinion that the only protein-independent lipid structural feature well documented in cell membranes is the simple bilayer in either α-structure (acyl chains mobile, with frequent gauche kinks in each chain but average chain orientation perpendicular to the plane of the bilayer), or β-structure (acyl chains all trans and rigidly extended perpendicular to the bilayer plane) (Fig. 1) (Tardieu *et al.*, 1973).

For several decades it has been recognized that certain isolated lipids can form nonbilayer micelles when hydrated. Pure cholesterol forms platelike

---

insert.] B3: Integral glycoprotein of MW 95,000 designed after *band 3* (Steck, 1978) to have 35,000- and 42,000-MW portions mushrooming from the outer and inner surfaces respectively; a 17,000-MW portion is buried in the membrane. With a single carbohydrate chain of some 40 sugar residues, it exists as a dimer with a second band 3 molecule to form an anion transport channel. G: integral glycoprotein described in Legend of Fig. 1. Each pair of band 3 molecules forms a lipid/protein cluster with one glycophorin; the other G in each basic unit is independent. Such clusters probably trap lipid (Davoust *et al.*, 1980). The associations between B3 and G were taken from Nigg *et al.*, 1980 but see also Haest, 1982. A: some band 3 molecules bind to a molecule (Ankyrin) of about 230,000 MW (Branton *et al.*, 1981) that may be analogous to intermediate filament proteins of other eukaryotic cells. The degree of lipid involvement of such species is unknown, but they represent suitable attachment points for cytoskeletal elements. AP: a peripheral protein having enzymatic activity (e.g., 150,000 MW alkaline phosphatase; see Section VII). It gives rise to its own phase separation via an affinity for phosphatidylinositol headgroups (here shown as 3 blackened structures). Note that it does not exclude cholesterol from its contacts. D: a cytoplasmic peripheral enzyme of no known lipid preference that might represent glucose-6-phosphate dehydrogenase. Ca^{2+}-mediated cluster of the acidic lipid, phosphatidylserine, immediately left of the B3 dimers at the cytoplasmic surface. Phosphatidylserine in the patch has black acyl chain borders; the headgroups are associated with Ca^{2+} (black circles). Part of the same patch is involved in binding to a hypothetical acidic lipid site on the integral protein (see Section IV). A rigid sphingomyelin/cholesterol patch (shaded with heavy dots) is shown at the outer surface to the right of each glycophorin that is not involved in the B3/G trimer. In these patches, the cholesterol has been drawn so closely apposed (in a 1:1 ratio) to five sphingomyelin molecules that only the outlines of the resultant sphingomyelin/cholesterol dimers are shown. Cholesterol has been shown as short, crosshatched cylinders inserted perpendicular to the bilayer; it is not found in close association with integral proteins.

GI: Glycosphingolipids (headgroups finely stippled) are shown with a finite tendency to cluster. At several points, Ca^{2+} (filled circles) is shown mediating sialic acid residue cross-links in the glycocalyx (Sharom and Grant, 1978; Peters *et al.*, 1982).

crystals in which it separates out from hydrated cholesterol–phospholipid mixtures when the molar ratio exceeds 1:1 (Oldfield and Chapman, 1972; Demel and de Kruijff, 1976). Pure glycosphingolipids in water form various nonbilayer phases (Tinker *et al.*, 1976; Bunow and Bunow, 1979), as do cardiolipin, phosphatidylethanolamine, and sphingomyelin under appropriate conditions (Luzzati, 1968; de Kruijff *et al.*, 1980). Such structures are not evident in homogeneous natural lipid mixtures. However, a number of workers have entertained the idea that given a tendency for local enrichment of certain lipids in cell membranes microregions of a form other than bilayer may be expected. If this is true it is very important and the efforts of Cullis and De Kruijff (1979) in this area are noteworthy, but to date no compelling experimental evidence exists for the actual existence of protein-independent nonbilayer structures in cell membranes. Nevertheless, it remains valid that any phase-separated domain enriched in a lipid with a propensity for existence in nonbilayer forms would be expected to reflect the fact in properties such as stability and ion permeability. Furthermore, regions of lipid/protein interface need not be bilamellar, and sites of membrane fusion or lipid flip-flop must be transiently disordered.

Lipids in membranes are, regardless of structural arrangement, driven together by entropic considerations; the single most important requirement is the avoidance of hydrocarbon chain contact with water. Given the resulting enforced tight proximity, maximizing the enthalpy of interaction through optimal packing of headgroups and acyl chains and optimal headgroup ionic or hydrogen bonding becomes a secondary consideration. Hence, the lipid bilayer has finite crystal properties even in "liquid crystal" form. Clearly this is most evident at low temperatures and when the lipids are homogeneous with regard to headgroup and acyl chain characteristics. In a pure, single phospholipid (e.g., dimyristoyl phosphatidylcholine), the hydrated bilayer form actually has a sharp "melting point" or *phase transition temperature.* When heated at a lower temperature, the rigidly extended, hexagonally packed acyl chains undergo a highly cooperative transition over a period of several degrees to the disordered (fluid) phase (Tardieu *et al.*, 1973; Chapman, 1975; Lee, 1977). That is, the favorable enthalpy of headgroup and acyl chain packing is eventually too weak to completely overcome the tendency to molecular motion. In most cases the bilayer structure itself is maintained with increased temperature, although the area per lipid molecule rises substantially and the bilayer thickness drops correspondingly. The melting point depends on the headgroup, on acyl chain length (higher with longer chains), and degree of chain unsaturation (lower with increasing unsaturation).

However, lipid melting is not a simple process. It has features of polymer behavior: measurable related rearrangements occur before the melting point is reached [e.g., premelting some 6–10°C below the transition temperature (Chapman, 1975; Lee, 1977; and references in Scott, 1981)], and well above the transition temperature [e.g., formation of prefreezing dissociable lipid clusters (Lee *et al.*, 1974; Lee, 1977; Lenaz, 1977)]. In other words the fluid–gel transition phenomenon in lipids has important structural consequences at temperatures very different from the actual lipid transition temperature.

The basic comparison of lipid bilayer thermal response with crystal melting has been extended by Chapman and McConnell and co-workers to explain certain characteristics of mixed lipids (Phillips *et al.*, 1970; Shimshick and McConnell, 1973; Shimshick *et al.*, 1973; Wu and McConnell, 1975). They suggested that a bilayer formed of two pure lipids, for example, may be treated in a fashion similar to a mixture of two pure solids, and that a phase diagram can be constructed to catalog the various phases. Of course lipid mixtures are essentially two-dimensional systems such that coexisting regions of different composition and properties are *laterally* separated. This phenomenon has now been demonstrated and characterized in an impressive array of model systems (reviewed in Lee, 1977), and in a less impressive array of real membranes: *E. coli* (Linden *et al.*, 1973; Schechter *et al.*, 1974; Kleemann *et al.*, 1974; Morrisett *et al.*, 1975; Linden *et al.*, 1977), *A. laidlawii* (Read and McElhaney, 1975; Wallace *et al.*, 1976; Bevers *et al.*, 1977), algae (Furtado *et al.*, 1979), and higher organisms (Wisnieski *et al.*, 1974; Petit and Ediden, 1974; Tanaka and Ohnishi, 1976; Marinetti, 1977; Sklar *et al.*, 1979; Klausner *et al.*, 1980). Typically, interpretations of the results from higher organisms are more debatable. In fact, given that the membranes of higher organisms contain a considerable mixture of different lipids, and that extensive regions of any pure component are unlikely to exist, it becomes nontrivial to apply the above logic to eukaryotes. For instance, although domains as small as 11 to 60 lipid molecules are capable of cooperative behavior, their melting takes place over a broader temperature range (Lee, 1975; Yellin and Levin, 1977), which probably explains in part the less dramatic temperature effects seen in intact cell membranes. The view taken in this chapter will be that treating cell membrane lipids as mixed melting solids is a very useful basic approach, but one that must be applied conservatively. Note that although Fig. 1 shows a fluid–rigid phase separation, fluid–fluid phase separations have also been documented (Recktenwald and McConnell, 1981).

An interesting corollary of lipid–lipid phase separations is that at phase boundaries one would expect increased lateral compressibility because small area increases would be absorbed at essentially zero energy cost at the expense of acyl chain flexibility (Linden *et al.*, 1973; Phillips *et al.*, 1975). Certainly this phenomenon is reflected in permeability and stability changes in simple lipid systems (Papahadjopoulos *et al.*, 1973; Lee, 1975; Marsh *et al.*, 1976; Thilo *et al.*, 1977), but its importance in eukaryotic cells remains to be demonstrated [see, however, Wisnieski *et al.*, (1974)].

Cation Effects

The monovalent cations Na^+ and K^+ apparently have a slight disordering effect at high concentration upon lipid bilayers (Trauble and Eibl, 1974), perhaps through their effect on water structure, but do not bind with high affinity. Rightly or wrongly, they will be largely ignored in this chapter. On the other hand, most workers will agree that Ca^{2+} and Mg^{2+} must not be neglected in a discussion of phase separations in membranes (Hauser *et al.*, 1976). Certainly chelation of Ca^{2+} leads to dissociation of numerous peripheral proteins and to considerable loss of membrane integrity. Ca^{2+} occurs at about 2 mM outside most eukaryotic cells and at less than 0.01 mM inside, Mg^{2+} is about 1–2 mM both inside and out. The molecular basis of their role is not clear, but it is generally agreed that they promote certain noncovalent associations involving proteins as well as lipids. Both Ca^{2+} and Mg^{2+} bind with appreciable affinity to negatively charged headgroups (e.g., K_D 10^3 M for Ca^{2+}/phosphatidylserine) resulting in charge neutralization and tighter packing (Behr and Lehn, 1973; Hauser *et al.*, 1976; van Dijck *et al.*, 1978). Ca^{2+} in particular is well known to cross-link phosphatidylserine headgroups leading to phase separation of the latter as indicated in Figs. 2, 3, 4 (Papahadjopoulos *et al.*, 1976). Berclaz and McConnell (1981) have demonstrated the production of specific lipid associations and fluid–fluid phase separations in cardiolipin–phosphatidylcholine bilayers upon Ca^{2+} addition. A lipid/protein interaction involving Ca^{2+} has been specifically indicated at the inner surface in Fig. 4. It is worth keeping in mind that membrane surfaces exposed to the cytoplasm may have to deal with fluctuating Ca^{2+} concentrations over a range such that the amount bound will be a sensitive function of the free cytoplasmic concentration. Thus one may expect certain Ca^{2+}-mediated associations within the cell to be under cellular control. This kind of argument has been invoked, for instance, in postulating a role for Ca^{2+}/phosphatidylserine in neurotransmitter release (Papahadjopoulos *et al.*, 1979). The outer surface of eukaryotic cells bears a net minus charge, largely because of glycolipid and glycoprotein sialic acid residues. At the high concentrations found in serum and interstitial fluid, divalent cation-induced cross-linking of such charged groups may contribute importantly to receptor distribution and glycocalyx integrity (Sharom and Grant, 1978) (Fig. 3).

Integral Proteins and the Phase Separation

Integral membrane proteins, which penetrate deeply into the hydrophobic interior, are thought to fold in such a way that predominantly hydrophilic amino acid residues are exposed to the lipid headgroup region and more hydrophobic ones are outermost at the level of the acyl chains. Thus, one must specifically consider headgroup lipid/protein forces as well as entropy-driven interactions with the acyl chains. It has been pointed out that the polypeptide hydrophobic surface presented to lipid need not be uniformly rigid, and may in fact be locally fluid depending upon the nature of the amino acid side chains involved (Wallach, 1979).

As outlined in Section III, even fluid, mixed lipid bilayer regions have a certain crystal nature. In general, a penetrating protein must be expected to fit less than optimally into such a lattice; it will be treated as an impurity in a zone-refined crystal, and there will be a tendency to extrusion (Grant and McConnell, 1974). Furthermore, because the (membrane) crystal is semi-fluid, extrusion can occur just as in zone refining of inorganic materials. However, protein "impurities" cannot be completely extruded vertically (see however Borochov *et al.*, 1979) as that would lead to exposure of hydrophobic amino acids to water; thus, extrusion in the plane of the bilayer, such that integral proteins occupy the more disordered (less crystalline) regions of membranes, occurs in a number of systems (Chen and Hubbell, 1973; Grant and McConnell, 1974; Kleemann *et al.*, 1974; Shechter *et al.*, 1974; Kleemann and McConnell, 1976; Wallace *et al.*, 1976; Sharom *et al.*, 1977; Lenaz, 1977). In the eukaryote, because gross protein distribution is regulated by cytoskeletal constraints into a more or less diffuse network, one may expect this phenomenon to operate over shorter distances and more subtly; instead of being grossly swept into large patches, proteins will tend to exist in small clusters surrounded locally by the least crystalline available lipids. Clearly one can envisage cases in which headgroup ionic or covalent attachments might override this nonspecific type of phase separation, but in general one can expect the enthalpy of an arrangement to be favorably optimized by keeping the most crystalline lipid patches away from integral proteins. This rule of thumb is illustrated in Fig. 1. The same sort of consideration may also have a bearing on distribution of other "oddly shaped" molecules such as cholesterol (Kleemann and McConnell, 1976; Lenaz, 1977; Demel *et al.*, 1977) and dolichol (McCloskey and Troy, 1980). It has a firm theoretical basis (Kleemann and McConnell, 1976; Marcelja, 1976; Owicki *et al.*, 1978).

Distinct from this phenomenon of integral proteins as "impurities" in a crystal lattice, there is the complex question of their potential for preferential interactions with certain lipids. Intuitively one might expect that a given integral protein might fit better among longer or shorter acyl chains, or might make more efficient ionic and hydrogen bonds with certain headgroups. Although acyl-chain preference has not been shown to be important where tested (Boggs and Moscarello, 1978; Caffrey and Feigenson, 1981), there is evidence of a measurable preference for association with certain negatively-charged headgroups in the case of rhodopsin (Watts *et al.*, 1979), lipophilin (Boggs *et al.*, 1977b), and (Na^+, K^+)-ATPase (Brotherus *et al.*, 1980), for example. Thus, one must expect at least some integral proteins to generate their own small phase separated regions of lipids via special affinity for various headgroups. Presumably such specific attachments will often be dynamic, with "bad fit" lipids simply occupying nearest neighbor positions less frequently and for shorter times than "best fit" lipids. Also, presumably different parts of a protein or protein cluster can have different requirements for lipid. At the other end of the spectrum, it is generally accepted that close association of cholesterol with integral proteins is universally improbable (Warren *et al.*, 1975; Wallach, 1979).

One is led inexorably to the thorny problem of annular lipid. A given type of integral protein is supposed by many to possess an annulus of 30 or more lipid molecules which behave, as a result of their nearest-neighbor association with protein, in a manner distinct from normal bilayer lipid. Certainly it is agreed that theoretically an integral protein should influence its nearest-neighbor lipids whereas lipids farther away will be considerably less affected (Kleemann and McConnell, 1976; Marcelja, 1976; Owicki *et al.*, 1978). Like cholesterol, integral proteins probably form a relatively rigid insertion along lipid acyl chains, resulting in reduced cooperativity in annular lipid melting; the amount of lipid so affected in membranes can be as much as 35%. As already pointed out, annular lipid should show a composition measurably different from that of the bulk lipid although in equilibrium with it. It seems that some annular lipid will be rigid, either through simple interaction with protein or by entrapment in clusters and crevices, and in other configurations it will be quite fluid (Jost and Griffith, 1980; Davoust *et al.*, 1980). Additional evidence (London and Feigenson, 1981) suggests that annular lipid composition may be a sensitive function of the degree of local lipid phase separation and rigidity as predicted theoretically by Owicki *et al.* (1978).

Cholesterol

The presence of cholesterol as a structural component of eukaryotic cell plasma membranes imposes some important constraints on the concepts so far presented (although internal organelle membranes contain relatively little cholesterol and virtually all prokaryote membranes are sterol free). Cholesterol, by far the most structurally significant steroid in the membrane, is a cylinder which fits with its long axis perpendicular to the membrane plane and its 3-OH group at the polar surface. Its locked ring structure forms a rigid insertion to a depth of about C-10 on the acyl chains surrounding it. The effect is to reduce the cooperativity of interaction among nearby acyl chains and to broaden phase transitions (Oldfield and Chapman, 1972; Kleemann and McConnell, 1976; Demel and de Kruijff, 1976; Lee, 1977). Cholesterol fluidizes gel phase lipid bilayers by perturbing the lattice and tends to rigidify fluid lipid bilayers. It has been claimed to form associations (1:1?) with lipids in the following order of affinity (van Dijck, 1979; Boggs, 1980, and refer therein; but see Barenholz and Thompson, 1980):

sphingomyelin ≫ phosphatidylserine, phosphatidylglycerol > phosphatidylcholine ≫ phosphatidylethanolamine

These preferences aside, cholesterol (like proteins) tends to occupy the more fluid phase (as pointed out in Section IV) and its phase behavior in bilayer membranes is poorly understood (Recktenwald and McConnell, 1981 and references therein). All cells have access to the exchangeable serum lipoprotein pool of cholesterol. Red blood cells and myelin are particularly rich with a 1:1 mol ratio of cholesterol/phospholipid; liver plasma membranes may have a ratio as low as 0.5:1. The values for plasma membranes of other tissues are less well known but are probably somewhat lower.

The significance of cholesterol to phase separations in cell membranes is (a) its tendency to reduce the cooperativity of acyl chain fluid–rigid transitions, which should in turn reduce the importance of acyl chain melting/rigidification as a source of phase separation (b) its probable general avoidance of contact with integral proteins (Section IV), and (c) the question of its distribution relative to other lipids. It is probably extreme to suggest that the presence of cholesterol in cell membranes abolishes the role of acyl chain melting in phase separations. On the other hand, it would be even more extreme to ignore cholesterol altogether. The question of how cholesterol fits into the picture of membrane structure is indeed a difficult one. It seems that cholesterol can form phase-separated regions with other lipids in 1:1 to 1:3 mol ratios (reviewed in Lee, 1977; Rectenwald and McConnell, 1981). Perhaps very importantly, cholesterol may also exist as pure patches in regions of the membrane so far unidentified (Wallach, 1979). In this regard there have been claims that the lateral distribution of cholesterol may be grossly inhomogeneous (Murphy, 1965). In Fig. 3, the quantity included is conservative relative to the amount known to exist somewhere in erythrocytes.

Temperature

It is often implied that although the effects of temperature variation on membrane lipids are interesting they have little relevance to the (homeothermic) mammal. Of course this is not true. In humans, for example, the majority of cells fail to be maintained at 37°C and frequently fluctuate widely in temperature over periods of hours or even minutes. It is only the core temperature that is closely regulated at 37°C (Burton and Edholm, 1955); virtually all cells in the legs, arms, superficial portions of torso, and head register a gradient down to ambient temperature. For an extreme, one might consider a lunchtime squash player who returns to his lab in winter lightly dressed; within minutes, his surface cells experiencing a drop from 40 to 5°C. One is reminded of a tongue-in-cheek article by Hershkowitz (1977) concerning the horrors of penile frostbite. For natural lipid mixtures, temperatures below 37°C greatly increase the tendency to phase separation. Furthermore, as already pointed out in Section I, the existence of acyl chain melting at a particular temperature is a determinant of lipid behavior even at fixed temperatures well above or below the actual transition.

Peripheral Proteins and the Phase Separation

By definition, peripheral proteins make less extensive contacts with hydrophobic lipid domains and relatively more with headgroup regions than do integral proteins. Thus, the basic premise in this section will be that peripheral proteins will not tend to be excluded from rigid lipid domains, nor will they exclude cholesterol from their lipid contacts (see London *et al.*,

1974). In practice, the distinction will not be as clear-cut and, although not specifically considered here, many peripheral proteins must be expected to make some degree of hydrophobic contact with lipid acyl chains (Boggs, 1980 and references therein).

At the whole cell and model membrane level, there is good evidence that lipid/peripheral protein interactions can induce phase separations. For instance, alkaline phosphatase, 5′-nucleotidase and acetyl cholinesterases seem, at least in some cases, to be associated with phosphatidylinositol (Michell, 1979; Low and Zilversmit, 1980), and myelin basic protein and cytochrome *c* show a strong tendency to bind to several negatively-charged lipids (Birrell and Griffith, 1976; Boggs *et al.*, 1977a; Stollery *et al.*, 1980). One must expect such interactions to be important determinants of membrane lipid architecture at sites of protein contact and to contribute to the separation of various lipids out of the bulk phase. An attempt has been made to include this factor in the model of membrane structure (Figs. 3 and 4).

There is also a class of peripheral protein, not usually considered, which probably forms a more or less continuous layer at all membrane surfaces. This is simply material nonspecifically adsorbed from the fluid in contact with a given membrane surface. For instance, albumins, globulins, and (nonactivated) complement proteins will coat the outer surfaces of all eukaryotic cells exposed to serum or interstitial fluid; all cytoplasmic proteins should be represented at the inner surface. The forces involved in holding adsorbed proteins to membranes can be sufficient to actually partially unfold the macromolecule involved. Hydrogen bonding via approximately 80 sites has been proposed in the case of serum albumin (Rehfeld *et al.*, 1975). Hence it is easy to imagine that the resultant forces exerted on the membrane will be sufficient to affect such properties as charge distribution, lipid spacing, membrane stability, and permeability (see also Boggs, 1980). Unfortunately, little is known about the role of this layer, although it has been recently shown to influence specific receptor binding properties (Ketis and Grant, 1982 and references therein).

Cell Surface Recognition Events

The plasma membrane is the first line of interaction between cellular machinery and the outside world. It is the site of key events in morphogenesis, attachment, immune response, and interactions with viruses, polypeptide hormones, and toxins. The outer surface of the plasma membrane is thickly populated with headgroups of glycolipids and glycoproteins, and these are highly mobile structures (Sharom and Grant, 1978; Aplin *et al.*, 1979; Lee and Grant, 1980; Cherry *et al.*, 1980). In fact, with the additional overlying mat of peripheral proteins and loosely attached species adsorbed from the surrounding medium (Figs. 3 and 4), clear patches of lipid bilayer are unlikely to be visible except intermittently. The whole surface layer will be in a state of constant motion and subject to a storm of adsorbing and desorbing macromolecules.

It is known that specific binding of certain lectins or antibodies to receptors at the cell periphery can set off a chain of events resulting in modification of cell behavior. One concrete observation from such experiments has been a correlation with lateral redistribution of the glycolipid and/or glycoprotein receptors involved. The suggestion has been made that, in some cases, lateral rearrangement of receptors may itself trigger a response by the cell via its effect on associated membrane structures (Edelmann, 1976). This view is supported by observations that multivalency of the exogenous ligand is a key factor (Shechter *et al.*, 1979). Given the existence of protein–protein and protein–lipid microdomains, and the possibility of oligosaccharide headgroup interactions (Figs. 3 and 4), it is obvious that any redistribution is likely to involve more than just the particular receptor bound by the external agent. Landsberger and associates have reported that specific binding by virus or concanavalin A (when valency is ≥ 4) leads to increased lipid fluidity in nucleated erythrocytes (Lyles and Landsberger, 1978) and cultured cells (F. R. Landsberger, *personal communication*). Such an observation might be explained by a release of relatively fluid annular lipid in exchange for increased protein–protein hydrophobic contacts.

It has alternatively been suggested that specific binding to a glycoprotein might affect other membrane components by *directly* inducing a conformational change in the receptor which would in turn lead to altered interaction with its surrounding lipid microdomain and/or a neighboring protein. Certainly, binding of a macromolecule to an oligosaccharide headgroup can induce immobilization of that strand (Lee and Grant, 1980a,b). However, existing data for carbohydrate headgroups of glycolipids and the integral membrane glycoprotein, glycophorin, indicate that their dynamics are largely and totally respectively independent of that portion which interacts with lipid (Lee and Grant, 1980a; Peters *et al.*, 1982).

Summary

The discussions in this chapter suggest that a quandary exists: on the one hand, there are well-characterized mechanisms for generation of phase separated regions within membranes; on the other hand, factors exist which can greatly reduce the importance of some of these mechanisms. It is easy to fantasize at great length about possible roles for phase-separated regions, such as provision of appropriate domains for special functions, control of fusion and local membrane stability, maintenance of protein complexes, and so forth. But do they really occur? There certainly is good evidence from a wide variety of sources that microdomains do exist. However, the extent of these domains is unclear; how large are they and what fraction of lipid is involved? Clearly, lipid heterogeneity will reduce crystallinity, integral proteins and cholesterol will reduce acyl chain cooperativity, and domains in general will be small. As a result, temperature effects in cell membranes typically occur over broader ranges than the 1–2°C specified in model systems. Exceptions occur among prokaryotes, which lack cholesterol and can often have simple lipid composition. For similar reasons, eukaryotic internal organelle membranes may under some conditions display more massive phase-separation behavior. But the most realistic approach is to apply phase-separation logic conservatively to help explain membrane structure and function (see legend, Figs. 3 and 4).

It should be clear from Figs. 3 and 4 that the term membrane fluidity must be used cautiously. For a given cell membrane, a drifting small molecule would experience great variation in fluidity as it traveled from annular lipid to sterol-free bilayer regions to domains rich in cholesterol to regions of ion- or protein-induced patching, and indeed, from one monolayer to the other.

References

Aplin, J. D., Bernstein, M. A., Culling, C. F. A., Hall, L. D., and Reid, P. E. (1979). *Carbohydr. Res.* **70,** C9–C12.

Barenholz, Y., and Thompson, T. E. (1980). *Biochim. Biophys. Acta* **604,** 129–158.

Behr, J., and Lehn, J. (1973). *FEBS Lett.* **31,** 297–300.

Berclaz, T., and McConnell, H. M. (1981). *Biochemistry* **20,** 6635–6640.

Bevers, E. M., Singal, S. A., Op den Kamp, J. A. F., and van Deenen, L. L. M. (1977). *Biochemistry* **16,** 1290–1295.

Birrell, G. B., and Griffith, O. H. (1976). *Biochemistry* **15,** 2925–2936.

Boggs, J. M. (1980). *Can. J. Biochem.* **58,** 755–770.

Boggs, J. M., and Moscarello, M. A. (1978). *Biochemistry* **17,** 5734–5739.

Boggs, J. M., Moscarello, M. A., and Papahadjopoulos, D. (1977a). *Biochemistry* **16,** 5420–5426.

Boggs, J. M., Wood, D. D., Moscarello, M. A., and Papahadjopoulos, D. (1977b). *Biochemistry* **16,** 2325–2329.

Borochov, H., Abbott, R. E., Schachter, D., and Shinitzky, M. (1979). *Biochemistry* **18,** 251–255.

Branton, D., Cohen, C. M., and Tyler, J. (1981). *Cell* **24,** 24–32.

Brotherus, J. R., Jost, P. C., Griffith, O. H., Keana, J. F., and Hokin, L. E. (1980). *Proc. Natl. Acad. Sci. U.S.A.* **77,** 272–276.
Bunow, M. R., and Bunow, B. (1979). *Biophys. J.* **27,** 325–337.
Burton, A. C., and Edholm, O. G. (1955). "Man in a Cold Environment." Arnold, London.
Caffrey, M., and Feigenson, G. W. (1981). *Biochemistry* **20,** 1949–1961.
Chap, H. J., Zwaal, R. F. A., and van Deenen, L. L. M. (1977). *Biochim. Biophys. Acta* **467,** 146–164.
Chapman, D. (1975). *Q. Rev. Biophys.* **8,** 185–235.
Chen, Y. S., and Hubbell, W. L. (1973). *Exp. Eye Res.* **17,** 517–532.
Cherry, R. J., Nigg, E. A., and Beddard, G. S. (1980). *Proc. Natl. Acad. Sci. U.S.A.* **77,** 5899–5903.
Cullis, P. R., and DeKruijff, B. (1979). *Biochim. Biophys. Acta* **559,** 399–420.
Davoust, J., Bienvenue, A., Fellmann, P., and Deavaux, P. F. (1980). *Biochim. Biophys. Acta* **596,** 28–42.
de Kruijff, B., Cullis, P. R., and Verkleij, A. J. (1980). *Trends Biochem. Sci., (Pers. Ed.)* **5,** 79–81.
Demel, R. A., and de Kruijff, B. (1976). *Biochem. Biophys. Acta* **457,** 109–130.
Demel, R. A., Jansen, J. W. C. M., van Dijck, P. W. M., and van Deenen, L. L. M. (1977). *Biochim. Biophys. Acta* **465,** 1–10.
Edelmann, G. M. (1976). *Science* **192,** 218–226.
Furtado, D., Williams, W. P., Brain, A. P. R., and Quinn, P. J. (1979). *Biochim. Biophys. Acta* **555,** 352–357.
Grant, C. W. M., and McConnell, H. M. (1974). *Proc. Natl. Acad. Sci. U.S.A.* **71,** 4653–4657.
Haest, C. W. M. (1982). *Biochim. Biophys. Acta* **694,** 331–352.
Hauser, H., Levine, B. A., and Williams, R. J. P. (1976). *Trends Biochem. Sci. (Pers. Ed.)* **1,** 278–281.
Hershkowitz, M. (1977). *New Engl. J. Med.* **296,** 178.
Jost, P. C., and Griffith, O. H. (1980). *Ann. N.Y. Acad. Sci.* **348,** 391–405.
Ketis, N. V., and Grant, C. W. M. (1982). *Biochim. Biophys. Acta* **689,** 194–202.
Klausner, R. D., Kleinfeld, A. M., Hoover, R. L., and Karnovsky, M. J. (1980). *J. Biol. Chem.* **255,** 1286–1295.
Kleemann, W., Grant, C. W. M., and McConnell, H. M. (1974). *J. Supramol. Struct.* **2,** 609–616.
Kleemann, W., and McConnell, H. M. (1976). *Biochim. Biophys. Acta* **419,** 206–222.
Lee, A. G. (1977). *Biochim. Biophys. Acta* **472,** 237–344.
Lee, A. G. (1975). *Prog. Biophys. Mol. Biol.* **29,** 3–56.
Lee, A. G., Birdsall, N. J. M., Metcalfe, J. C., Toon, P. A., and Warren, G. B. (1974). *Biochemistry* **13,** 3699–3705.
Lee, P. M., and Grant, C. W. M. (1980a). *Can. J. Biochem.* **58,** 1197–1205.
Lee, P. M., and Grant, C. W. M. (1980b). *Biochim. Biophys. Acta* **601,** 302–314.
Lenaz, G. (1977). *In* "Membrane Proteins and Their Interactions with Lipids" (R. A. Capaldi, ed.), pp. 47–150. Dekker, New York.
Linden, C. D., Blasie, J. K., and Fox, C. F. (1977). *Biochemistry* **16,** 1621–1625.
Linden, C. D., Wright, K. L., McConnell, H. M., and Fox, C. F. (1973). *Proc. Natl. Acad. Sci. U.S.A.* **70,** 2271–2275.
London, E., and Feigenson, G. W. (1981). *Biochemistry* **20,** 1939–1948.
London, Y., Demel, R. A., Geurts van Kessel, W. S. M., Zahler, P., and van Deenen, L. L. M. (1974). *Biochim. Biophys. Acta* **332,** 69–84.
Low, M. G., and Zilversmit, D. B. (1980). *Biochemistry* **19,** 3913–3918.

Luzzati, V. (1968). *In* "Biological Membranes" (D. Chapman, ed.), Vol. 1, pp. 71–123. Academic Press, New York.
Lyles, D. S., and Landsberger, F. R. (1978). *Virology* **88,** 25–32.
Marcelja, S. (1976). *Biochim. Biophys. Acta* **455,** 1–7.
Marinetti, G. V. (1977). *Biochim. Biophys. Acta* **465,** 198–209.
Marsh, D., Watts, A., and Knowles, P. F. (1976). *Biochemistry* **15,** 3570–3578.
McCloskey, M. A., and Troy, F. A. (1980). *Biochemistry* **19,** 2061–2066.
Michell, R. H. (1979). *Trends Biochem. Sci. (Pers. Ed.)* **4,** 128–131.
Morrisett, J. D., Pownall, H. J., Plumlee, R. T., Smith, L. C., Zehner, Z. E., Esfhani, M., and Wakil, S. J. (1975). *J. Biol. Chem.* **250,** 6969–6976.
Murphy, J. R. (1965). *J. Lab. Clin. Med.* **65,** 756–763.
Nigg, E. A., Bron, C., Girardet, M., and Cherry, R. J. (1980). *Biochemistry* **19,** 1887–1893.
Oldfield, E., and Chapman, D. (1972). *FEBS Lett.* **23,** 285–296.
Owicki, J. C., Springgate, M. W., and McConnell, H. M. (1978). *Proc. Natl. Acad. Sci. U.S.A.* **75,** 1616–1619.
Papahadjopoulos, D., Poste, G., and Vail, W. J. (1979). *In* "Methods in Membrane Biology" (E. D. Korn, ed.), Vol. 10, pp. 1–122. Plenum, New York.
Papahadjopoulos, D., Jacobson, K., Nir, S., and Isac, T. (1973). *Biochim. Biophys. Acta* **311,** 330–348.
Papahadjopoulos, D., Vail, W. J., Pangborn, W. A., and Poste, G. (1976). *Biochim. Biophys. Acta* **448,** 265–283.
Peters, M. W., Barber, K. R., and Grant, C. W. M. (1982). *Biochim. Biophys. Acta* **693,** 417–424.
Petit, V. A., and Ediden, M. (1974). *Science* **184,** 1183–1184.
Phillips, M. C., Ladbrooke, B. D., and Chapman, D. (1970). *Biochim. Biophys. Acta* **196,** 35–44.
Phillips, M. C., Graham, D. E., and Hauser, H. (1975). *Nature (London)* **254,** 154–155.
Read, B. D., and McElhaney, R. M. (1975). *J. Bacteriol.* **123,** 47–55.
Recktenwald, D. J., and McConnell, H. M. (1981). *Biochemistry* **20,** 4505–4510.
Rehfeld, S. J., Eatough, D. J., and Hansen, L. D. (1975). *Biochem. Biophys. Res. Commun.* **66,** 568–591.
Scott, H. L. (1981). *Biochim. Biophys. Acta* **643,** 161–167.
Sharom, F. J., and Grant, C. W. M. (1978). *Biochim. Biophys. Acta* **507,** 280–293.
Sharom, F. J., Barratt, D. G., and Grant, C. W. M. (1977). *Proc. Natl. Acad. Sci. U.S.A.* **74,** 2751–2755.
Shechter, E., Letellier, L., and Gulik-Krzywicki, T. (1974). *Eur. J. Biochem.* **49,** 61–76.
Shechter, Y., Chang, K., Jacobs, S., and Cuatrecasas, P. (1979). *Proc. Natl. Acad. Sci. U.S.A.* **76,** 2720–2724.
Shimshick, E. J., and McConnell, H. M. (1973). *Biochemistry* **12,** 2351–2360.
Shimshick, E. J., Kleemann, W., Hubbell, W. L., and McConnell, H. M. (1973). *J. Supramol. Struct.* **1,** 285–294.
Sklar, L. A., Miljanich, G. P., Bursten, S. L., and Dratz, E. A. (1979). *J. Biol. Chem.* **254,** 9583–9591.
Steck, T. L. (1978). *J. Supramol. Struct.* **8,** 311–324.
Stollery, J. G., Boggs, J. M., and Moscarello, M. A. (1980). *Biochemistry* **19,** 1219–1226.
Tanaka, K., and Ohnishi, S. (1976). *Biochim. Biophys. Acta* **426,** 218–231.
Tardieu, A., Luzzati, V., and Reman, F. C. (1973). *J. Mol. Biol.* **75,** 711–733.
Thilo, L., Trauble, H., and Overath, P. (1977). *Biochemistry* **16,** 1283–1289.
Tinker, D. O., Pinteric, L., Hsia, J. C., and Rand, R. P. (1976). *Can. J. Biochem.* **54,** 209–218.

Tomita, M., Furthmayr, H., and Marchesi, V. T. (1978). *Biochemistry* **17,** 4756–4770.
Trauble, H., and Eibl, H. (1974). *Proc. Natl. Acad. Sci. U.S.A.* **71,** 214–219.
van Dijck, P. W. M. (1979). *Biochim. Biophys. Acta* **555,** 89–101.
van Dijck, P. W. M., de Kruiff, B., Verkleij, A. J., van Deenen, L. L. M., and de Gier, J. (1978). *Biochim. Biophys. Acta* **512,** 84–96.
Wallace, B. A., Richards, F. M., and Engleman, D. M. (1976). *J. Mol. Biol.* **107,** 255–269.
Wallach, D. F. (1979). "Plasma Membranes and Disease." Academic Press, New York.
Warren, G. B., Houslay, M. D., Metcalfe, J. C., and Birdsall, N. J. M. (1975). *Nature (London)* **255,** 684–687.
Watts, A., Volotovski, I. D., and Marsh, D. (1979). *Biochemistry* **18,** 5006–5013.
Wisnieski, B. J., Parkes, J. G., Huang, Y. O., and Fox, C. F. (1974). *Proc. Natl. Acad. Sci. U.S.A.* **71,** 4381–4385.
Wu, S. H., and McConnell, H. M. (1975). *Biochemistry* **14,** 847–854.
Yellin, N., and Levin, I. W. (1977). *Biochim. Biophys. Acta* **468,** 490–494.

Chapter 6

Phospholipid Transfer Proteins and Membrane Fluidity[1]

George M. Helmkamp, Jr.

Introduction

In 1966, Dawson proposed, based upon numerous observations that phospholipid biosynthesis was localized on the endoplasmic reticulum, that newly synthesized phospholipids should be transported intracellularly to other

[1]The personal research cited in this chapter has been supported by the National Institutes of Health, United States Public Health Service (Grant No. GM 24035).

ISBN 0-12-053002-3

membranes by soluble cytoplasmic proteins or through a continuous matrix of membrane-bound lipoproteins. Shortly thereafter, several research groups independently provided experimental evidence in support of the former process (Wirtz and Zilversmit, 1968; McMurray and Dawson, 1969; Akiyama and Sakagami, 1969). In monitoring the transfer of phospholipid molecules between microsomes and mitochondria, addition of a membrane-free supernate preparation gave stimulations which were entirely consistent with the presence of a catalytic factor in that supernate. Further fractionation of the liver homogenates firmly established the protein nature of the factor (Wirtz *et al.*, 1972). Phospholipid transfer proteins have been identified in a large number of animal and plant tissues; they characteristically exhibit specificity for one or several phospholipid classes.

In recent years, the detailed structural analysis of biological and artificial membranes has been advanced by the use of phospholipid transfer proteins (Zilversmit, 1978). Among the applications which have been reported are the insertion of spectroscopic (spin-labeled, fluorescent) probes into membranes, the determination of membrane phospholipid topography, and the measurement of transbilayer mobility of phospholipids. These applications, as well as the proposed role of protein-catalyzed transfer in intracellular phospholipid movement, depend upon selective associations between phospholipid transfer proteins, membrane interfaces, and bound phospholipid molecules. The chemical and physical nature of such associations is the focus of this chapter. More general aspects of phospholipid transfer proteins have been reviewed by Wirtz (1974, 1982), Zilversmit and Hughes (1976), and Kader (1977).

Survey of Phospholipid Transfer Proteins

Protein Nomenclature

Historically, the process of protein-catalyzed movement of phospholipids between membranes has been described as primarily one of exchange rather than transfer. Such a designation was used although the typical measurement considered only a unidirectional flux and often assumed that the total phospholipid mass of the participating membranes remained unchanged. As will be described in greater detail in Section III,C, appreciation for the difference between a strict one-for-one phospholipid exchange and a less restrictive one-way transfer into or out of a membrane interface has developed in the last few years. The early observations that certain classes of phospholipids, particularly phosphatidylinositol, could be transported to membranes initially devoid of that phospholipid suggested that *transfer* was a general mechanism and *exchange* a special case (Harvey *et al.*, 1974; Zborowski and Wojtczak, 1975; Demel *et al.*, 1977). Indeed, a protein may function as a exchange vehicle under one set of experimental conditions and may effect net transfer under other circumstances; the mechanism is dependent upon the chemical and physical properties of not only the membrane or membrane-like particle but also the protein itself.

In keeping with the broadest definition possible, the protein catalysts discussed in this chapter will be referred to as phospholipid transfer proteins. The preference toward a specific class of phospholipid, whether absolute or significantly greater than other phospholipids, will also be affixed to the protein, for example, bovine phosphatidylinositol transfer protein. Finally, a pronounced lack of specificity will be identified by the term *nonspecific*, as done by Zilversmit and co-workers (Bloj and Zilversmit, 1977; Crain and Zilversmit, 1980a).

Distribution of Phospholipid Transfer Activities

Homogenates of rat liver provided the first conclusive evidence that a protein factor could facilitate and accelerate the transfer of phospholipids between microsomes and mitochondria. Early successful purifications of phospholipid transfer proteins were achieved, however, from tissues of bovine origin, namely liver, heart, and brain (Wirtz *et al.*, 1972; Ehnholm and Zilversmit, 1973; Helmkamp *et al.*, 1974). More recently, transfer proteins have been isolated and characterized from potato tuber, rat liver, and sheep lung (Kader, 1975; Lumb *et al.*, 1976; Robinson *et al.*, 1978). The major exchange proteins from these tissues have been extensively studied for characteristics of molecular weight, isoelectric point, amino acid composition, and substrate specificity, as summarized in Table I. In the case of each protein, a membrane-free supernate obtained by either high-speed centrifugation or adjustment of a postmitochondrial fraction to pH 5.1 was used as the starting material for further purification, which emphasizes the soluble nature of these catalysts. Liver, whether bovine or rat, is a major source of a transfer protein with a high degree of specificity for phosphatidylcholine. Bovine heart and brain, on the other hand, contain phosphatidylinostiol transfer protein as the principal catalytic entity and, accordingly, serve as the major source of this protein.

TABLE I

Properties of Purified Phospholipid Transfer Proteins

Protein	Principal source	Charge isomer	Molecular weight	Isoelectric point	Phospholipid(s) transferred[a]	Reference
Phosphatidylcholine	Bovine liver	—	24,681[b]	5.8	PtdCho	Kamp *et al.* (1973); Akeroyd *et al.* (1981)
Phosphatidylinositol	Bovine heart	I	33,500[c]	5.3	PtdIns PtdCho	DiCorleto *et al.* (1979)
		II	33,500[c]	5.6		
Phosphatidylinositol	Bovine brain	I	32,300[c]	5.2		Helmkamp *et al.* (1974)
		II	32,500[c]	5.5		
Nonspecific	Bovine liver	I	14,500[c]	9.6	PtdCho PtdOH	Crain and Zilversmit (1980a)
					PtdEtn Sphingomyelin	
		II	14,500[c]	9.8	PtdIns Cholesterol	Bloj and Zilversmit (1981)
					PtdSer Glycosylceramide	
					PtdGro Ganglioside	
Phosphatidylcholine	Rat liver	—	28,000[d]	8.4	PtdCho	Poorthuis *et al.* (1980)

Nonspecific	Rat liver	I	12,500[c]	8.6	PtdCho PtdSer PtdEtn Sphingomyelin PtdIns Cholesterol	Bloj and Zilversmit (1977)
		II	12,500[c]	9.0		
Nonspecific	Rat hepatoma	—	11,200[d]	5.2	PtdCho PtdIns PtdEtn Sphingomyelin	Dyatlovitskaya *et al.* (1978)
Phosphatidylcholine	Sheep lung	—	21,000[d]	7.1	PtdCho	Robinson *et al.* (1978)
Phospholipid	Sheep lung	—	22,000[d]	5.8	PtdCho PtdIns	Robinson *et al.* (1978)
Phospholipid	Potato tuber	—	22,000[d]	N.D.[e]	PtdCho PtdEtn	Kader (1975)

[a]PtdCho, phosphatidylcholine; PtdIns, phosphatidylinositol; PtdEtn, phosphatidylethanolamine; PtdSer, phosphatidylserine; PtdGro, phosphatidylglycerol; PtdOH, phosphatidic acid.

[b]Amino acid sequence.

[c]Polyacrylamide gel electrophoresis in presence of sodium dodecylsulfate.

[d]Gel filtration chromatography.

[e]Not determined.

Although isolated from different tissues and purified by radically different procedures, the major phosphatidylinositol transfer proteins from bovine brain and heart appear to be identical (Helmkamp *et al.*, 1974; DiCorleto *et al.*, 1979). Evidence in support of this distribution includes molecular weight, amino acid composition, and isoelectric points of the two isomeric forms which occur in each tissue; the substrate specificity and behavior toward membranes of different lipid composition are also in agreement. The physical and chemical similarities of these proteins confirm the earlier observation that rabbit antibody raised against brain protein could precipitate most of the phosphatidylinositol and phosphatidylcholine transfer activities in unfractionated heart homogenates (Helmkamp *et al.*, 1976a).

In addition to those proteins, which exhibit a narrow specificity toward one or two phospholipid classes, nonspecific lipid transfer proteins have been identified in bovine liver, rat liver, and rat lung (Bloj and Zilversmit, 1977; Crain and Zilversmit, 1980a; Post *et al.*, 1980). These nonspecific proteins have been shown to accelerate the intermembrane movement of phosphatidylcholine, phosphatidylinositol, phosphatidylethanolamine, phosphatidylserine, phosphatidic acid, phosphatidylglycerol, sphingomyelin, and cholesterol. This broad specificity has been extended recently to include neutral glycosphingolipid and ganglioside (Bloj and Zilversmit, 1981). Those lipids for which there was no demonstrable transfer activity are bisphosphatidylglycerol, cholesteryl ester, and triacylglycerol. It is now apparent that the nonspecific lipid transfer protein in rat liver is identical to one of the soluble sterol carrier proteins which activate some of the steps in the microsomal conversion of lanosterol to cholesterol as well as the subsequent esterification of cholesterol by fatty acid (Noland *et al.*, 1980; Gavey *et al.*, 1981; Trzaskos and Gaylor, 1983). Whether the two diverse functions of this protein, namely, intermembrane lipid transport and membrane enzyme activation are related remains to be established.

Phospholipid transfer activity has been demonstrated in a number of other mammalian tissues, including kidney, intestine, lung, and thyroid, several plant tissues, and the yeast *Saccharomyces cerevisiae* (Kader, 1975; Lumb *et al.*, 1976; Zilversmit and Hughes, 1976; Cobon *et al.*, 1976). Cell-free extracts of the photosynthetic bacterium *Rhodopseudomonas sphaeroides* have been shown to transfer phosphatidylglycerol between phospholipid vesicles and intracytoplasmic membrane vesicles isolated from the same organism, demonstrating exchange activity in a prokaryote for the first time (Cohen *et al.*, 1979). Protein-mediated transfer of phosphatidylethanolamine, phosphatidylglycerol, and bisphosphatidylglycerol has been described by Lemaresquier *et al.* (1982) in a system composed of mesosomes, protoplasts, and soluble protein isolated from *Bacillus subtilis*.

Mechanism of Protein-Catalyzed Phospholipid Transfer

Measurement of Phospholipid Transfer Activity

In a typical intermembrane phospholipid transfer system, minimum requirements are two membrane populations, an ability to distinguish or separate those membranes, and a recognition of the transferred phospholipid. The inclusion of a protein catalyst is optional. Many different combinations of membranes both biological and artificial have been employed in the measurement of transfer protein activity, including microsome–mitochondrion, microsome–single bilayer vesicle, mitochondrion–single bilayer vesicle, single bilayer vesicle–multilamellar vesicle, single bilayer vesicle–monolayer, and monolayer–monolayer. For many of these membrane pairs, separation at the completion of an experiment can be achieved by differential or density gradient centrifugation; for others, particularly the monolayer–monolayer system, actual physical separation can be maintained throughout the experiment. In recent years, the vesicle–vesicle systems have received the greatest attention. These membranes are easily prepared and readily characterized for size, concentration, and lipid composition. Separation can be effected in a number of ways: (1) the incorporation of sufficient phosphatidic acid (7–9 mol%) into one of the vesicle populations to permit those vesicles to bind to columns of DEAE-cellulose (Hellings *et al.*, 1974); (2) the incorporation of Forssman antigen into the bilayer and the subsequent precipitation by a specific immunoglobulin (Ehnholm and Zilversmit, 1973); (3) the use of dimannosyldiacylglycerol in vesicle preparations and the formation of precipitable complexes with concanavalin A (Sasaki and Sakagami, 1978); and (4) the inclusion of lactosylceramide (8–10 mol%) in one of the bilayers to enable the aggregation and precipitation of those membranes in the presence of *Ricinus communis* agglutinin (Kasper and Helmkamp, 1981a). The principal advantages of the vesicle–vesicle assay systems described above are nearly complete recovery of both membrane populations and a reduction of the amount of exchange protein needed to obtain meaningful rates of transfer. The common means of identifying the transferred phospholipid is radioisotopic labeling. Selected experiments, however, have utilized appropriate phospholipid analogs for detection by electron spin resonance, nuclear magnetic resonance, or fluorescence spectroscopy. As pointed out by Zilversmit (1978), it is highly desirable, if not essential, to incorporate a nontransferable lipid as an internal standard in one of the membrane populations to permit estimations of membrane recovery and nonspecific fusion or adsorption between membranes under incubation conditions. Examples of such markers are radiolabeled cholesteryl oleate and trioleoylglycerol.

Kinetic Analysis

Several lines of experimentation support the hypothesis that phospholipid transfer proteins function as intermembrane carriers of phospholipid molecules through the aqueous phase. The current working model includes the following events: (1) formation of transient dissociable complexes between protein and membrane and between protein and phospholipid molecule and (2) deposition of a protein-bound phospholipid into the membrane and/or acquisition of a membrane-bound phospholipid by the protein. Both the bovine phosphatidylcholine transfer protein and the bovine phosphatidylinositol transfer protein display similar catalytic properties. Protein–phospholipid and protein–vesicle complexes from which appropriate dissociation constants can be calculated have been demonstrated (Kamp *et al.*, 1975, 1977; Johnson and Zilversmit, 1975; Helmkamp *et al.*, 1976b; Wirtz *et al.*, 1979). Kinetic analyses, in which the concentration of one or both membrane populations is varied, yield results consistent with a mechanism such that the protein oscillates between the two membranes (van den Besselaer *et al.*, 1975; Helmkamp *et al.*, 1976b). Using enzyme terminology, this mechanism is analogous to a two-substrate reaction in which an enzyme (transfer protein) interacts first with one substrate (donor membrane) and then, independently, with the other (acceptor membrane); in other words, a classical ping-pong mechanism. Measurable fluxes of phospholipid between physically separated membranes, for example, two surface monolayers sharing a common subphase, exemplify the ability of the proteins to diffuse freely in an aqueous environment (Demel *et al.*, 1973, 1977). Interestingly, no bound phospholipid could be demonstrated for the bovine nonspecific phospholipid transfer protein, suggesting that this catalyst may operate by a different mechanism (Crain and Zilversmit, 1980a).

Net Transfer versus Exchange

Although the carrier nature of the transfer proteins is well documented, the molecular events at the membrane surface remain unresolved. Two mechanistic possibilities exist; (1) *exchange:* a protein-bound phospholipid is always replaced by a membrane-bound phospholipid, such that the flux of phospholipid in one direction (donor to acceptor) is balanced by an equivalent flux in the opposite direction (acceptor to donor); (2) *net transfer:* a one-for-one replacement of protein-bound and membrane-bound phospholipids is not necessary. Consequently, for the latter case the bidirectional fluxes need not be equivalent, and changes in the phospholipid mass of the participating membranes may result. It should be stressed that neither of these mechanisms excludes the unidirectional movement of a specific class of phospholipid to a membrane initially deficient in that phospholipid. Indeed, such movements of phosphatidylinositol or phosphatidylcholine catalyzed by bovine phosphatidylinositol transfer protein have been observed under a wide variety of experimental conditions. DiCorleto *et al.* (1979), using single bilayer donor vesicles composed of phosphatidylethanolamine/phosphatidylcholine/phosphatidylinositol (80:10:10 mol%), showed a unidirectional flux of phosphatidylcholine to multilamellar acceptor vesicles of phosphatidylethanolamine/phosphatidylinositol (90:10 mol%), and a similar flux of phosphatidylinositol to acceptors of phosphatidylethanolamine/phosphatidylcholine/bisphosphatidylglycerol (80:10:10 mol%). Kasper and Helmkamp (1981b) extended these findings by quantitating phospholipid fluxes in both directions between two populations of single bilayer vesicles. When both membranes contained phosphatidylinositol, the flux of that phospholipid in one direction was balanced by an equivalent flux in the opposite direction. This was also observed when phosphatidylcholine was a component of both membranes. If one population of vesicles (I) contained phosphatidylcholine, phosphatidylinositol, and lactosylceramide, and the other (II) contained phosphatidylcholine and phosphatidic acid, the expected net flux of phosphatidylinositol would be from I to II, but the flux of phosphatidylcholine was greater in the direction II to I. These results suggest a compensatory response to the movement of phosphatidylinositol. In fact, the *total* phospholipid flux in one direction equalled that in the opposite. If, however, the acceptor membrane was comprised of phosphatidylethanolamine, phosphatidic acid, and lactosylceramide and lacked a transferable phospholipid, no detectable phospholipid transfer could be measured. These results strongly imply an exchange mechanism for the major phospholipid transfer protein of bovine brain and heart.

Somewhat more complex is the mode of action of bovine liver phosphatidylcholine transfer protein at membrane interfaces. By monitoring the movement of spin-labeled phosphatidylcholine molecules, Wirtz *et al.* (1980) showed limited transfers from phosphatidylcholine/phosphatidic acid (75:25 mol%) donor vesicles to acceptor membranes composed of phosphatidylethanolamine/phosphatidic acid (81:19 mol%). Because of the pronounced specificity of this protein for phosphatidylcholine, it is clear that only a unidirectional, net transfer could have occurred. Nichols and Pagano (1983) confirmed this net-transfer capability of bovine liver phosphatidylcholine transfer protein, using a fluorescent resonance energy transfer assay with phosphatidylcholine/phosphatidic acid (1:99, mol%) donor vesicles and phosphatidylethanolamine/phosphatidic acid (80:20, mol%) acceptor vesicles. In another series of qualitative observations, Wilson *et al.* (1980) noted a unidirectional movement of phosphatidylcholine catalyzed by bovine phosphatidylcholine transfer protein. Donor membranes were either phosphatidylcholine vesicles or human low density lipoprotein; acceptors were particles prepared from sphingomyelin and apolipoprotein A-II. An absence of a reverse flux of sphingomyelin to the donor pool confirmed the net transfer of phospholipid. These demonstrations of net transfer required relatively large amounts of protein (12–20 μg) and/or a 10- to 20-fold molar excess of acceptor membranes. Under radically different conditions, namely, low levels of protein (0.10 μg), and approximately equal quantities of donor and acceptor membranes, bidirectional rates of phosphatidylcholine transfer between single bilayer vesicles were found to be identical (Helmkamp, 1980c). Thus, no net accumulation of phosphatidylcholine by either membrane population took place. A strict exchange process has also been noted for the phosphatidylcholine transfer protein-catalyzed movement of lipid between two monolayers of equivalent surface area and composition (Demel *et al.*, 1973).

Several other experimental systems have employed unusual acceptor particles. Phosphatidylcholine transfer protein-bound phosphatidylcholine could be released into micelles of sodium deoxycholate or lysophosphatidylcholine (Kamp *et al.*, 1975; Wirtz *et al.*, 1980). In a broad sense, these results have been interpreted as examples of net transfers of phospholipid.

In an extensive comparative study, Crain and Zilversmit (1980b) measured lipid transfer from multilamellar vesicles to intact human high density lipoprotein or ethanol–diethylether-delipidated human high-density lipoprotein in the presence and absence of the three major bovine phospholipid transfer proteins. With intact lipoprotein as acceptor, phosphatidylinositol transfer protein and the nonspecific lipid transfer protein facilitated the transfers of phosphatidylcholine and phosphatidylinositol, and phosphatidylcholine transfer protein catalyzed a flux of phosphatidylcholine. With the delipidated acceptors, only the nonspecific protein provided a significant transfer of either phospholipid, albeit at rates which were markedly less than those with the intact lipoprotein. Calculation of bidirectional fluxes suggested that the phosphatidylinositol and phosphatidylcholine transfer proteins operated by a strict exchange mechanism with respect to phosphatidylcholine movement between multilamellar vesicles and intact high density lipoprotein. On the other hand, the nonspecific lipid transfer protein catalyzed a threefold greater flux of phosphatidylcholine in the vesicle-to-lipoprotein direction. Under these conditions, net transfer is the preferred mechanism for the nonspecific protein. Net transfers of phospholipid have also been observed between single bilayer vesicles and rat liver mitochondria from which the outer membrane had been stripped away by digitonin.

Relationship between Membrane Lipid Composition and Phospholipid Transfer

Variation in Polar Head Groups

The first extensive survey of vesicle lipid composition and phospholipid transfer protein activity was reported by Hellings *et al.* (1974). Using an assay system of two vesicle populations, phosphatidylcholine transfer protein responded to increasing amounts of phosphatidic acid or phosphatidylinositol in the donor membrane. At a level of 20 mol% of either of these anionic lipids in phosphatidylcholine vesicles, all tansfer activity was lost. Reference donor vesicles contained 7 mol% phosphatidic acid or 9 mol% phosphatidylinositol. This apparent inhibition by negatively charged lipids was subsequently extended to include phosphatidylserine and phosphatidylglycerol which, at 20 mol%, led to 50% and 60% reductions in transfer activity, respectively (Wirtz *et al.*, 1976). A similar proportion of sphingomyelin (20 mol%) had no effect on the movement of phosphatidylcholine between vesicles, but at a level of 30 mol%, inhibition was approximately 35%. Like sphingomyelin, the zwitterionic phosphatidylethanolamine was only moderately inhibitory, yielding 80% of the normal activity for vesicles containing 20 mol%. The introduction of a positive charge to the vesicle surface, in the form of 10 mol% stearylamine, produced phosphatidylcholine transfers which were 150% of normal. The above lipid compositions refer to the single bilayer acceptor vesicles in a mitochondrion–vesicle system.

In contrast to the above analysis, DiCorleto and Zilversmit (1977) observed that phosphatidylcholine transfer protein activity was stimulated by certain anionic phospholipids. Thus, phosphatidylinositol (10 mol%) or bisphosphatidylglycerol (5 mol%) increased the activity by three- to four-fold when incorporated into multilamellar vesicle donors. The reference membranes in this study contained phosphatidylcholine/phosphatidylethanolamine (70:30 mol%), and the anionic species replaced the latter membrane component. However, the comparison between negatively charged membranes (phosphatidylcholine/bisphosphatidylglycerol, 95:5 mol%) and neutral, phosphatidylethanolamine-containing membranes (phosphatidylcholine/phosphatidylethanolamine, 70:30 mol%) gave results qualitatively similar to those obtained earlier. Zwitterionic membranes were significantly less active than slightly acidic membranes. Some rather striking results were reported by Machida and Ohnishi (1978) with a vesicle–vesicle system. As little as 2 mol% phosphatidylserine in phosphatidylcholine acceptor membranes caused a 40% decrease in activity, and a decrease of nearly 90% was found at a level of 10 mol%. Interestingly, the transfer system could be protected from inhibition by this anionic phospholipid upon addition of 15 m*M* Ca^{2+} or Mg^{2+} to the medium, but monovalent ions were not effective.

The phosphatidylinositol transfer protein, isolated from bovine brain and heart, has also been investigated with respect to membrane polar head group variations. In an early study, DiCorleto *et al.* (1977) prepared mixed phospholipid single bilayer vesicles from phosphatidylcholine and either phosphatidic acid or phosphatidylinositol. The inclusion of these acidic lipids at 6–17 mol% levels gave phosphatidylcholine transfer activities 50–60% greater than those with vesicles containing only phosphatidylcholine. Thus, while the inclusion of a few mol% of an acidic phospholipid in a membrane generated a much more favorable interaction between that membrane and phosphatidylinositol transfer protein, increases up to 20 mol% did not benefit the reaction further. This phenomenon was also noted by Helmkamp (1980a) for the phosphatidylinositol transfer protein isolated from bovine brain. Compared to acceptor single bilayer vesicles containing phosphatidylcholine/phosphatidic acid (98:2 mol%), those which contained 5–20 mol% phosphatidic acid, phosphatidylserine, or phosphatidylglycerol supported essentially unchanged levels of phosphatidylinositol and phosphatidylcholine transfer activity.

A striking exception to the effect of anionic lipids on phosphatidylinositol transfer protein activity is that of the preferred substrate, phosphatidylinositol. Harvey *et al.* (1974) observed a sharp decrease in transfer activity for the series of acceptor vesicles containing 2, 4, 8, or 12 mol% phosphatidylinositol. This apparent inhibition was further increased as the vesicle concentration approached 0.8 m*M* (in phospholipid). In the concentration range 20 to 80 μM, however, the opposite was seen. Vesicles containing 8 or 20 mol% phosphatidylinositol were, in fact, more active than those with only 2 mol%. A detailed kinetic analysis provided a possible resolution of this paradox (Helmkamp *et al.*, 1976b). The affinity of the protein for a membrane surface increases dramatically if phosphatidylinositol is a structural component of that membrane. At low vesicle concentrations, this is observed as a stimulus to transfer activity. But as the concentration of the phosphatidylinositol-containing vesicle increases, the strong association between protein and membrane becomes limiting, and the decline in the intermembrane movement of the protein is seen as an inhibition of transfer activity. The extreme sensitivity of phosphatidylinositol transfer protein to this phospholipid, not only as a bound substrate but as a membrane constituent, is a critical feature of its catalytic activity.

The replacement of membrane phosphatidylcholine by other zwitterionic phospholipids and the effect on phosphatidylinositol transfer protein activity were investigated by DiCorleto *et al.* (1979) and Helmkamp (1980a). Both groups found that the addition of phosphatidylethanolamine to a level of 25–30 mol% produced 2- to 10-fold increases in transfer activity. This stimulation was independent of the participating membrane; that is, phosphatidylethanolamine could be added to multilamellar vesicle donors in a multilamellar vesicle–small unilamellar vesicle system for phosphatidylcholine transfer, or to small unilamellar vesicle acceptors in a microsome–vesicle assay system for phosphatidylinositol or phosphatidylcholine transfer. Significant enhancement of transfer activity occurred at 2 mol% phosphatidylethanolamine and continued at concentrations as high as 40 mol%. It is important to recall that the structure of membranes containing more than 60–70 mol% phosphatidylethanolamine is not well defined and may involve cylindrical morphology and hexagonal phases (Cullis and de Kruijff, 1979).

Sphingomyelin presents a more complex picture. At low levels (2–6 mol%) of incorporation into acceptor vesicles, a marked increase in phosphatidylinositol and phosphatidylcholine transfer was measured, analogous to that described above for phosphatidylethanolamine (Helmkamp, 1980a). But with further increases, sphingomyelin became strongly inhibitory. The ratio of phosphatidylinositol transfer activity for multilamellar vesicle acceptors composed of phosphatidylethanolamine/phosphatidylcholine/phosphatidylinositol (80:10:10 mol%) or phosphatidylethanolamine/sphingomyelin/phosphatidylinositol (80:10:10 mol%) was 3.4 (DiCorleto *et al.*, 1979). The similar ratio for small unilamellar vesicle acceptors composed of phosphatidylcholine/phosphatidic acid (98:2 mol%) or phosphatidylcholine/sphingomyelin/phosphatidic acid (58:40:2 mol%) was 3.6 (Helmkamp, 1980a).

As discussed by Helmkamp (1980a), the transfer rate enhancement observed with egg phosphatidylcholine vesicles containing phosphatidylethanolamine or sphingomyelin may be attributed to additional electrostatic and hydrogen bonding interactions (Hertz and Barenholz, 1975; Yeagle *et al.*, 1977) which, by reducing surface hydration, could facilitate membrane interface–transfer protein association.

The construction of model membranes with positively charged surfaces is difficult to achieve since the availability of naturally occurring cationic amphiphilic lipids is very limited. However, long-chain alkylamines can be used in their place. As increasing proportions of stearylamine were incorporated into phosphatidylcholine vesicles, these membranes became less competent in accepting phosphatidylinositol from rat liver microsomes in the presence of phosphatidylinositol transfer protein (Fig. 1). A nearly linear decline in transfer activity to less than 10% of the control rate was seen between 0 and 40 mol% stearylamine (Helmkamp, 1980a). This inhibitory effect of stearylamine on the activity of phosphatidylinositol transfer protein should be contrasted to the stimulatory effect on the activity of phosphatidylcholine transfer protein (Wirtz *et al.*, 1976).

In addition to imparting a net positive charge to the surfaces of phosphatidylcholine vesicles, stearylamine may also influence the molecular organization of the lipid bilayer. That this indeed occurred was demonstrated by the fluorescence behavior of diphenylhexatriene in a series of stear-

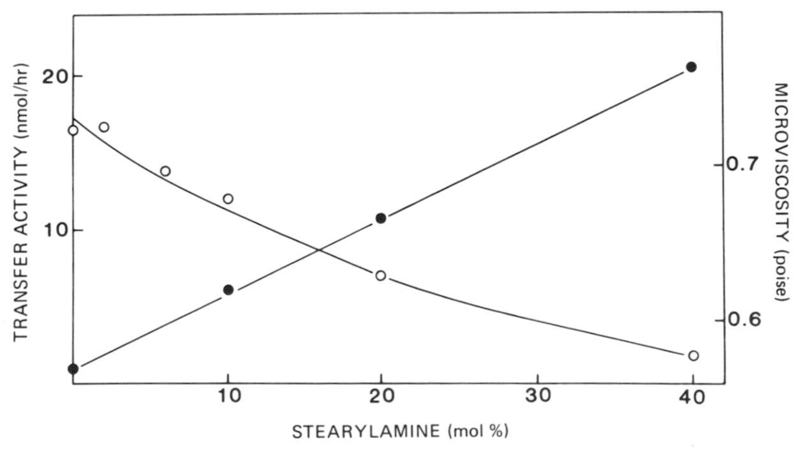

Fig. 1. Effect of stearylamine on the activity of bovine phosphatidylinositol transfer protein. The indicated molar proportion of stearylamine is mixed with egg phosphatidylcholine and 2 mol% egg phosphatidic acid. After sonication and centrifugation these vesicles are tested at 37°C as acceptors of [^{3}H]phosphatidylinositol from rat liver microsomes in the presence and absence of 2.2 μg transfer protein. Vesicle microviscosity at 37°C was determined from the polarization of diphenylhexatriene in the membrane preparation, according to Shinizky and Barenholz (1978); in these measurements, the membrane lipid concentration was 0.4 m*M* and the probe concentration 0.5 μ*M*. Transfer activity (○); microviscosity (●). Adapted with permission from Helmkamp (1980a).

ylamine-phosphatidylcholine mixed vesicles. Microviscosity increased in direct proportion to the amount of the saturated alkylamine in the membranes (Fig. 1). Concurrently, the ability of these membranes to function as acceptors of microsomal phosphatidylinositol in the presence of phosphatidylinositol transfer protein declined. Further experiments using other cationic amphiphiles are needed to distinguish between altered surface charge or hydrocarbon fluidity as the underlying cause of stearylamine inhibition.

Similar arguments may be made for the inhibitory effect of high levels of bovine spinal cord sphingomyelin. The fatty acid composition of this zwitterionic phospholipid is rich in C_{18}-saturated and C_{24}-monounsaturated species (White, 1973), both of which would be expected to reduce the fluidity of egg phosphatidylcholine bilayers. Surprisingly, however, the effect of even 40 mol% sphingomyelin on vesicle microviscosity, as monitored by diphenylhexatriene polarization, was very modest. An increase from 0.52 to 0.77 was measured. In comparison to the data plotted in Fig. 3, the increase in microviscosity alone was insufficient to explain the decrease in phosphatidylcholine and phosphatidylinositol transfer activity.

Variation in Phospholipid Fatty Acyl Groups

Advances in chemical synthetic techniques in the last decade have permitted the preparation of phosphatidylcholines of known fatty acid composition. Positionally specific derivatives at the *sn*-1 and *sn*-2 glycerol carbons may also be synthesized. Single bilayer vesicles prepared from these lipids can, therefore, be tested as donor or acceptor membranes in protein-catalyzed phospholipid transfer systems.

For bovine phosphatidylcholine transfer protein, the earliest comparison of phospholipid fatty acid composition was carried out by Kamp (1975). Donor vesicles were prepared from ^{14}C-labeled egg or [^{14}C]dipalmitoylphosphatidylcholine, containing 2 mol% egg phosphatidic acid; acceptor vesicles consisted of egg phosphatidylcholine/egg phosphatidic acid (92:8 mol%). At 37°C, transfers of egg phosphatidylcholine and dipalmitoylphosphatidylcholine were calculated to an activity ratio of almost 25. Interpretation of this difference in activity is difficult because the egg phosphatidylcholine vesicles contained unsaturated fatty acid residues in a liquid-crystalline state, whereas the dipalmitoylphosphatidylcholine vesicles were fully saturated and in a gel state.

The interaction of bovine phosphatidylcholine transfer protein with phospholipid membranes of different fatty acid composition was investigated in greater detail by Kasper and Helmkamp (1981a) by measuring phospholipid transfer from egg phosphatidylcholine donor vesicles to egg or dimyristoylphosphatidylcholine acceptor vesicles. At 37°C, the unsaturated egg phosphatidylcholine membranes supported a 2.5-fold greater rate of transfer per unit protein than did the saturated dimyristoylphosphatidylcholine membranes. Since both these lipids undergo gel to liquid-crystalline phase transitions at temperatures considerably less than that of the assay (Ladbrooke *et al.*, 1968; Phillips *et al.*, 1969), a clear preference for the unsaturated species is evident. The marked difference between these two vesicle populations in functioning as acceptors of egg phosphatidylcholine was confirmed in experiments where donor membranes were prepared from [^{14}C]dimyristoylphosphatidylcholine. With the unsaturated acceptors, transfer rates were approximately three times greater.

Similar observations were made by Schulze *et al.* (1977) using unfractionated rat liver supernate as a source of transfer protein in a vesicle–mitochondrion assay system. Donor vesicles composed of dioleoylphosphatidylcholine supported 15–20 times greater phosphatidylcholine transfer at 25°C than vesicles of dipalmitoylphosphatidylcholine or distearoylphosphatidylcholine. Noncatalyzed transfer rates differed only by a factor of two under these conditions. Comparable experiments were performed at 25, 37, and 43°C to assess lipid phase transitions. Although trans-

TABLE II

Characteristics of Phosphatidylcholine Vesicles of Various Fatty Acid Composition

Membrane phosphatidylcholine[a]	Transfer activity[b] (nmol/hr)	Fluorescence polarization[c]	Vesicle radius[d] (nm)	Transbilayer distribution[d] (mol (outer)/mol (inner))
Egg	15.4	0.102	12.0	2.0
Dioleoyl	15.2	0.100	13.1	1.8
Dielaidoyl	9.6	0.137	12.3	2.0
Dimyristoyl	1.6	0.155	8.4	2.6

[a]Membranes also contained 2 mol% egg phosphatidic acid.

[b]Phosphatidylinositol transfer was measured at 37°C from microsomes to vesicles in presence of bovine phosphatidylinositol transfer protein (Helmkamp, 1980b).

[c]Values represent corrected determination of diphenylhexatriene polarization at 37°C (Helmkamp, 1980b).

[d]Data are taken from de Kruijff *et al.* (1976).

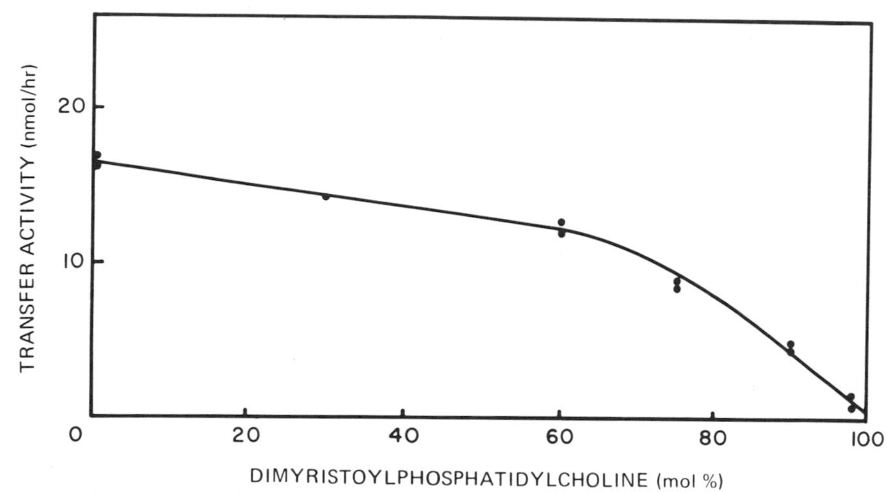

Fig. 2. Effect of dimyristoylphosphatidylcholine on the activity of bovine phosphatidylinositol transfer protein. Mixtures of egg and dimyristoylphosphatidylcholines are prepared as single bilayer vesicles and tested as acceptors of microsomal phosphatidylinositol. See Fig. 1 for other details. Reproduced with permission from Helmkamp (1980a).

fer activity increased with increasing temperature, the authors concluded that no correlation between phosphatidylcholine exchange and phase transition could be established. It should be emphasized that these data were limited to only three temperatures.

A more extensive series of phosphatidylcholines was investigated in the presence of phosphatidylinositol transfer protein (Helmkamp, 1980b). With rat liver microsomes containing [^{3}H]phosphatidylinositol as donor membranes, single bilayer acceptor vesicles were prepared from egg, dioleoyl-, dielaidoyl-, or dimyristoylphosphatidylcholine. Initial rates of transfer were highest with the two cis-unsaturated species, intermediate with the trans-unsaturated analog of oleic acid, and lowest with the saturated lipid (Table II). These values were recorded at 37°C, at which temperature all phospholipid vesicles are in a liquid-crystalline state. Furthermore, for acceptor vesicles containing mixtures of egg and dimyristoylphosphatidylcholines, transfer activity decreased dramatically as the proportion of the saturated species increased (Fig. 2).

Thus, for liquid–crystalline vesicles, the following order of reactivity in phospholipid transfer systems may be established: cis-unsaturated > trans-unsaturated > saturated. That this same order is correct for the relative fluidity of these vesicles has been verified by the polarization of 1,6-di-

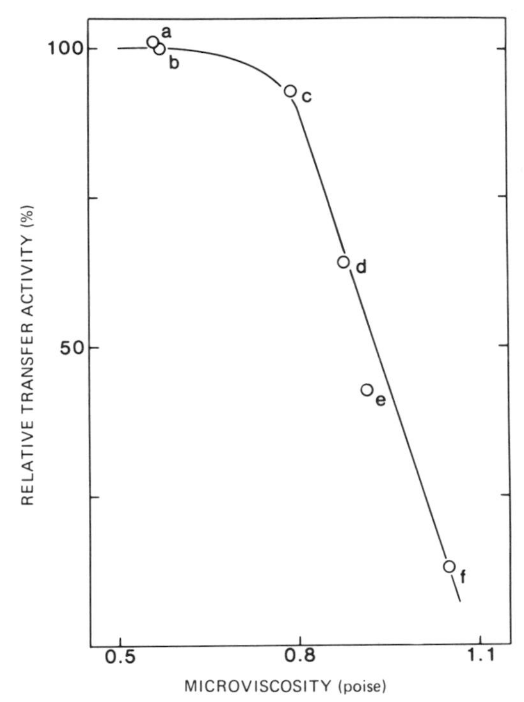

Fig. 3. Correlation between bilayer microviscosity and rates of phosphatidylinositol transfer activity. Single bilayer vesicles serve as acceptors of phosphatidylinositol from either rat liver microsomes (a,b,d,f) or egg phosphatidylcholine-phosphatidylinositol vesicles (b,c,e) catalyzed by bovine phosphatidylinositol transfer protein. Transfer activity is relative to that found with egg phosphatidylcholine (b); other vesicles are prepared from dioleoylphosphatidylcholine (a), 1-stearoyl-2-oleoylphosphatidylcholine (c), dielaidoylphosphatidylcholine (d), 1-stearoyl-2-elaidoylphosphatidylcholine (3), and dimyristoylphosphatidylcholine (f). Microviscosity was determined as described in Fig. 1. Adapted with permission from Helmkamp (1980b). Copyright (1980) American Chemical Society.

phenyl-1,3,5-hexatriene (DPH), a fluorescent molecule which associates with the hydrocarbon interior of a lipid bilayer (Lentz *et al.*, 1976). These values, calculated by Shinitzky and Barenholz (1978), can be found in Table II. Indeed, an excellent correlation between phosphatidylinositol transfer protein activity and acceptor vesicle microviscosity is apparent from these data (Fig. 3). Also included in Fig. 3 are several additional acceptor vesicles of defined lipid composition (1-stearoyl-2-oleoyl-, and 1-stearoyl-2-elaidoylphosphatidylcholine) which were evaluated in a vesicle–vesicle assay system. The chemistry of the membrane phosphatidylcholine fatty acyl groups plays a critical role in the fluidity of the membranes and, in turn, in the interaction between the membranes and phospholipid transfer protein.

Bilayer Phase Transitions and Phospholipid Transfer Protein Activity

The reversible transition of oriented, hydrated phospholipids between gel and liquid–crystalline phases is most readily effected by changes in temperature and leads to dramatic changes in the physical properties of the membranes (Lee, 1977). Among the changes that occur at temperatures above the thermal transition are decreased bilayer thickness, increased vesicle radius, increased amount of bound water, and reorganization of the fatty acyl hydrocarbon chains from a rigid, all-trans configuration to a more relaxed gauche configuration (Watts *et al.*, 1978). The resulting alteration in membrane fluidity can be monitored, with minimum ambiguity, by the polarization characteristics of appropriate fluorescent compounds. Diphenylhexatriene is particularly useful in this regard since it partitions to the hydrophobic interior of a lipid bilayer and exhibits equal preference for the gel and liquid–crystalline phases (Lentz *et al.*, 1976; Shinitsky and Barenholz, 1978).

Typical fluorescence polarization results are shown in Fig. 4 for various single bilayer phosphatidylcholine vesicles, each containing 2 mol% egg phosphatidic acid. The discontinuity represents the gel to liquid-crystalline phase transition, which for these vesicles occurs over a range of 12 to 15 degrees. The transition midpoints, at temperatures a and b in Fig. 4 (b), compare favorably with those obtained by other physical techniques. The thermotropic phase transition of the dielaidoylphosphatidylcholine vesicles occurred between 6–15°C with a midpoint of 11°C. Transition temperatures have been reported to be −5 to −15°C for egg phosphatidylcholine, 12–13°C for dielaidoylphosphatidylcholine, and −22°C for dioleoylphosphatidylcholine (Ladbrooke *et al.*, 1968; Phillips *et al.*, 1972; Norman *et al.*, 1972; Wu and McConnell, 1975). The principal molecular species in egg phosphatidylcholine, 1-palmitoyl-2-oleoylphosphatidylcholine, has a phase transition temperature of −3 to −5°C (Davis *et al.*, 1980). The inclusion of 8 mol% lactosylceramide in egg phosphatidylcholine vesicles had no measurable effect on the fluidity of those bilayers (Kasper and Helmkamp, 1981a).

The spectroscopic description of changes in bilayer fluidity and phase behavior of phospholipid vesicles permits a straightforward comparison of these parameters and the activity of phospholipid transfer protein. To this end, phospholipid transfer and fluorescence polarization measurements were performed over the temperature range 15–45°C, with acceptor vesicles prepared from different phosphatidylcholines. Phosphatidylinositol transfer protein-catalyzed transfer of phosphatidylinositol from rat liver microsomes to vesicles prepared from egg, dioleoyl-, or dielaidoylphosphatidylcholine yielded Arrhenius plots of transfer activity (Fig. 5) which were linear and

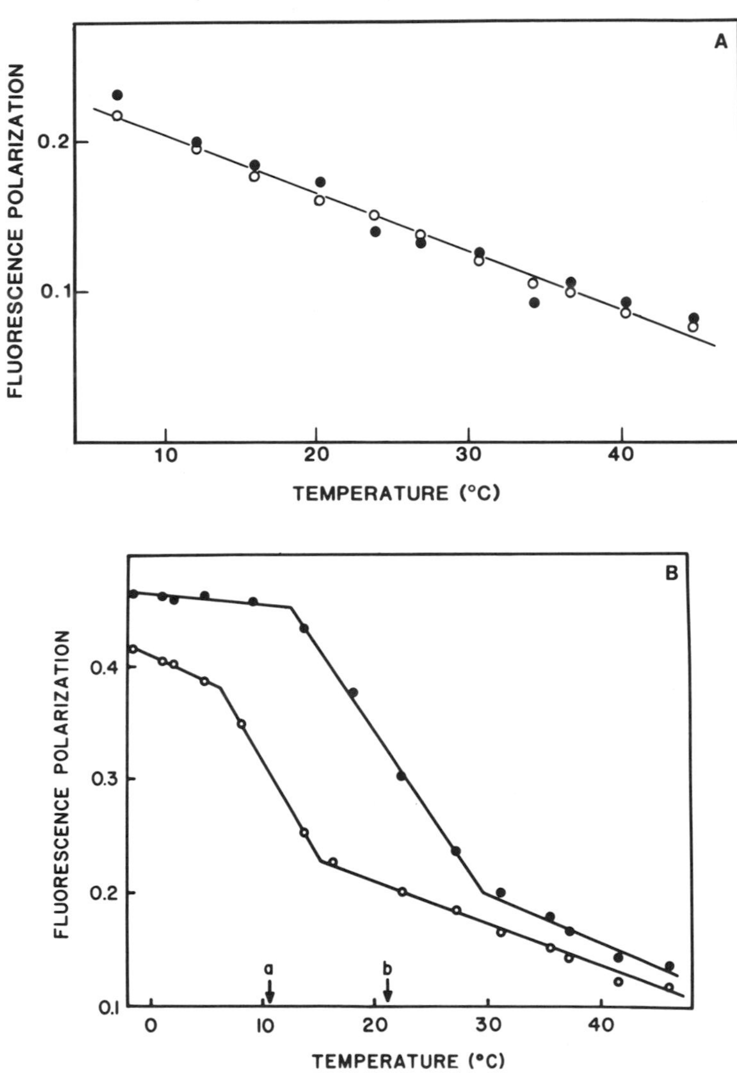

Fig. 4. Polarization of diphenylhexatriene fluorescence in phospholipid vesicles. Phosphatidylcholine vesicles, containing 2 mol% egg phosphatidic acid, are prepared as described in Fig. 1. Samples are heated to new temperatures and allowed to equilibrate completely before subsequent polarization measurements; appropriate corrections for light scattering are made when necessary. (a) Egg phosphatidylcholine (●); dioleoylphosphatidylcholine (○). (b) dielaidoylphosphatidylcholine (○); dimyristoylphosphatidylcholine (●). Gel to liquid-crystalline phase transition midpoint temperatures are indicated by arrows a and b.

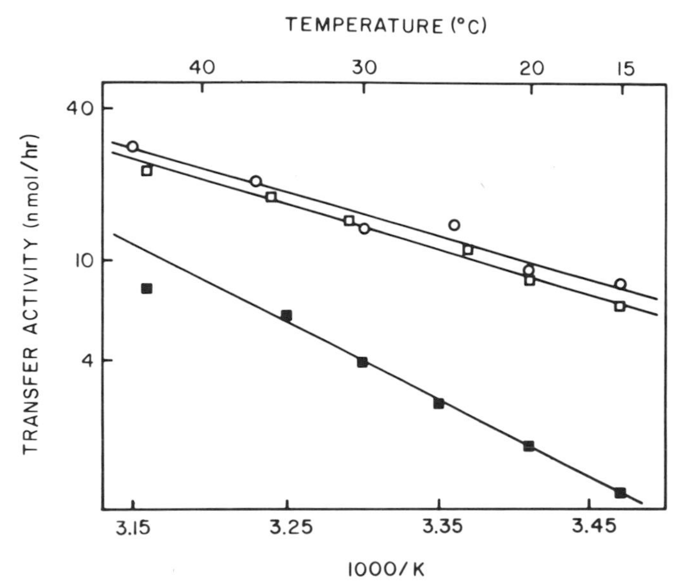

Fig. 5. Arrhenius plots of rates of phosphatidylinositol transfer. The catalyst was bovine phosphatidylinositol transfer protein; assays were carried out as described in Fig. 1. Vesicles were prepared from egg phosphatidylcholine (□), dioleoylphosphatidylcholine (○), or dielaidoylphosphatidylcholine (■). Reprinted with permission from Helmkamp (1980b). Copyright (1980) American Chemical Society.

without discontinuity (Helmkamp, 1980b). The shape of these curves can be compared with those obtained by fluorescence polarization, all of which were linear in the same temperature range (Fig. 4).

It was evident from the Arrhenius plots of phosphatidylinositol transfer activity that the apparent activation energies of the process differed with the various acceptor membranes. Values for egg phosphatidylcholine and dioleoylphosphatidylcholine vesicles were 35 and 45 kJ mol^{-1}, respectively; with dielaidoylphosphatidylcholine vesicles, it was 60 kJ mol^{-1} (Table III). The nearly twofold increase in activation energy for the trans-unsaturated membrane suggests that the configurational geometry of membrane phospholipid fatty acyl residues is a significant determinant in the transfer process.

The low but measurable amount of activity of phosphatidylcholine transfer protein toward the fully saturated dimyristoylphosphatidylcholine vesicles provided an opportunity to investigate the effect of the gel to liquid–crystalline phase transition on transfer activity. Egg phosphatidylcholine vesicles exhibited a linear, continuous activity/temperature Arrhenius relationship over the range 11–45°C (Fig. 6) with an apparent

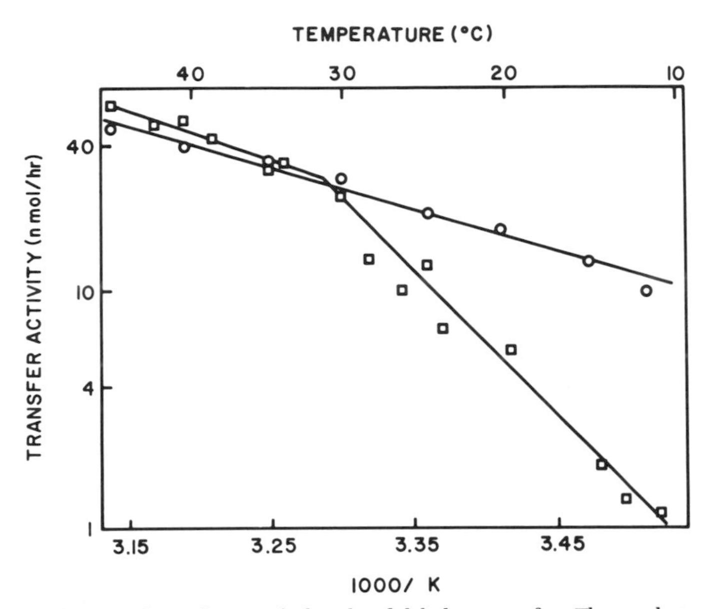

Fig. 6. Arrhenius plots of rates of phosphatidylcholine transfer. The catalyst was bovine phosphatidylcholine transfer protein; activity was measured between egg phosphatidylcholine donor vesicles and the following acceptor vesicle and quantity of transfer protein: egg phosphatidylcholine and 0.12 μg protein (○); dimyristoylphosphatidylcholine and 0.30 μg protein (□). Adapted with permission from Kasper and Helmkamp (1981a). Copyright (1981) American Chemical Society.

of 33 kJ/mol (Table III). Throughout this temperature range, egg phosphatidylcholine vesicles were in the liquid-crystalline phase, as indicated graphically in Fig. 7 by the Arrhenius plot of the anisotropy parameter $[(r_0/r)-1]^{-1}$.

In contrast, the Arrhenius plot of transfer activity with dimyristoylphosphatidylcholine acceptors displayed a discontinuity at 31°C. Above that temperature, the apparent activation energy was 35 kJ/mol, essentially identical to that observed for egg phosphatidylcholine membranes. The temperature of this discontinuity corresponded precisely to the upper end of the region in which these vesicles changed from gel to liquid-crystalline phases. Below 31°C, phosphatidylcholine transfer rates were characterized by an apparent activation energy of 115 kJ/mol. Thus, phospholipid bilayers in which the gel and liquid-crystalline phases coexist participate readily in protein-catalyzed lipid transfer, but with significantly altered kinetic and thermodynamic parameters. Kasper and Helmkamp (1981a) also found a similarly large apparent activation energy, 141 kJ/mol, for transfer of egg phosphatidylcholine to vesicles prepared from dipalmitoylphosphatidyl-

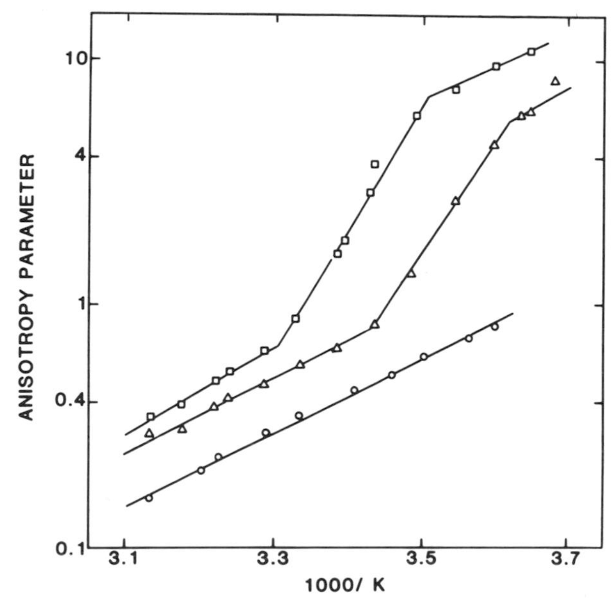

Fig. 7. Arrhenius plots of anisotropy parameter of diphenylhexatriene fluorescence. The polarization values (P) of Fig. 4 are related to fluorescence anisotropy (r) by the relationship, $r = [2P/(3-P)]$. The anisotropy parameter is derived from the Perrin equation and is given by the function $[(r_0/r)-1]^{-1} = \bar{\eta}/C(r)T\tau$ where r_0 is the fluorescence anisotropy in the absence of rotational motion ($r_0 = 0.395$ for diphenylhexatriene), $\bar{\eta}$ is microviscosity, $C(r)$ is a parameter relating to the molecular shape and orientation of the rotating probe, and τ is the reciprocal lifetime of the excited state (Kawato *et al.*, 1977; Shinitzky and Barenholz, 1978). Vesicles were prepared from egg phosphatidylcholine (○), dielaidoylphosphatidylcholine (△), or dimyristoylphosphatidylcholine (□).

choline between 28 and 39°C. This saturated phospholipid has a phase transition midpoint at 42°C (Phillips *et al.*, 1969). The Arrhenius plot was monophasic, and it was assumed that the vesicles remained in the fully ordered state.

An assay system of single bilayer donor vesicles of egg phosphatidylcholine/egg phosphatidylglycerol (95:5 mol%) and multilamellar acceptor vesicles of dipalmitoylphosphatidylcholine/dipalmitoylphosphatidylglycerol (95:5 mol%) provided the framework for a detailed kinetic analysis of bovine phosphatidylcholine transfer protein activity (Bozzato and Tinker, 1982). Initial transfer rates were measured at 37 and 45°C, below and above the phase transition temperature of dipalmitoylphosphatidylcholine. In proceeding from the gel to the liquid-crystalline state, the calculated maximum velocity increased nearly fivefold. This change reflected a comparable increase in the dissociation rate constant of the protein–acceptor membrane complex. Multilamellar acceptor vesicles prepared for egg phosphatidylcholine exhibited no such change in the rate-determining step between these two temperatures. Thus, the increase fluidity of the liquid-crystalline state markedly facilitates the dissociation step of the transfer protein catalytic cycle. Whether these kinetic differences derive from the altered physical state of the acceptor measure or from an altered affinity of the transfer protein for dipalmitoylphosphatidylcholine remains to be established.

Thermodynamic Considerations

Arrhenius plots of phospholipid transfer rates yield apparent activation energies of this process (Table III). Activation energies for the bovine phosphatidylcholine and phosphatidylinositol transfer proteins were 33–35 kJ/mol for transfer to bilayers composed of egg or dioleoylphosphatidylcholine. A similar value was also found for phosphatidylcholine. A similar value was also found for phosphatidylcholine transfer protein with liquid–crystalline dimyristoylphosphatidylcholine vesicles, suggesting that for this protein the actual chemistry of the phospholipid is of minor importance compared to the phase behavior of the organized bilayers. By contrast, the shift from a cis-unsaturated to a trans-unsaturated liquid-crystalline bilayer results in a nearly twofold increase in the apparent activation energy for the phosphatidylinositol transfer protein. Thus, for this protein catalyst, the chemical properties of membrane phospholipids play a critical role in the transfer event.

It is instructive to compare the above activation energies with those for intermembrane phospholipid transfer which takes place in the absence of protein. Although such measurements have not been carried out for unmodified phospholipids in microsome–vesicle or vesicle–vesicle membrane systems, transfers of spin-labeled or ^{3}H-labeled phosphatidylcholine between a variety of biological and artificial membranes are characterized by apparent activation energies of 43–95 kJ/mol (Table III). These data clearly indicate that a significant aspect of the protein-catalyzed transfer mechanism involves a lowering of the apparent activation energy. Just how this is achieved is not yet known. Some information is available on the thermodynamics of lipid–water and lipid–protein interactions. Recalling that the available kinetic evidence is consistent with a mechanism in which a transfer protein–phospholipid complex is free to traverse the aqueous space between outer surfaces of different membranes, a rather formidable energy barrier must still exist for the transfer of a phospholipid molecule between an accommodating membrane bilayer environment and even the most hydro-

TABLE III

Arrhenius Activation Energies for Intermembrane Phospholipid Transfer[a]

Transfer protein	Transferred phospholipid	Donor membrane	Acceptor membrane	Apparent E_a (kJ/mol)	Reference
Bovine phosphatidylinositol	PtdIns	Rat liver microsomes	Egg PtdCho vesicles	35	Helmkamp (1980a)
			Ole_2PtdCho vesicles	34	
			Ela_2PtdCho vesicles	60	
Bovine phosphatidylcholine	PtdCho	Egg PtdCho vesicles	Egg PtdCho vesicles	33	Kasper and Helmkamp (1981a)
			Egg PtdCho vesicles containing 33 mol% chol	33	
			Myr_2PtdCho vesicles (>31°C)	35	
			Myr_2PtdCho vesicles (<31°C)	115	
			Myr_2PtdCho vesicles containing 33 mol% chol	81	
			Pal_2PtdCho (<39°C)	141	
None	PtdCho	PtdCho vesicles	Egg PtdCho vesicles	82	Iida *et al.* (1978)
			Pal_2PtdCho vesicles (>41°C)	95	
None	PtdCho	*Tetrahymena pyriformis* microsomes	*Tetrahymena pyriformis* microsomes	75	
None	PtdCho	Egg PtdCho vesicles	Bovine plasma high density lipoprotein	52	Jonas and Maine (1979)
None	PtdCho	Sendai virus	Human erythrocyte ghosts	43	Kuroda *et al.* (1980)

[a]Lipid abbreviations are those used in Table I, and the following: Myr_2PtdCho, dimyristoyl-PtdCho; Pal_2PtdCho, dipalmitoyl-PtdCho; Ole_2PtdCho, dioleoyl-PtdCho; Ela_2PtdCho, dielaidoyl-PtdCho.

phobic protein environment. The free energy of transfer of dipalmitoylphosphatidylcholine from a micelle (bilayer) to water is -63 kJ/mol, based on an observed critical micelle concentration of 4.7×10^{-10} *M* (Smith and Tanford, 1972). The free energy of transfer of that same molecule from a phospholipid transfer protein to water must account for the zwitterionic charges and two hydrocarbon chains; a value in the range of -70 to -90 kJ/mol would be anticipated (Tanford, 1973). This calculation is based upon a doubling of the free energy of association between a long-chain alkyl sulfate and bovine plasma albumin. Although the analogy between albumin–alkyl sulfate and transfer protein–phospholipid complexes is tenuous at best, it does suggest that only a small change in the protein–water and micelle–water transfer free energies is needed to permit the formation of highly specific protein–lipid complexes. Of course, the exclusion or reduction of water at the transfer protein–membrane interface would greatly facilitate the formation and dissolution of stable protein–phospholipid complexes.

The change in the apparent activation energy for protein-catalyzed phospholipid transfers to gel-state vesicles, compared to liquid–crystalline vesicles, must take into consideration the enthalpy change for the thermotropic phase transition. For hydrated dimyristoylphosphatidylcholine bilayers, this parameter has a value of 23–27 kJ/mol (Phillips *et al.*, 1969). However, the activation energy for phosphatidylcholine transfer protein increases 80 kJ/mol as dimyristoylphosphatidylcholine acceptor vesicles become more ordered in the gel state and bilayer fluidity decreases. Therefore, the higher activation energy reflects more than the change in transition enthalpy and possibly indicates a major difference in the physical interaction between the phosphatidylcholine transfer protein and gel–phase phospholipid bilayers.

Effect of Cholesterol

Cholesterol is a common constituent of many biological membranes; its ability to fluidize or rigidify phospholipid hydrocarbon chains is well documented (Hinz and Sturtevant, 1972; Oldfield and Chapman, 1972). The influence of this sterol on phospholipid protein activity has been investigated with mixed phospholipid–cholesterol vesicles (Helmkamp, 1980a; Kasper and Helmkamp, 1981a). The transfer of phosphatidylinositol by bovine brain phosphatidylinositol transfer protein was unaltered with egg phosphatidylcholine acceptor vesicles containing 15 or 30 mol% cholesterol; at 45 mol% the activity decreased 20% (Fig. 8). Similar results were observed for phosphatidylcholine transfer protein. For acceptor vesicles prepared from

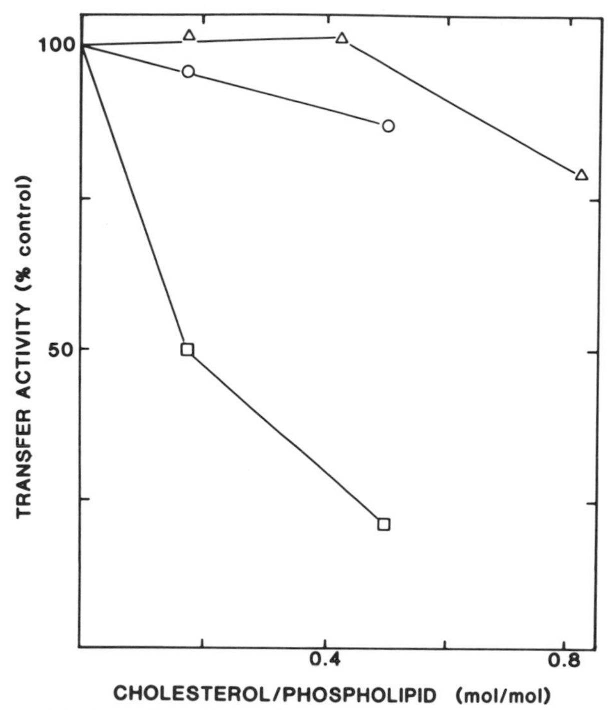

Fig. 8. Effect of cholesterol on phospholipid transfer activity. Acceptor vesicles contained the indicated molar ratio of cholesterol and egg phosphatidylcholine (△, ○) or dimyristoylphosphatidylcholine (□). Control experiments used vesicles that did not contain cholesterol. Catalysts and donor membranes were bovine phosphatidylinositol transfer protein and rat liver microsomes (△) or bovine phosphatidylcholine transfer protein and egg phosphatidylcholine vesicles (○, □).

egg phosphatidylcholine/cholesterol (67:33 mol%), transfer activity decreased to 88% of control. Thus, for the highly unsaturated, fluid egg phosphatidylcholine bilayers, cholesterol had only a minimal effect. On the other hand, the incorporation of cholesterol in dimyristoylphosphatidylcholine membranes led to a striking loss in phosphatidylcholine transfer protein activity at 37°C. Only one-half the control activity remained at 15 mol% cholesterol, and less than 25% was measured at 33 mol% cholesterol. Johnson and Zilversmit (1975) had earlier noted some decrease in the activity of bovine heart phosphatidylinositol transfer protein when the cholesterol content of mixed cholesterol–egg phosphatidylcholine vesicles exceeded a molar ratio of 1. These workers were investigating the transfer of phosphatidylcholine from vesicles to beef heart mitochondria. Thus, in keeping with the lipid-ordering effect of cholesterol on liquid–crystalline phospholipid bilayers and the decrease in membrane fluidity, lower rates of protein-catalyzed phospholipid transfer are generally observed.

The effect of cholesterol on phospholipid transfer may be interpreted in terms of decreased membrane fluidity. Molecular motion within a phospholipid bilayer is sensitive to the presence of cholesterol. Thus, under conditions where dimyristoylphosphatidylcholine–cholesterol bilayers assumed organizational and fluidity states intermediate between pure gel and pure liquid-crystalline phases (Rubenstein *et al.*, 1979), the rates of phosphatidylinositol transfer accordingly decreased. The smaller although significant reduction in the activity of phosphatidylinositol transfer protein toward egg phosphatidylcholine–cholesterol acceptor vesicles also reflects the expected decrease in fluidity of these membranes. It is obvious that cholesterol had a greater effect with saturated phosphatidylcholine. Experiments utilizing mixed phosphatidylcholine–cholesterol vesicles, at temperatures below the phospholipid transition temperature, would be helpful in further defining the relationship between phospholipid transfer proteins and membrane fluidity. Rate enhancements are predicted, because cholesterol will render such lipid bilayers more fluid.

Another important consequence of cholesterol incorporation into bilayers is in the alteration of selected physical parameters of the unilamellar vesicles (Johnson, 1973; de Kruijff *et al.*, 1976). The addition of cholesterol to approximately 33 mol% causes an increase in vesicle radius (from 10.2 to 13.3 nm for egg phosphatidylcholine) and a corresponding decrease in the transbilayer lipid distribution or the outer layer to inner layer ratio (from 2.6 to 2.1 for dimyristoylphosphatidylcholine). Even though this would cause approximately a 10% decrease in the pool of transferable phospholipid in the vesicle–vesicle assay system, such a small change cannot possibly account for the dramatic changes in transfer activity noted for dimyristoylphosphatidylcholine acceptor vesicles.

Membranes containing 33 mol% cholesterol were used as acceptors in the presence of phosphatidylcholine transfer protein over the temperature range 20–41°C (Kasper and Helmkamp, 1981a). Arrhenius plots for egg phosphatidylcholine–cholesterol vesicles remain linear and continuous; the apparent activation energy, 33 kJ/mol, was identical to that for cholesterol-free membranes. Also yielding a continuous activity–temperature Arrhenius relationship were dimyristoylphosphatidylcholine–cholesterol vesicles. The monotonic curve gave a slope of 81 kJ/mol, an activation energy intermediate between those calculated in the absence of sterol above and below 31°C. These data are summarized in Fig. 9 and Table III. Therefore, by abolishing the gel to liquid-crystalline phase transition, cholesterol led to an intermediate fluid state in these membranes. Reduced rates of phosphatidylcholine transfer are probably a direct consequence of decreased membrane fluidity.

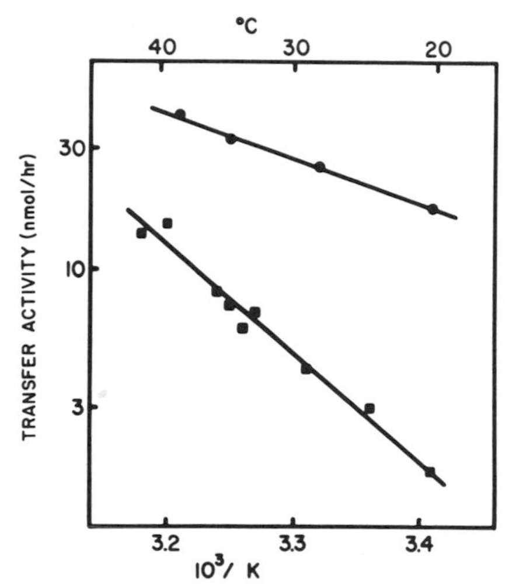

Fig. 9. Arrhenius plots of rates of phosphatidylcholine transfer to cholesterol-containing vesicles. Experimental conditions are similar to those described in Fig. 6, with the following acceptor vesicle and quantity of transfer protein: egg phosphatidylcholine and 0.12 μg protein (●); dimyristoylphosphatidylcholine and 0.50 μg protein (■). Reprinted with permission from Kasper and Helmkamp (1981a). Copyright (1981) American Chemical Society.

Although cholesterol spontaneously redistributes among single bilayer vesicles ($t_{1/2}$ = 1.5 hr, 37°C) and displays a rapid transmembrane movement ($t_{1/2}$ = 1 min, 37°C) (Backer and Dawidowicz, 1979, 1981), phospholipid transfer measurements were carried out under the conditions which minimized these contributions to the assay system.

Spontaneous Phospholipid Transfer

A spontaneous, protein-independent transfer of lipid molecules occurs between various membranes. Such movement has been observed for cholesterol, fatty acids, phospholipids, and other amphiphilic compounds in membrane systems consisting of micelles, single bilayer vesicles, liposomes, or lipoproteins. As in the case of catalyzed transfer, two membrane populations are required, as well as a means of identifying the transferred molecules. Several systems took advantage of initially different gel to liquid–crystalline transition temperatures in order to monitor the phase transition by calorimetric or turbidimetric techniques during the course of lipid transfer (Martin and McDonald, 1976, Kremer *et al.*, 1977; Duckwitz-

Peterlein *et al.*, 1977). Other systems employed radiolabeled or fluorescent phospholipids to establish the kinetics of intermembrane transfer (Sengupta *et al.*, 1976; Galla *et al.*, 1979; Jonas and Maine, 1979; Roseman and Thompson, 1980; Nichols and Pagano, 1981, 1983). Certain spectroscopic properties of the fluorescent analogs, namely eximer formation of pyrenedecanoyl-substituted phosphatidylcholine, self-quenching of (7-nitro-2,1,3-benzoxadiazol-4-yl)aminoacyl-substituted phosphatidylcholine, and resonance energy transfer between different chromophores, have made them especially attractive for initial rate measurements.

The mixing of phospholipids from different membranes may be accomplished by several distinct methods: (1) transfer through the aqueous phase of monomeric or micellar species, (2) transfer upon collision of the membranes, and (3) transfer following fusion of the membranes. Attempts to distinguish among these mechanisms have failed to arrive at a general consensus, even for experimental systems which are essentially identical. Nevertheless, recent investigations of lipid transfer between unilamellar phospholipid vesicles support a mechanism of monomer diffusion rather than collision (Roseman and Thompson, 1980; Nichols and Pagano, 1981; McLean and Phillips, 1981). Although the diffusion-mediated model is usually based on an observed independence of initial transfer rates on the concentration of the acceptor vesicle, Nichols and Pagano (1981) stress that this is not a necessary condition. The rates of noncatalyzed phospholipid transfer exhibited half times of 2–48 hr. As for the catalyzed process, this parameter is a complex function of the structure of the lipid molecule, the characteristics of the bilayer membranes, and the temperature and ionic strength of the medium.

For example, Martin and MacDonald (1976) followed phospholipid transfer between dimyristoylphosphatidylcholine vesicles and dipalmitoylphosphatidylcholine vesicles by changes in solution turbidity. The rate at which equilibrium was approached, that is, the generation of vesicles composed of equimolar amounts of each phospholipid, increased at 24 and 41°C. These values correspond to the phase transition temperatures of the pure vesicles. The authors interpreted these observations in terms of changes in the rate of dissociation of a dimyristoylphosphatidylcholine molecule from a dimyristoylphosphatidylcholine vesicle. With vesicles prepared from the same phospholipids, Kremer *et al.* (1977) observed virtually complete phosphatidylcholine exchange at 50°C, but no change in phase transition or vesicle composition at 20°C. Transfers of spin-labeled phosphatidylcholine from vesicles to dipalmitoylphosphatidylcholine vesicles occurred readily above 41°C; below the phase transition temperature, the transfer rate was small and essentially independent of temperature (Iida *et al.*, 1978). These investigators also noted a most interesting effect of membrane fluidity on spontaneous phospholipid transfer. By growing cultures of *Tetrahymena pyriformis* at 15 or 39.5°C, membranes of different fluidity are formed, the more fluid at the lower temperature. At 37°C, transfer of spin-labeled phosphatidylcholine to membrane fractions (microsome, pellicle, or cilia) of the cells grown at 15°C was approximately twice as rapid as that seen for the cells grown at 39.5°C. A similar difference was evident for vesicle–vesicle transfers using extracted microsomal lipids from the two cell populations.

These results demonstrate that bilayer phase behavior and membrane fluidity strongly influence selected noncatalyzed movements of phospholipids between membranes.

Phospholipid Transfer Protein–Lipid Interactions

It is generally accepted that for many protein–lipid interactions the chemical properties and physical organization of the lipids play important roles (Sandermann, 1978; Chapman *et al.*, 1979). Factors such as the character of the polar head group of a phospholipid and the chain length and degree of unsaturation of its constituent fatty acyl groups can profoundly influence protein–lipid interactions. Similarly, the amino acid composition, secondary and tertiary conformations, and hydrophobic nature of a protein will determine the extent of interaction with individual or multidimensional arrays of lipid molecules. Phospholipid transfer proteins have provided an excellent opportunity to investigate protein–lipid interactions at two very distinct levels. One level involves the membrane-dissociated, water-soluble complex between phospholipid transfer protein and phospholipid. Examples of this interaction occupy the spectrum from bovine phosphatidylcholine transfer protein, with its near absolute specificity for the phosphorylcholine head group and relaxed specificity toward the fatty acyl moieties (Kamp *et al.*, 1977), to bovine or rat nonspecific phospholipid transfer proteins which can bind and transport virtually all amphiphilic molecules (Bloj and Zilversmit, 1977, 1981; Crain and Zilversmit, 1980a). Between these extremes is bovine phosphatidylinositol transfer protein, which has a binding domain or domains that accommodates an anionic, highly polar phosphatidylinositol molecule and a zwitterionic phosphatidylcholine molecule, yet discriminates against a similarly zwitterionic phosphatidylethanolamine molecule (Helmkamp *et al.*, 1974).

The other level of protein–lipid interaction concerns the adsorption complexes between phospholipid transfer protein and various membrane interfaces. The existence of electrostatic as well as hydrophobic components to these interactions has been discussed earlier. Several general conclusions may be drawn.

(1.) Membrane surface charge is an important determinant. Both anionic and cationic lipids modify the rates of protein-catalyzed phospholipid transfer. Among zwitterionic membrane phospholipids, there are striking differences in the quantity of ion-pair bonds and bound water which, in turn, may alter the affinity of a transfer protein for a bilayer surface.

(2.) Preference is exhibited toward liquid-crystalline membranes. As bilayer membranes proceed through a thermotropic transition from gel to liquid-crystalline phase, phospholipid transfer protein activity is characterized by a sharp decrease in apparent activation energy. A discontinuity in the Arrhenius plot of transfer activity is observed at the upper end of temperature range over which the phase transition occurs. If penetration of the bilayer by phospholipid transfer protein is a critical step in the catalytic process, this could be facilitated by the physicochemical properties of the liquid-crystalline phase.

(3.) Membrane fluidity and transfer activity are highly correlated. For a series of liquid-crystalline phosphatidylcholine bilayers, phospholipid transfer protein activity increases in direct proportion to the fluidity of the hydrocarbon interior of the membrane. Again, this relationship may reflect the intimate association between phospholipid transfer proteins and the membranes with which they interact. It is likely that the inhibition of intermembrane phospholipid flux by certain amphiphilic molecules is caused, at least in part, by alterations of membrane fluidity. Thus, the incorporation of cholesterol, stearylamine, or sphingomyelin into phospholipid bilayers leads to decreased hydrocarbon fluidity and diminished phospholipid protein activity.

Although the experimental results discussed in this review have been obtained by *in vitro* measurements, indeed, most by using artificial membranes, certain analogies to the biological world may be drawn. It is assumed, although by no means proven, that phospholipid transfer proteins catalyze intracellular intermembrane lipid transport. Such transport could enhance the flux of phospholipid from a synthetic site on one membrane to sites on another membrane where the phospholipid is required for structural, regulatory, or metabolic activities. Obviously, a strict relationship between phospholipid synthesis and degradation, although teleologically attractive, is not a necessary condition for protein-catalyzed phospholipid transfer. Phospholipid transfer proteins could participate in (a) the biogenesis of new membrane, (b) the repair, maintenance, and modification of existing membranes, and (c) the formation and dissociation of lipid or lipoprotein aggregates. Each of these roles involves an interaction between phospholipid transfer protein and a (membrane) lipid interface. The current body of knowledge provides a useful foundation upon which further insight into the structure and function of phospholipid transfer proteins may be pursued.

Acknowledgments

The personal research cited in this chapter has been supported by the expert technical assistance of C. R. Badger, R. H. Baldridge, and A. M. Kasper. The skillful assistance of Elaine Davis during the preparation of this manuscript is also appreciated.

References

Akeroyd, R., Moonen, P., Westerman, J., Puyk, W. C., and Wirtz, K. W. A. (1981). *Eur. J. Biochem.* **114,** 385–391.

Akiyama, M., and Sakagami, T. (1969). *Biochim. Biophys. Acta* **187,** 105–112.

Backer, J. M., and Dawidowicz, E. A. (1979). *Biochim. Biophys. Acta* **551,** 260–270.

Backer, J. M., and Dawidowicz, E. A. (1981). *J. Biol. Chem.* **256,** 586–588.

Bloj, B., and Zilversmit, D. B. (1977). *J. Biol. Chem.* **252,** 1613–1619.

Bloj, B., and Zilversmit, D. B. (1981). *J. Biol. Chem.* **256,** 5988–5991.

Bozzato, R. P., and Tinker, D. O. (1982). *Can. J. Biochem.* **60,** 409–418.

Chapman, D., Gómez-Fernández, J. C., and Goñi, F. M. (1979). *FEBS Lett.* **98,** 211–223.

Cobon, G. S., Crowfoot, P. D., Murphy, M., and Linnane, A. W. (1976). *Biochim. Biophys. Acta* **441,** 255–259.

Cohen, L. K., Leuking, D. R., and Kaplan, S. (1979). *J. Biol. Chem.* **254,** 721–728.

Crain, R. C., and Zilversmit, D. B. (1980a). *Biochemistry* **19,** 1433–1439.

Crain, R. C., and Zilversmit, D. B. (1980b). *Biochim. Biophys. Acta* **620,** 37–48.

Cullis, P. R., and de Kruijff, B. (1979). *Biochim. Biophys. Acta* **559,** 399–420.

Davis, P. J., Coolbear, K. P., and Keough, K. M. (1980). *Can. J. Biochem.* **58,** 851–858.

Dawson, R. M. C. (1966). In "Essays in Biochemistry" (P. N. Campbell and G. D. Greville, eds.), Vol. 2, pp. 69–115. Academic Press, New York.

de Kruijff, B., Cullis, P. R., and Radda, G. K. (1976). *Biochim. Biophys. Acta* **436,** 729–740.

Demel, R. A., Wirtz, K. W. A., Kamp, H. H., Geurts van Kessel, W. S. M., and van Deenen, L. L. M. (1973). *Nature (London)* **246,** 102–105.

Demel, R. A., Kalsbeek, R., Wirtz, K. W. A., and van Deenen, L. L. M. (1977). *Biochim. Biophys. Acta* **466,** 10–22.

DiCorleto, P. E., and Zilversmit, D. B. (1977). *Biochemistry* **16,** 2145–2150.

DiCorleto, P. E., Fakharzadeh, F. F., Searles, L. L., and Zilversmit, D. B. (1977). *Biochim. Biophys. Acta* **468,** 296–304.

DiCorleto, P. E., Warach, J. B., and Zilversmit, D. B. (1979). *J. Biol. Chem.* **254,** 7795–7802.
Duckwitz-Peterlein, G., Eilenberger, G., and Overath, P. (1977). *Biochim. Biophys. Acta* **469,** 311–325.
Dyatlovitskaya, E. V., Timofeeva, N. G., and Bergelson, L. D. (1978). *Eur. J. Biochem.* **82,** 463–471.
Ehnholm, C., and Zilversmit, D. B. (1973). *J. Biol. Chem.* **248,** 1719–1724.
Galla, H.-J., Theilen, U., and Hartman, W. (1979). *Chem. Phys. Lipids* **23,** 239–251.
Gavey, K. L., Noland, B. J., and Scallen, T. J. (1981). *J. Biol. Chem.* **256,** 2993–2999.
Harvey, M. S., Helmkamp, G. M., Jr., Wirtz, K. W. A., and van Deenen, L. L. M. (1974). *FEBS Lett.* **46,** 260–262.
Hellings, J. A., Kamp, H. H., Wirtz, K. W. A., and van Deenen, L. L. M. (1974). *Eur. J. Biochem.* **47,** 601–605.
Helmkamp, G. M., Jr. (1980a). *Biochim. Biophys. Acta* **595,** 222–234.
Helmkamp, G. M., Jr. (1980b). *Biochemistry* **19,** 2050–2056.
Helmkamp, G. M., Jr. (1980c). *Biochem. Biophys. Res. Commun.* **97,** 1091–1096.
Helmkamp, G. M., Jr., Harvey, M. S., Wirtz, K. W. A., and van Deenen, L. L. M. (1974). *J. Biol. Chem.* **249,** 6382–6389.
Helmkamp, G. M., Jr., Nelemans, S. A., and Wirtz, K. W. A. (1976a). *Biochim. Biophys. Acta* **424,** 168–182.
Helmkamp, G. M., Jr., Wirtz, K. W. A., and van Deenen, L. L. M. (1976b). *Arch. Biochem. Biophys.* **174,** 592–602.
Hertz, R., and Barenholz, Y. (1975). *Chem. Phys. Lipids* **15,** 138–156.
Hinz, H. J., and Sturtevant, J. M. (1972). *J. Biol. Chem.* **247,** 3679–3700.
Iida, H., Maeda, T., Ohki, K., Nozawa, Y., and Ohnishi, S. (1978). *Biochim. Biophys. Acta* **508,** 55–64.
Johnson, L. W., and Zilversmit, D. B. (1975). *Biochim. Biophys. Acta* **375,** 165–175.
Johnson, S. M. (1973). *Biochim. Biophys. Acta* **307,** 27–41.
Jonas, A., and Maine, G. T. (1979). *Biochemistry* **18,** 1722–1728.
Kader, J. C. (1975). *Biochim. Biophys. Acta* **380,** 31–44.
Kader, J. C. (1977). In "Dynamic Aspects of Cell Surface Organization" (G. Poste and G. L. Nicolson, eds.), pp. 127–204. Elsevier, Amsterdam.
Kamp, H. H. (1975). Doctoral Dissertation, pp. 74–76. Rijksuniv. Utrecht, Utrecht, Netherlands.
Kamp, H. H., Wirtz, K. W. A., and van Deenen, L. L. M. (1973). *Biochim. Biophys. Acta* **318,** 313–325.
Kamp, H. H., Wirtz, K. W. A., and van Deenen, L. L. M. (1975). *Biochim. Biophys. Acta* **398,** 401–414.
Kamp, H. H., Wirtz, K. W. A., Baer, P. R., Slotboom, A. M., Rosenthal, A. F., Paltauf, F., and van Deenen, L. L. M. (1977). *Biochemistry* **16,** 1310–1316.
Kasper, A. M., and Helmkamp, G. M., Jr. (1981a). *Biochemistry* **20,** 146–151.
Kasper, A. M., and Helmkamp, G. M., Jr. (1981b). *Biochim. Biophys. Acta* **664,** 22–32.
Kawato, S., Kinosita, K., Jr., and Ikegami, A. (1977). *Biochemistry* **16,** 2319–2324.
Kremer, J. M. H., Kops-Werkhoven, M. M., Pathmamanoharan, C., Gizeman, O. L. J., and Wiersema, P. H. (1977). *Biochim. Biophys. Acta* **471,** 177–188.
Kuroda, K., Maeda, T., and Ohnishi, S. (1980). *Proc. Natl. Acad. Sci. U.S.A.* **77,** 804–807.
Ladbrooke, B. D., Williams, R. M., and Chapman, D. (1968). *Biochim. Biophys. Acta* **150,** 333–340.
Lee, A. G. (1977). *Biochim. Biophys. Acta* **472,** 237–281.
Lemaresquier, H., Bureau, G., Mazliak, P., and Kader, J. C. (1982). *Int. J. Biochem.* **14,** 71–74.

Lentz, B. R., Barenholz, Y., and Thompson, T. E. (1976). *Biochemistry* **15,** 4521–4528.

Lumb, R. H., Kloosterman, A. D., Wirtz, K. W. A., and van Deenen, L. L. M. (1976). *Eur. J. Biochem.* **69,** 15–22.

McLean, L. R., and Phillips, M. C. (1981). *Biochemistry* **20,** 2893–2900.

McMurray, W. C., and Dawson, R. M. C. (1969). *Biochem. J.* **112,** 91–108.

Maeda, T., and Ohnishi, S. (1974). *Biochem. Biophys. Res. Commun.* **60,** 1509–1516.

Martin, F. J., and MacDonald, R. C. (1976). *Biochemistry* **15,** 321–327.

Nichols, J. W., and Pagano, R. E. (1981). *Biochemistry* **20,** 2783–2789.

Nichols, J. W., and Pagano, R. E. (1983). *J. Biol. Chem.* **258,** 5368–5371.

Noland, B. J., Arebalo, R. E., Hansbury, E., and Scallen, T. J. (1980). *J. Biol. Chem.* **255,** 4282–4289.

Norman, A. W., Demel, R. A., de Kruijff, B., Geurts van Kessel, W. S. M., and van Deenen, L. L. M. (1972). *Biochim. Biophys. Acta* **290,** 1–14.

Oldfield, E., and Chapman, D. (1972). *FEBS Lett.* **23,** 285–297.

Phillips, M. C., Williams, R. M., and Chapman, D. (1969). *Chem. Phys. Lipids* **3,** 234–244.

Phillips, M. C., Hauser, H., and Paltauf, F. (1972). *Chem. Phys. Lipids* **8,** 127–133.

Poorthuis, B. H. J. M., van der Krift, T. P., Teerlink, T., Akeroyd, R., Hostetler, K. Y., and Wirtz, K. W. A. (1980). *Biochim. Biophys. Acta* **600,** 376–386.

Post, M., Batenburg, J. J., Schuurmans, E. A. J. M., and van Golde, L. M. G. (1980). *Biochim. Biophys. Acta* **620,** 317–321.

Robinson, M. E., Wu, L. N. Y., Brumley, G. W., and Lumb, R. H. (1978). *FEBS Lett.* **87,** 41–44.

Roseman, M. A., and Thompson, T. E. (1980). *Biochemistry* **19,** 439–444.

Rubenstein, J. L. R., Smith, B. A., and McConnell, H. M. (1979). *Proc. Natl. Acad. Sci. U.S.A.* **76,** 15–18.

Sandermann, H., Jr. (1978). *Biochim. Biophys. Acta* **515,** 209–237.

Sasaki, T., and Sakagami, T. (1978). *Biochim. Biophys. Acta* **512,** 461–471.

Schulze, G., Jung, K., Kunze, D., and Egger, E. (1977). *FEBS Lett.* **74,** 220–224.

Sengupta, P., Sackmann, E., Kühnle, W., and Scholz, H. P. (1976). *Biochim. Biophys. Acta* **436,** 869–878.

Shinitzky, M., and Barenholz, Y. (1978). *Biochim. Biophys. Acta* **515,** 367–394.

Smith, R., and Tanford, C. (1972). *J. Mol. Biol.* **67,** 75–83.

Tanford, C. (1973). In "The Hydrophobic Effect," pp. 12–15, 49–53, 126–151. Wiley, New York.

Trzaskos, J. M., and Gaylor, J. L. (1983). *Biochim. Biophys. Acta* **751,** 52–65.

van den Besselaar, A. M. H. P., Helmkamp, G. M., Jr., and Wirtz, K. W. A. (1975). *Biochemistry* **14,** 1852–1858.

Watts, A., Marsh, D., and Knowles, P. F. (1978). *Biochemistry* **17,** 1792–1801.

White, D. A. (1973). In "Form and Function of Phospholipids" (G. B. Ansell, R. M. C. Dawson, and Hawthorne, J. N., eds.), 2nd edition, pp. 441–482. Elsevier, Amsterdam.

Wilson, D. B., Ellsworth, J. L., and Jackson, R. L. (1980). *Biochim. Biophys. Acta* **620,** 550–561.

Wirtz, K. W. A. (1974). *Biochim. Biophys. Acta* **344,** 95–117.

Wirtz, K. W. A. (1982). In "Lipid-Protein Interactions" (P. C. Jost and O. H. Griffith, eds.), Vol. 1, pp. 151–231. J. Wiley and Sons, Inc., New York.

Wirtz, K. W. A., and Zilversmit, D. B. (1968). *J. Biol. Chem.* **243,** 3596–3602.

Wirtz, K. W. A., Kamp, H. H., and van Deenen, L. L. M. (1972). *Biochim. Biophys. Acta* **274,** 606–617.

Wirtz, K. W. A., Geurts van Kessel, W. S. M., Kamp, H. H., and Demel, R. A. (1976). *Eur. J. Biochem.* **61,** 515–523.

Wirtz, K. W. A., Vriend, G., and Westerman, J. (1979). *Eur. J. Biochem.* **94,** 215–221.
Wirtz, K. W. A., Devaux, P. F., and Bienvenue, A. (1980). *Biochemistry* **19,** 3395–3399.
Wu, S. H. W., and McConnell, H. M. (1975). *Biochemistry* **14,** 847–854.
Yeagle, P. L., Hutton, W. C., Huang, C., and Martin, R. B. (1977). *Biochemistry* **16,** 4344–4349.
Zborowski, J., and Wojtczak, L. (1975). *FEBS Lett.* **51,** 317–320.
Zilversmit, D. B. (1978). *Ann. N.Y. Acad. Sci.* **308,** 149–163.
Zilversmit, D. B., and Hughes, M. E. (1976). *Methods Membrane Biol.* **7,** 211–259.

Chapter 7

Ionotropic Effects on Phospholipid Membranes: Calcium/Magnesium Specificity in Binding, Fluidity, and Fusion[1]

Nejat Düzgüneş and Demetrios Papahadjopoulos

Introduction[2]

The phospholipid bilayer is an integral component of cellular membranes, acting as a permeability barrier and a matrix in which membrane proteins

[1]This work was supported by NIH grant GM-28117 (Demetrios Papahadjopoulos) and fellowship CA-06190 (Nejat Düzgüneş).

[2]Abbreviations: PS, phosphatidylserine; PC, phosphatidylcholine; PE, phosphatidylethanolamine; PI, phosphatidylinositol; PG, phosphatidylglycerol; PA, phosphatidic acid; CL, cardiolipin; DSPC, distearoylphosphatidylcholine; DPPC, dipalmitoylphosphatidylcholine; DPPA, dipalmitoylphosphatidic acid; DPPS, dipalmitoylphosphatidylserine; DMPS, dimyristoylphosphatidylserine; DMPG, dimyristoylphosphatidylglycerol; DMPC, dimyristoylphosphatidylcholine; DPPG, dipalmitoylphosphatidylglycerol.

ISBN 0-12-053002-3

are embedded. The functional role of individual phospholipid species in membranes is not well understood at present and requires further exploration. The studies outlined in the present review indicate that Ca^{2+} and Mg^{2+} alter the fluidity of phospholipid bilayers and facilitate interactions between phospholipid vesicles, depending on the phospholipid composition of the bilayers. These effects are likely to be manifested in several membrane phenomena in biological systems.

The lateral distribution and functions of intrinsic membrane proteins are expected to be closely related to the physical and chemical nature of the lipid bilayer. Therefore, molecular factors that influence the fluidity and structure of the lipid bilayer and the formation of microdomains of specific lipid components should exhibit a corresponding influence on membrane proteins. Furthermore, these molecular factors may play a more direct role by regulating the permeability properties of the plasma membrane or the fusion and interconversion of subcellular membranes. In this chapter, we will examine the effects of two such regulators, calcium and magnesium ions, on membrane fluidity and intermembrane interactions. Calcium is involved in the regulation of numerous cellular functions such as stimulus–contraction coupling, protoplasmic motility, intercellular interactions, and stimulus–secretion coupling. Although Ca^{2+} mediates these effects largely via its interaction with proteins, the involvement of membrane lipids is still largely unexplored. High concentrations of Ca^{2+} are maintained in mitochondria, secretory granules, and the extracellular milieu. The membranes of these organelles and the plasma membrane are certainly affected by Ca^{2+}. On the other hand, Ca^{2+} levels in the cytoplasm are estimated to be quite low, but the Mg^{2+} level is in the millimolar range. In this chapter, we focus on the phospholipid bilayer part of cellular membranes and discuss the differential effects of Ca^{2+} and Mg^{2+} on the structure, phase behavior, and fusion of phospholipid membranes.

Thermotropic and Ionotropic Phase Transitions in Phospholipid Membranes

Fully hydrated phospholipid bilayers undergo a transition from the gel to the liquid–crystalline phase at a characteristic temperature T_c. The thermotropic transitions of synthetic phospholipid membranes, particularly dipalmitoylphosphatidylcholine (DPPC) and dimyristoylphosphatidylcholine (DMPC) in multilamellar vesicles, have been studied extensively by means of calorimetry (Chapman *et al.*, 1967; Ladbrooke and Chapman, 1969; Hinz and Sturtevant, 1972; Chapman, 1975; Jacobson and Papahadjopoulos, 1975; Suurkuusk *et al.*, 1976; Thompson *et al.*, 1977), X-ray diffraction (Ranck *et al.*, 1974; Luzzati and Tardieu, 1974; Janiak *et al.*, 1976), ^{1}H and ^{31}P NMR (Sheetz and Chan, 1972; Cullis and deKruijff, 1976), dilatometry (Nagle and Wilkinson, 1978), fluorescence polarization (Jacobson and Papahadjopoulos, 1975; Lentz *et al.*, 1976; Thompson *et al.*, 1977; Shinitzky and Barenholz, 1978), and Raman spectroscopy (Spiker and Levin, 1976; Gaber *et al.*, 1978). When formed into small (approximately 300 Å diameter), single bilayer vesicles, these lipids exhibit different transition characteristics, such as a broader endothermic transition whose peak occurs at a considerably lower temperature than the T_c of the multilamellar dispersions (Papahadjopoulos *et al.*, 1976; Suurkuusk *etal.*, 1976; van Dijck *et al.*, 1978a). Large unilamellar vesicles of 1000–2000 Å diameter composed of DPPC or DMPC have thermotropic properties closer to those of multilamellar vesicles (Düzgüneş *et al.*, 1983a). The transition temperatures for various phospholipids have been reported in recent reviews (Szoka and Papahadjopoulos, 1980; Boggs, 1980; Lee, 1977).

Phase transitions can be induced in phospholipid bilayers not only by changes in temperature but also by changes in the ionic environment. For example, a decrease in pH from 7.4 to 3.5 causes phosphatidylserine (PS) bilayers, which are in the liquid crystalline state at 12°C at neutral pH, to change into the gel phase (Jacobson and Papahadjopoulos, 1975). The main transition temperature of acidic phospholipids increases as the pH is decreased (Watts *et al.*, 1978; Träuble, 1977; Sacré *et al.*, 1979; MacDonald *et al.*, 1976; van Dijck *et al.*, 1978b).

The transition temperature of acidic phospholipid membranes is also shifted to higher temperatures in the presence of divalent metal ions such as Ca or Mg. The T_c of dipalmitoylphosphatidylglycerol (DPPG) increases from 42°C in 100 m*M* NaCl to 57°C in 100 m*M* NaCl containing 1 m*M* Ca^{2+} (Table I; Papahadjopoulos, 1977). The interaction of Ca^{2+} with phosphatidylglycerol (PG) or PS vesicles results in a transformation of the membrane into a multilamellar cylindrical structure (cochleate) as visualized by freeze-fracture electron microscopy (Tocanne *et al.*, 1974; Papahadjopoulos *et al.*, 1975). Monolayer surface pressure measurements indicate that both Mg^{2+} and Ca^{2+} condense the membrane; that is, the area occupied by each molecule at a constant surface pressure decreases (Papahadjopoulos, 1968; Tocanne *et al.*, 1974; Ohki, 1982). The increase in T_c of acidic phospholipid membranes caused by divalent cations has been explained in terms of the reduction of the (negative) surface charge density, which influences the electrostatic free energy of the interface (Träuble and Eibl, 1974; Träuble, 1977). Besides the electrostatic free energy, the isothermal phase transition

TABLE I

Midpoint of the Main Transition Endotherm T_c of Dipalmitoylphosphatidylglycerol Multilayers in the Presence of Na^+, Mg^{2+}, or Ca^{2+} [a]

	Cations (m*M*)				
	Na^+	Mg^{2+}		Ca^{2+}	
	100	1	5	1	5
T_c (°C)	42	51	53	57	68

[a] The divalent cation concentrations are in addition to 100 m*M* Na^+. (Data from Papahadjopoulos, 1977.)

could be effected by changes in hydrogen bonding, water of hydration, or packing of the polar head groups upon divalent cation binding (Jacobson and Papahadjopoulos, 1975).

The upward shift in T_c induced by divalent cations has been studied in detail with PS membranes. The broad endothermic transition at 5–8°C of bovine brain PS in 100 m*M* NaCl is shifted to about 18°C when 0.5 m*M* Ca^{2+} is present and is considerably broader (Jacobson and Papahadjopoulos, 1975). This transition completely disappears when the Ca^{2+} concentration is raised to 1.0 m*M*. When multilamellar vesicles of PS are prepared in the same salt medium containing only 0.15 m*M* Ca^{2+} (in contrast to dialyzing the preformed vesicles against a particular Ca^{2+} concentration), the phase transition is also not apparent up to 70°C (Newton *et al.*, 1978). When the vesicles are prepared in NaCl buffer containing Mg^{2+} above 3 m*M*, the midpoint of the transition is approximately 18°C regardless of the Mg^{2+} concentration (Jacobson and Papahadjopoulos, 1975; Newton *et al.*, 1978). At lower concentrations of Mg^{2+}, the T_c of PS shifts upward as a function of the divalent ion concentration (Newton, 1978). It should be noted that the effect of Ca^{2+} on the phase transition temperature of PS appears to depend on the composition of the acyl chains. Sklar *et al.* (1979) have reported that retinal rod outer segment PS, which contains highly unsaturated fatty acids (Daemen, 1973), does not undergo as drastic a transition in the presence of Ca^{2+} as does brain PS which contains less unsaturated fatty acids.

X-ray diffraction studies have revealed important structural differences between the Ca^{2+} and Mg^{2+} complexes of PS. In the presence of 100 m*M* Na^+ at 25°C, PS membranes have a lamellar repeat distance of 78 Å and a diffuse 4.6 Å high-angle reflection, which indicates that the acyl chains are in a fluid state and that there is considerable distance between the bilayer surfaces (Papahadjopoulos and Miller, 1967; Shipley, 1973). When Ca^{2+}

interacts with PS bilayers, the lamellar spacing is reduced to 53 Å, and sharp 4.1 and 4.5 Å high-angle diffraction spacings, which reflect an ordered packing of the hydrocarbon chains are observed (Jacobson and Papahadjopoulos, 1975). It is interesting that this is a different crystal lattice compared to the chain packing of phospholipid membranes below their T_c which give sharp 4.2 Å high-angle diffraction patterns. The short lamellar spacing is not altered after high vacuum drying, suggesting that free water between the bilayers is absent (Newton *et al.*, 1978; Papahadjopoulos *et al.*, 1978; Portis *et al.*, 1979). ^{2}H-NMR measurements also indicate that less than three water molecules are associated with each lipid molecule in the Ca/PS complex (Hauser *et al.*, 1977). The Mg^{2+} complex of PS membranes, however, allows water to remain between bilayers (Portis *et al.*, 1979). This complex gives rise to the 4.2 Å sharp high-angle spacing below the phase transition temperature and the 4.6 Å diffuse spacing above it, which are characteristic for phospholipid membranes in the gel and liquid–crystalline phases, respectively (Luzzati and Tardieu, 1974; Shipley, 1973). A summary of the X-ray diffraction spacings of Ca^{2+} or Mg^{2+} complexes of PS bilayers is given in Table II.

Raman scattering from Ca^{2+} and Mg^{2+} complexes of PS have essentially confirmed these observations on the difference between the acyl chain packing in the two complexes (Hark and Ho, 1979). The ratio of the trans/gauche

TABLE II

X-ray Diffraction Spacing of Phosphatidylserine (PS) Membranes in the Presence of Various Cations[a]

Phosphatidylserine/ metal ion complex	X-ray diffraction spacings (Å)	
	Lamellar repeat	High angle
Na/PS[b] (25°C)	78 diffuse	4.6 diffuse
Na/PS (dry[c]; 25°C)	60	4.2
Mg/PS[d] (25°C)	53	4.6 diffuse
Mg/PS[d] (5°C)	67	4.2
Mg/PS[e] (dry; 5° and 25°C)	60	4.2
Ca/PS[f] (5°, 25° and 60°C)	53	4.1 and 4.5
Ca/PS[e] (dry; 5° and 25°C)	53	4.1 and 4.5

[a]Adapted from Portis *et al.*, 1979.
[b]100 m*M* NaCl, pH 7.4.
[c]Sodium salt of PS was dried from chloroform.
[d]5–20 m*M* $MgCl_2$ in 100 m*M* NaCl.
[e]Vesicles hydrated in 100 m*M* NaCl containing divalent cations followed by high vacuum drying.
[f]5–20 m*M* $CaCl_2$ in 100 m*M* NaCl.

bonds is similar for PS membranes and the Mg^{2+} complex below or above their respective T_c. The Ca^{2+} complex however, appears to be more rigid than the Mg^{2+} complex below 20°C. Hark and Ho (1979) have also observed a temperature dependence of the trans–gauche bond distribution which is not apparent from X-ray diffraction experiments.

The isothermal crystallization of PS acyl chains induced by Ca^{2+} may be observed by batch or titration microcalorimetry (Papahadjopoulos *et al.*, 1978; Portis *et al.*, 1979; Rehfeld *et al.*, 1981a,b). Addition of 5 m*M* Mg^{2+} or 0.7 m*M* Ca^{2+} to a suspension of PS vesicles in a batch microcalorimeter results in a small (0.2 kcal/mol lipid) endothermic reaction. When the Ca^{2+} concentration is increased to 1.3 m*M*, an exothermic reaction is observed with an enthalpy of 5.5 ± 0.5 kcal/mol. The rate of this process increases when the Ca^{2+} concentration is increased to 5 m*M* (Portis *et al.*, 1979). In titration microcalorimetry, the heat released by small incremental additions of Ca^{2+} can be determined (Rehfeld *et al.*, 1981a). Figure 1 shows the thermometric titration of small unilamellar PS vesicles by Ca^{2+} or Mg^{2+} at 25°C. If the titration of Ca^{2+} is stopped before the onset of the exothermic reaction, no further heat change is observed. When the titration is stopped

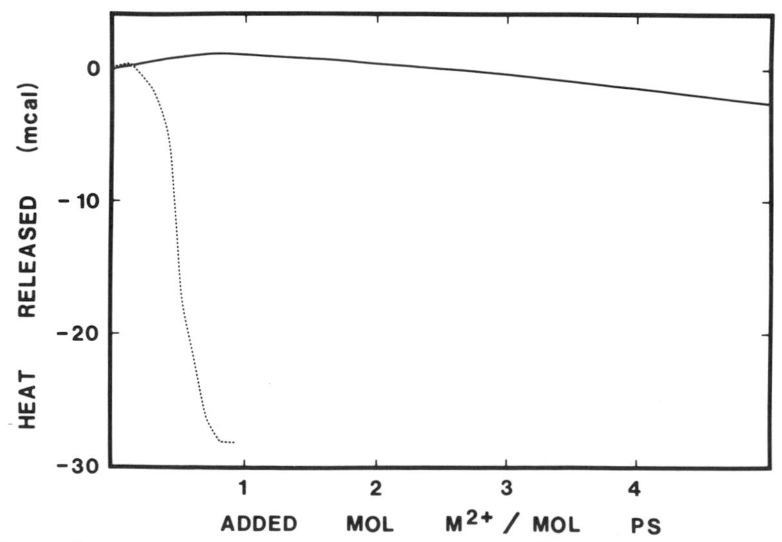

Fig. 1. Thermometric titration of small unilamellar phosphatidylserine (PS) vesicles with Ca^{2+} (- - - -) or Mg^{2+} (———). The vesicles were prepared by sonication in 100 m*M* NaCl, 2 m*M* L-histidine, 2 m*M* *N*-trishydroxymethyl methyl-2-aminoethane sulfonic acid, pH 7.4; any multilamellar vesicles were separated by centrifugation at 115000 *g* for 1 hr. The divalent cations were titrated from 0.1 *M* solutions in the NaCl buffer at a rate of 4.7 μl/min into a 2.5 ml suspension of vesicles at 25°C. Exothermic reactions are indicated by negative values of the enthalpy, which is given in units of mcal. (Data from Rehfeld *et al.*, 1981b.)

beyond the threshold concentration of Ca^{2+} necessary for the exothermic reaction, however, the heat release continues spontaneously. The reaction does not go to completion, presumably because the free Ca^{2+} concentration drops below the threshold level as it binds to the inner monolayer phospholipids and also forms a complex with a higher binding constant (see Sections III and V). If the same sample is titrated further, the total heat release is identical to the value obtained in continuous titrations. It is clear that Mg^{2+} induces a very small amount of heat release compared to Ca^{2+} at this temperature, as would be expected from the known T_c of the Mg/PS complex. At 25°C this complex is still in a liquid–crystalline state, and the chain packing is similar to the Na/PS complex (Table II). We would predict that if the Mg^{2+} titration were performed at 12°C, a larger exothermic component would result; the system would undergo an isothermal phase transition and would be below the T_c of the complex. Membrane fusion in acidic phospholipid vesicles induced by Ca^{2+}, and to a limited extent by Mg^{2+}, has been related to the divalent cations' ability to induce such isothermal phase transitions (Papahadjopoulos *et al.*, 1977; Papahadjopoulos, 1978); this will be discussed further on p. 207. Strehlow and Jähnig (1981) have measured the kinetics of the phase transition induced by Ca^{2+} in dimyristoylmethylphosphatidic acid bilayers by stopped-flow measurements of the fluorescence anisotropy of diphenylhexatriene embedded in the bilayer as a probe of lipid order. They found a relaxation time of about 10 msec that increased with temperature and concluded that the relaxation process is dependent on a nucleation step.

In cardiolipin (CL) bilayers, Ca^{2+} and Mg^{2+} induce an isothermal phase transition from the lamellar phase into the hexagonal H_{II} phase (Rand and Sengupta, 1972; Cullis *et al.*, 1978; Vail and Stollery, 1979). A similar transition occurs in phosphatidic acid membranes in the presence of Mg^{2+} at pH 6.0 (Papahadjopoulos *et al.*, 1976). The T_c of the Mg^{2+} or Ca^{2+} complexes of palmitoyl CL is higher than that of the Na^+ complex (Rainier *et al.*, 1979) in agreement with observations on brain PS (Jacobson and Papahadjopoulos, 1975; Newton *et al.*, 1978) and DPPS (MacDonald *et al.*, 1976). The phase transition of zwitterionic dihexadecylphosphatidylethanolamine (at pH 8) does not change in the presence of 10 m*M* Ca^{2+}, but that of the deprotonated lipid (at pH 12) is shifted to higher temperatures and exhibits complex phase behavior (Harlos and Eibl, 1980). The effect of Ca^{2+} and Mg^{2+} on phosphatidylcholine (PC) bilayers has been investigated by Raman spectroscopy. Lis *et al.* (1975) have shown that these ions decrease the proportion of gauche bonds to trans bonds in the hydrocarbon chain. The T_c of DPPC shifts from 41°C in NaCl (1 *M*) to 45°C in $CaCl_2$ and 43°C in $MgCl_2$ (both 1 *M* and without any NaCl) but at lower, physiological concentrations of Ca^{2+} the T_c is about 39°C (Simon *et al.*, 1975). Similarly, DMPC bilayers undergo the gel–liquid crystalline transition at higher temperatures in the presence of divalent cations; the increase is 2°C for 1 *M* $MgCl_2$ and 4°C for $CaCl_2$ (Chapman *et al.*, 1977). These findings indicate that the effect of Ca^{2+} and Mg^{2+} on PC bilayers is considerably less than their effect on acidic phospholipid bilayers. It is probable that the effect of these divalent cations on phospholipid bilayer fluidity is closely related to their binding to the polar head groups of the phospholipids.

Binding of Calcium and Magnesium to Phospholipid Membranes

The differences in the effects of Ca^{2+} and Mg^{2+} on PS bilayers, such as the structure and molecular motion within the hydrocarbon interior of the bilayer (Hauser *et al.*, 1977; Newton *et al.*, 1978; Portis *et al.*, 1979), the water of hydration and the mobility of the polar headgroups (Kurland *et al.*, 1979a; Portis *et al.*, 1979), the endothermic transition of the metal ion/phospholipid complex (Jacobson and Papahadjopoulos, 1975; Newton *et al.*, 1978), and membrane fusion (Papahadjopoulos *et al.*, 1977; Portis *et al.*, 1979; Wilschut *et al.*, 1981), may be manifestations of the differential binding characteristics of the ions to the membrane surface. The effect of ions on membranes has been related to the binding constant of the ion to ligands on the membrane surface, the concentration of the ion, the partition of the ion in the membrane, and a rate constant relating the binding to a time-dependent effect (Hauser *et al.*, 1976a). The binding of Ca^{2+} and Mg^{2+} to PS membranes has been studied by a number of techniques. Equilibrium dialysis of PS vesicles, utilizing $^{45}Ca^{2+}$ or atomic absorption spectroscopy to detect bound ions, has yielded apparent binding constants K_a of 3.9×10^3 M^{-1} for Ca^{2+} and 1.7×10^3 M^{-1} for Mg^{2+} (Portis *et al.*, 1979). It should be noted, however, that apparent binding constants do not take into account the effect of the negative surface potential, which concentrates the divalent ions at the membrane surface according to a Boltzmann distribution. Thus, the concentration of a cation of valency z at the surface is given by

$$C_z(0) = C_z Y(0)^z = C_z \exp[-ze\psi(0)/kT] \tag{1}$$

where $\psi(0)$ is the surface potential (negative for PS membranes), k is the Boltzmann constant, e is the charge of a proton, T is the absolute temperature, C_z is the bulk concentration of the cation, and $Y(0)^z$ is the Boltzmann factor $Y(x)^z = \exp[-ze\psi(x)/kT]$ (McLaughlin *et al.*, 1971; Nir *et al.*, 1978;

Düzgüneş *et al.*, 1981a). For the interaction of divalent cations with two negatively charged phospholipids we may write (in the case of Ca^{2+}, for example)

$$Ca^{2+} + P_2^{2-} \rightleftharpoons P_2Ca \tag{2}$$

with an intrinsic association constant K_2

$$K_2 = \frac{[P_2\mathrm{Ca}]}{[\mathrm{Ca}^{2+}(0)][\mathrm{P}_2^{2-}]} \tag{3}$$

where Ca^{2+} (0) = Ca^{2+} $Y(0)^2$ is the surface concentration of calcium and P_2^{2-} implies a doubly charged site of interaction for the divalent cation (McLaughlin *et al.*, 1971; Nir *et al.*, 1978; Ohki and Sauvé, 1978). Here the interaction is assumed to be 2:1 (phospholipid:cation). McLaughlin *et al.* (1981) have assumed that divalent cations form 1:1 complexes with negatively charged phospholipids. In this case, we would have

$$K_2 = \frac{[\mathrm{CaP}^{+}]}{[\mathrm{Ca}^{2+}(0)][\mathrm{P}^{-}]} \tag{4}$$

Similar equations may be written for the monovalent cation (here Na^+) which also binds to PS (Newton *et al.*, 1978; Nir *et al.*, 1978; Eisenberg *et al.*, 1979; Kurland *et al.*, 1979b):

$$Na^{+} + P^{-} \rightleftharpoons PNa \tag{5}$$

and

$$K_1 = \frac{[\mathrm{P\ Na}]}{[\mathrm{P}^{-}][\mathrm{Na}^{+}(0)]} \tag{6}$$

where K_1 is the intrinsic association (or binding) constant of Na^+ or monovalent cations in general.

The surface charge density of the membrane, σ, is altered as a result of charge neutralization upon ion binding, and is related to the initial surface charge density in the absence of any ions, $\sigma_{\mathrm{initial}}$, by

$$\sigma = \frac{\sigma_{\mathrm{initial}}}{1 + K_2[\mathrm{Ca}^{2+}]\mathrm{Y}(0)^2 + K_1[\mathrm{Na}^{+}]\mathrm{Y}(0)} \tag{7}$$

where K_1 and K_2 are the intrinsic binding constants of monovalent and divalent cations, respectively (McLaughlin *et al.*, 1971; Nir *et al.*, 1978; Düzgüneş *et al.*, 1981a). This treatment may be extended to include trivalent cations as well as mono- and divalent cations (Bentz and Nir, 1980). The K_2 for Ca^{2+} and Mg^{2+} binding to PS bilayers have been calculated to be 35 M^{-1} and 20 M^{-1}, respectively (Portis *et al.*, 1979), by utilizing a modified Gouy-Chapman equation which takes into account the influence of ion binding on the surface potential (Nir *et al.*, 1978). From the observation that Ca^{2+} or Mg^{2+} binding saturates at 0.5 Ca or Mg bound per PS molecule, the mode of binding has been inferred to be a 2:1 PS/metal ion complex (Newton *et al.*, 1978; Nir *et al.*, 1978). Utilizing the modified Gouy-Chapman theory and the assumption of 2:1 binding, it has also been possible to predict the divalent cation bound per PS in mixed PS/PC membranes (Düzgüneş *et al.*, 1981a).

Binding of divalent cations to multilamellar liposomes composed of pure PS or mixtures of PS with PC or phosphatidylethanolamine (PE) has also been determined by electrophoretic mobility measurements which yield values of the zeta potential. McLaughlin *et al.* (1981) have determined the intrinsic 1:1 binding constants of the divalent cations from the reciprocal of the bulk divalent cation concentration which leads to charge reversal on the liposome surface. They have found values of 12 M^{-1} for Ca^{2+} and 8 M^{-1} for Mg^{2+}. Ohki and Kurland (1981) have measured the variation of the surface potential of PS monolayers by divalent cations in the subphase and have determined 2:1 binding constants of 30 and 10 M^{-1} for Ca^{2+} and Mg^{2+}, respectively. Ohki *et al.* (1982) have calculated the binding constants of various divalent and monovalent cations from the threshold concentrations of these ions which induce PS vesicle aggregation.

The definition of the association of divalent cations with anionic phospholipids as 2:1 or 1:1 cannot be distinguished by these methods of analysis. Cohen and Cohen (1981) have addressed this question from a theoretical treatment which considers both modes of binding; at physiological concentrations of Ca^{2+} it is difficult to distinguish between 1:1 or 2:1 binding. Bentz *et al.* (1983) have shown that under given ionic conditions the same amount of divalent cation bound per PS molecule can be predicted by using either the 2:1 or 1:1 binding constants.

Another unsolved problem in the association of Ca^{2+} and Mg^{2+} with PS membranes is the location of the bound cation. Seimiya and Ohki (1972, 1973) have suggested that Ca^{2+} binds more strongly to the carboxyl group than to the phosphate group. Hauser and Dawson (1967) have proposed that the phosphate group is the binding site. ^{31}P NMR studies have shown that the phosphate group is immobilized by Ca^{2+} and therefore supports this proposal (Kurland *et al.*, 1979a). It is possible, however, that this immobilization is partially a result of Ca^{2+} binding to the carboxyl group, which would be expected to immobilize the head group as a whole. The coordination of Ca^{2+} between phosphate, carboxyl, and amino groups in adjacent phospholipid molecules has also been proposed (Papahadjopoulos, 1968). Hauser *et al.* (1976b) have concluded that surface chemistry techniques for studying Ca^{2+} adsorption to phospholipid monolayers cannot distinguish between the different possibilities. It is important to note at this point that whatever the exact nature of the chemical association of Ca^{2+} and Mg^{2+}

with phospholipid head groups, the coordination chemistry of Ca^{2+} and Mg^{2+} are very different. Ca^{2+} has a coordination number of seven or eight, and the coordination geometry with respect to bond angle and bond length is not uniform; Mg^{2+} has six coordination bonds shaped as a regular octahedron (Williams, 1976). The flexible geometry of Ca^{2+} makes it a very suitable ion for cross-linking molecules in a reversible manner. This difference between Ca^{2+} and Mg^{2+} may also underlie some of the differential effects of these ions on phospholipid structure and intermembrane interactions such as membrane fusion.

Binding of Ca^{2+} and Mg^{2+} to phospholipids other than PS has been studied by several investigators using different techniques. McLaughlin *et al.* (1981) have determined that the intrinsic 1:1 binding constants of these ions to PC are 3 and 2 M^{-1} for Ca^{2+} and Mg^{2+}, respectively. Lau *et al.* (1981) have found 1:1 binding constants of 8.5 for Ca^{2+} and 6 M^{-1} for Mg^{2+} binding to PG. Binding of these divalent cations to PE has been reported to be similar to their binding to PC (McLaughlin *et al.*, 1981). This comparison is based on measurements of electrophoretic mobility at high (0.1 *M*) divalent cation concentrations. At lower concentrations of Ca^{2+} (0.1 m*M*), surface radioactivity measurements have indicated differences in Ca^{2+} binding between the two phospholipids (Rojas and Tobias, 1965). Preferential adsorption of Ca^{2+} over Mg^{2+} to phosphatidylinositol (PI), triphosphoinositide, and brain ganglioside monolayer membranes has also been observed (Hauser and Dawson, 1967).

We must emphasize that the binding of divalent cations to single membranes such as monolayers or individual liposomes may be quite different from their binding to suspensions of liposomes in which aggregation may occur at a threshold concentration of the divalent ions. A discontinuous increase in the amount of Ca^{2+} bound to PS is observed during the titration of PS vesicles with Ca^{2+} at the point at which the turbidity of the suspension increases sharply, regardless of the size of the vesicles, both in the presence and absence of a Ca^{2+} ionophore in the membrane (Rehfeld *et al.*, 1981a; Düzgüneş *et al.*, 1981b; Ekerdt and Papahadjopoulos, 1982). The significance of this result will be discussed on pp. 200–204.

Ionotropic Phase Separations in Phospholipid Mixtures

Lateral phase separations of membrane components have been observed in biological membranes (Kleeman and McConnell, 1974), reconstituted protein–lipid membranes (Chen and Hubbell, 1973), and binary mixtures of various phospholipids in liposomes (Phillips *et al.*, 1972; Shimshick and McConnell, 1973; Papahadjopoulos *et al.*, 1974; Wu and McConnell, 1975; Luna and McConnell, 1977; Findlay and Barton, 1978; Stewart *et al.*, 1979). In certain mixtures of acidic and neutral phospholipids phase separations may be induced by divalent cations. In PS/PC membranes, for example, Ca^{2+} causes the phase separation of the Ca/PS complex, which forms a solid domain, from the fluid PC domains (Ohnishi and Ito, 1974; Papahadjopoulos *et al.*, 1974; Ohnishi, 1975; Papahadjopoulos, 1978), as predicted in previous studies (Ohki and Papahadjopoulos, 1970).

Phase separations in phospholipid mixtures have been detected by electron spin resonance, differential scanning calorimetry, and freeze-fracture electron microscopy. In PS/PC membranes containing spin-labeled PC, the ESR spectrum of the membrane is broadened in the presence of Ca^{2+} due to intermolecular spin–spin exchange interactions, indicating the formation of PC clusters separated from the Ca^{2+}-complexed PS molecules (Ohnishi and Ito, 1974). This phase separation does not occur in the presence of Mg^{2+} or when the membrane contains only PC in addition to the spin-labeled PC. Similar Ca^{2+}-induced phase separations have been observed in PA/PC (Ito and Ohnishi, 1974; Galla and Sackmann, 1975) and PS/PE membranes (Tokutomi *et al.*, 1981). No phase separation occurs in PE/PC, PI/PC or CL/PC membranes as detected by the spin-label technique (Ohnishi and Ito, 1974).

The Ca^{2+}-induced phase separation occurs in a time scale of minutes, depending on the particular PS/PC mixture, and is reversible by EDTA (Ohnishi and Ito, 1974; Ito *et al.*, 1975). Barium and strontium are also effective in inducing phase separation with the order of effectiveness Ca $>$ Ba $>$ Sr. Magnesium causes a slight motional freezing of the acyl chains of the phospholipids although no phase separation, and retards the effectiveness of Ca^{2+} when they are added together. The local anesthetic tetracaine is able to disperse phase-separated PS domains by replacing the bound Ca^{2+} (Ohnishi and Ito, 1975). Studies with spin-labeled PS in PS/PC mixtures have shown that some PS molecules are still present in the fluid PC domains in the presence of Ca^{2+}; more PS is dissolved in PC as the mol fraction of PC is increased (Ito *et al.*, 1975). Similarly to the effect of tetracaine, protons appear to reverse the Ca^{2+}-induced phase separation if the latter has not proceeded to completion, that is, at shorter times of preincubation with Ca^{2+} (Tokutomi *et al.*, 1979). It has been suggested that protons (pH 2.5) replace Ca^{2+} but themselves induce a phase separation which can then be reversed by raising the pH in a buffered salt solution. Further studies have shown that lowering the pH at constant ionic strength or lowering the ionic strength at constant pH also induces lateral phase separation as a result of the protonation of the PS molecules and their subsequent solidification (Tokutomi *et al.*, 1980).

The rate and extent of Ca^{2+}-induced PS cluster formation in PS/PC membranes depends on the Ca^{2+} concentration as well as the time of incubation and the mol fraction of PS. The rate of aggregation of PS molecules into domains increases with higher Ca^{2+} concentrations and with membranes containing higher mol fractions of PS (Ito and Ohnishi, 1974; Ito *et al.*, 1975). PS/PE membranes are more susceptible to phase separation by Ca^{2+} than are PS/PC membranes with respect to both the extent of the solid PS fraction formed and the threshold concentration of Ca^{2+}. Half-maximal phase separation is obtained at 1.4×10^{-7} *M* Ca^{2+} for PS/PE and 1.2×10^{-6} *M* for PS/PC membranes (Tokutomi *et al.*, 1981). It should be noted that Ca^{2+} concentrations reported in this study are calculated from its binding constant to a Ca^{2+} buffer and are unusually low. Negligible binding of Ca^{2+} to PS would be expected at such low free Ca^{2+} concentrations (Portis *et al.*, 1979; Ekerdt and Papahadjopoulos, 1982).

The Ca^{2+} concentration dependence of phase separation in PS/PC mixtures has also been shown by differential scanning calorimetry. 5 m*M* Ca^{2+} causes an upward shift of the endothermic phase transition of sonicated PS/DSPC (3:2) vesicles. When the Ca^{2+} concentration is increased to 10 m*M*, however, in addition to the main transition a new double peak appears at temperatures corresponding to the main and pretransition of DSPC. When the PS content of the mixture is increased (PS/DSPC, 2:1), these new peaks are more pronounced. The appearance of these new transition endotherms has been ascribed to the molecular segregation of the DSPC from the Ca/PS complex (Papahadjopoulos *et al.*, 1974). Mg^{2+} is ineffective in inducing a shift in the transition temperature of the PS/DSPC mixture or molecular segregation. Similar observations have been made with PS/DPPC vesicles (Jacobson and Papahadjopoulos, 1975). In agreement with the observations of Ohnishi and Ito (1974), the effect of Ca^{2+} is reversible. In contrast to the experiments with spin-labeled phospholipids, differential scanning calorimetry indicates no phase separation in mixtures containing less than 50% PS (Papahadjopoulos *et al.*, 1974). Mixtures of DPPA/DPPC (2:1) also exhibit phase separation in the presence of Ca^{2+} at pH 8, as seen with PA/PC mixtures studied by Ito and Ohnishi (1974), but not at pH 6, indicating a possible requirement for two negative charges per molecule (Jacobson and Papahadjopoulos, 1975). Phase separation in phospholipid mixtures have also been observed by freeze-fracture electron microscopy (van Dijck *et al.*, 1978b) and Raman spectroscopy (Hark and Ho, 1980). Van Dijck *et al.* (1978b) have demonstrated that with membranes composed of DMPS/DMPC in the presence of Ca^{2+}, the fracture plane consists of a complex mixture of smooth and banded surfaces as well as cylindrical bilayers. Coexisting smooth and banded domains are also observed in DMPC/DPPG mixtures in excess Ca^{2+}. With Raman spectroscopy, it is possible to study the thermotropic behavior of individual components in a mixture. Hark and Ho (1980) have shown the

TABLE III

Phase Separation of Phospholipid Mixtures in the Presence of Divalent Cations

Phospholipid mixture	Divalent cation	Phase separation
PS/PC[a,b,c,d]	Ca^{2+}	+
PS/PC[a,b,c]	Mg^{2+}	−
CL/PC[a]	Ca^{2+}	−
PI/PC[a]	Ca^{2+}	−
PE/PC[a]	Ca^{2+}	−
PA/PC[c,d,e,f]	Ca^{2+}	+
PA/PC[e]	Mg^{2+}	+
PS/PE[g,h]	Ca^{2+}	+
PS/PE[h]	Mg^{2+}	−
PG/PC[d,i]	Ca^{2+}	+

[a]Ohnishi and Ito, 1974; Ito *et al.*, 1975.
[b]Papahadjopoulos *et al.*, 1974.
[c]Jacobson and Papahadjopoulos, 1975.
[d]van Dijck *et al.*, 1978b.
[e]Ito and Ohnishi, 1974.
[f]Galla and Sackmann, 1975.
[g]Tokutomi *et al.*, 1981.
[h]Düzgünes, *et al.*, 1983c.
[i]Phase separation was observed with DPPG/DMPC mixtures but not with DMPG/DMPC.

partial immiscibility of a mixture of PS and deuterated DMPC and the enhancement of this immiscibility in the presence of Ca^{2+}. Observations on phase separation in various phospholipid mixtures are summarized in Table III.

Intermembrane Interactions

The interaction of Ca^{2+} with phospholipid membranes varies substantially depending on whether the membranes are in close contact or in a dispersed state. This is not a generally recognized phenomenon, but a striking example can be found in a comparison of the binding of Ca^{2+} to single membranes and its binding to aggregating and fusing vesicles. Figure 2 shows the zeta potential of single multilamellar PS liposomes in the presence of varying concentrations of Ca^{2+} in the medium. McLaughlin *et al.* (1981) have utilized this data to estimate binding constants of Ca^{2+} to PS. The potential increases uniformly over two orders of magnitude of Ca^{2+} concentration and reaches zero at about 80 m*M* (not shown). In contrast, the Ca^{2+} bound per PS in large unilamellar vesicles containing a Ca^{2+} ionophore increases sharply at a threshold Ca^{2+} concentration at which the vesicle membranes begin to interact and reaches a plateau at 0.5 bound Ca/PS (Rehfeld *et al.*, 1981a; Ekerdt and Papahadjopoulos, 1982).

In PS vesicles containing globoside, which has a bulky head group with four carbohydrate residues, the bound Ca/PS remains at about 0.2 at Ca^{2+} concentrations, at which a sharp increase in binding occurs in PS vesicles. It is known that the Ca^{2+}-induced fusion of these vesicles is inhibited considerably and requires higher threshold concentrations of Ca^{2+} (Düzgüneş *et al.*, 1981c). However, when galactosylcerebroside (which has a single sugar residue as the head group) is included in PS vesicles instead of the globoside, the binding ratio increases at the threshold Ca^{2+} concentration (Ekerdt and Papahadjopoulos, 1982). Similarly, in large PS/PC (1:1) vesicles which do not fuse but aggregate (Düzgüneş *et al.*, 1981d), the bound Ca remains at the lower level; in PS/PE vesicles which undergo fusion, Ca^{2+} binding increases at a slightly higher free Ca^{2+} concentration than with pure PS, as

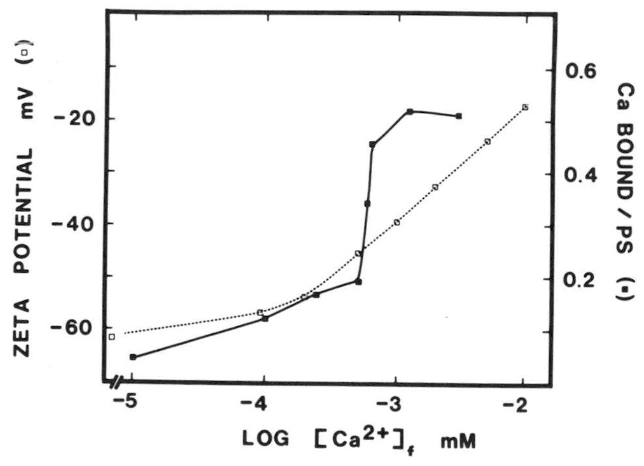

Fig. 2. Ca^{2+}-binding to single multilamellar PS vesicles and a suspension of large unilamellar vesicles. The zeta potential of multilamellar PS vesicles is given on the left hand ordinate (open symbols). (Data from McLaughlin *et al.*, 1981.) The Ca^{2+} bound per PS in large (0.1 μm diameter) unilamellar vesicles containing the ionophore A23187 is given on the right hand ordinate (closed symbols). (Data from Ekerdt and Papahadjopoulos, 1982.)

would be expected from the higher fusion threshold for the PS/PE vesicles. Thus, it appears that when the membranes are sufficiently close for fusion to occur, the binding of Ca^{2+} increases drastically.

Another indication of the importance of intermembrane contact in Ca^{2+}-induced phase changes is the higher threshold concentration of Ca^{2+} required for the exothermic reaction of Ca^{2+} and PS in large unilamellar vesicles compared to small unilamellar vesicles (Fig. 3). We have shown elsewhere that the threshold Ca^{2+} concentration for fusion is higher for the large vesicles than the small vesicles (Wilschut *et al.*, 1980). It is apparent from the data shown in Fig. 3 that the heat release associated with the crystallization of the acyl chains of PS is dependent on contact and fusion of the membranes rather than the free Ca^{2+} concentration in the medium. The experiment in Fig. 3 also shows that the total heat released from large unilamellar vesicles is less than that obtained by the complete titration of small vesicles of PS (Fig. 1). The enthalpy of this transition is 4.3 ± 0.4 kcal/mol PS compared to 5.6 ± 0.8 kcal/mol PS in the small vesicles. This difference may be a reflection of the expected differences in packing of the phospholipids in the two vesicle types.

Experiments employing the extrinsic membrane protein spectrin have indicated that spectrin can act as a spacer which prevents Ca^{2+}-induced

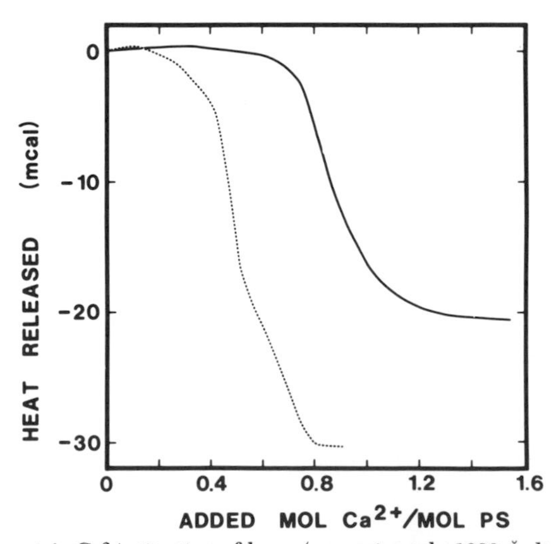

Fig. 3. Thermometric Ca^{2+} titration of large (approximately 1000 Å diameter) unilamellar PS vesicles made by reverse phase evaporation (———). Other conditions were the same as in Fig. 1. Titration of the small unilamellar vesicles is also shown for comparison (- - -). (Data from Rehfeld *et al.*, 1981b.)

contact between PS vesicles, and thereby can prevent the formation of an intermembrane complex. PS vesicles in the presence of spectrin and 1 m*M* Ca^{2+} exhibit an endothermic phase transition with a peak at 32°C instead of the shift to $T_c > 100$°C in the absence of the protein (Portis *et al.*, 1979). This effect is similar to the limited upward shift of the transition temperature observed at low Ca^{2+} concentrations (Jacobson and Papahadjopoulos, 1975). The packing of the hydrocarbon chains of PS in the presence of Ca^{2+} and spectrin at 22°C is characteristic of hexagonally close-packed acyl chains as observed below the T_c for phospholipids in general. This is in contrast to the highly ordered orthorhombic–perpendicular packing of PS/Ca complexes in the absence of spectrin. Spectrin also inhibits the release of aqueous contents of PS vesicles in the presence of Ca^{2+} and the fusion of the vesicles (Portis *et al.*, 1979; N. Düzgüneş, J. Wyatt, and D. Papahadjopoulos, unpublished observations). These results have been interpreted as evidence for two types of Ca^{2+} binding to PS membranes. When membranes are allowed to form close contact (less than 10 Å between the head groups), Ca^{2+} forms a trans complex which eventually results in the crystallization of the hydrocarbon chains in a unique packing geometry. Below the threshold concentration for aggregation of the vesicles, or if the vesicles are kept apart physically as in the case of spectrin (or globoside), Ca^{2+} forms a cis complex on a single membrane (Fig. 4). Similarly, Mg^{2+} has been proposed to form a cis complex with PS both at low and high concentrations of the ion, with a layer of water remaining between the apposed membranes (Portis *et al.*, 1979; Fig. 4).

If intermembrane contact is an important determinant of the mode of

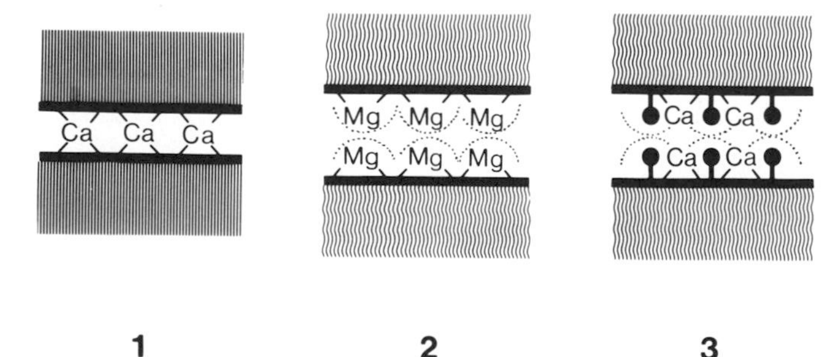

Fig. 4. Schematic representation of the divalent cation complexes of phosphatidylserine. The shaded areas represent the lipid bilayers and the dotted spheres show the hydration layer. (1) The anhydrous *trans* complex of Ca^{2+} and phosphatidylserine. (2) The hydrated *cis* Mg/phosphatidylserine complex. (3) The *cis* Ca/phosphatidylserine complex which is presumed to form when the bilayers are not allowed to establish close contact by spacer molecules such as globoside or spectrin.

Ca^{2+} binding to PS membranes, the higher binding of Ca^{2+} induced by the aggregation and fusion of the vesicles (Fig. 2) should also be observed if aggregation is induced by Mg^{2+} in the presence of low (subthreshold) concentrations of Ca^{2+}. Indeed, Ca^{2+} binding is enhanced in the presence of even 5–25-fold excess Mg^{2+}, but Mg^{2+} binding is decreased (Portis *et al.*, 1979). This enhanced binding upon introduction of an aggregating concentration of Mg^{2+} can also be monitored continuously by a Ca^{2+}-selective electrode (Ekerdt and Papahadjopoulos, 1982; McLaughlin *et al.*, 1981).

These observations suggest that the interaction of Ca^{2+} with membranes may be determined by factors which control the distance between membranes such as proteins, the lateral distribution of particular phospholipids in the membrane, and other divalent cations. In addition, the structural changes induced by Ca^{2+} in phospholipid bilayers will depend on the mode of Ca^{2+}-binding to the membrane.

Calcium/Magnesium Specificity in Membrane Fusion

The differences between Ca^{2+} and Mg^{2+} in their interactions with acidic phospholipid membranes and in the structural transformations they induce is also reflected in their ability to induce fusion between these membranes. The fusion of phospholipid vesicles by divalent cations has been reviewed in detail (Papahadjopoulos *et al.*, 1979; Düzgüneş *et al.*, 1980, 1983b; Nir *et al.*, 1983). In this section, we will consider vesicle systems where a considerable, and in some cases absolute, Ca^{2+}/Mg^{2+} specificity is observed. Fusion of PS vesicles is of particular interest here, since extensive studies have been carried out on Ca^{2+} and Mg^{2+} binding to PS, as well as on phase transitions induced by these ions in PS bilayers.

Large unilamellar vesicles (approximately 1000 Å diameter) of pure PS undergo extensive fusion above a threshold Ca^{2+} concentration of about 2 m*M*, as determined by intermixing of aqueous contents of the vesicles and by freeze-fracture electron microscopy (Wilschut *et al.*, 1980). In the presence of 5–20 m*M* Mg^{2+}, however, the vesicles aggregate but do not fuse. They also retain their contents, unlike the situation with Ca^{2+} (Wilschut *et al.*, 1981). The light scattering change also reflects the retention of the intact vesicular structure, again unlike the case with Ca^{2+} where the vesicles eventually collapse and produce an anhydrous Ca/PS complex with a different refractive index than that of the initial vesicle suspension. Hence, in this phospholipid vesicle system, Ca^{2+} has an absolute specificity over Mg^{2+} in the induction of fusion.

Although the difference in the binding affinity of Ca^{2+} and Mg^{2+} to PS (Newton *et al.*, 1978; Nir *et al.*, 1978; Portis *et al.*, 1979) may explain the difference in their threshold concentrations necessary to induce aggregation, the physicochemical nature of the interaction of these ions with PS must govern the drastic specificity of Ca^{2+} over Mg^{2+} in inducing fusion. The degree of dehyration of the membrane surface by Ca^{2+} may be one crucial criterion (Papahadjopoulos *et al.*, 1978; Portis *et al.*, 1979; Hoekstra, 1982). The role of interfacial water and the nature of the membrane surface, such as its hydrophobicity, has also been emphasized as important factors in divalent cation-induced membrane fusion (Ohki and Düzgüneş, 1979; Ohki, 1982; McIver, 1979). Divalent cations would become partially dehydrated as they bound to the negatively charged groups on the membrane and also interfere with the stabilizing hydration properties of the phospholipid–water interface (Wilschut *et al.*, 1981; Hauser *et al.*, 1975). Since repulsive forces between phospholipid bilayers are dominated by the hydration forces at short (30 Å) distances of separation (Cowley *et al.*, 1978; Rand, 1981), changes in the hydration layer caused by divalent cations would result in altered energies of interaction between the bilayers. Ca^{2+} is known to be more readily dehydrated than Mg^{2+} (Nancollas, 1966; Gresh, 1980) and to form a more anhydrous complex with PS (Portis *et al.*, 1979). We envisage several steps in the interaction of Ca^{2+} with the surface of PS vesicles:

1. Before the vesicles aggregate, Ca^{2+} would cause partial dehydration as a result of its binding to the functional groups on the surface and form a coordination complex (Papahadjopoulos, 1968). This process, which would condense the outer monolayer (Rojas and Tobias, 1965; Papahadjopoulos, 1968) and increase the lateral compressibility of the membrane, could have a destabilizing effect (Papahadjopoulos *et al.*, 1977; Nir *et al.*, 1980a).
2. At concentrations of Ca^{2+} above a threshold value, the vesicles aggregate as a result of charge neutralization (Lansman and Haynes, 1975; Düzgüneş and Ohki, 1977; Papahadjopoulos *et al.*, 1977; Nir *et al.*, 1980a; Wilschut *et al.*, 1981). If the membrane surfaces can achieve a close contact, Ca^{2+} would then be able to interact with the polar groups of the apposed membrane, thereby losing the remainder of its water of hydration and forming an anhydrous trans complex (Portis *et al.*, 1979).
3. While a few PS molecules are in the trans complex, which probably involves condensation in packing, several neighboring molecules may still be in the cis complex. This could be one of the factors contributing to the

formation of defects localized at the region of contact between two membranes. Thus, the alteration of membrane structure caused by the formation of an interbilayer Ca/PS complex may initiate fusion (Portis *et al.*, 1979; Düzgüneş and Ohki, 1981). Another factor could be the lateral fluctuations of the condensed PS/Ca domains in the membrane. These domains would have packing discontinuities at their boundaries (Papahadjopoulos *et al.*, 1977).

4. These defects and discontinuities could be nucleation points for the intermixing of the hydrocarbon phases of the two bilayers and the merging of the membranes into a continuous structure.

In the presence of Mg^{2+}, the large PS vesicles may aggregate preferentially in a secondary minimum of the potential energy (Nir *et al.*, 1981). It is apparent that in this case there is no sufficient disruption of the hydration layer and destabilization of the bilayer to alter the interaction energy between the vesicles (Wilschut *et al.*, 1981).

In contrast to the large vesicles, Mg^{2+} can induce the fusion of small (~250 Å in diameter) unilamellar PS vesicles, although only to a limited extent. When the vesicles have undergone a few rounds of fusion and grown in size, fusion ceases, presumably because the strain of the highly curved bilayer is relieved (Wilschut *et al.*, 1981). The rate of fusion at saturation concentrations of divalent cation, that is, about 15 m*M*, is approximately 10-fold lower for Mg^{2+} than for Ca^{2+}. This may reflect a lower rate of either aggregation or fusion. Even at these high concentrations of divalent cation, the rate of vesicle aggregation would be affected by the residual electrostatic repulsion between the vesicles; because of the higher affinity of Ca^{2+} for PS compared to Mg^{2+}, the repulsion would be higher in the presence of Mg^{2+} than of Ca^{2+} (Nir *et al.*, 1980b). Nevertheless, studies on the kinetic order of fusion show that whereas an order of 2.0 is obtained with Ca^{2+}, a value of 1.6 is obtained with Mg^{2+}, indicating that the initial rate of Mg^{2+}-induced fusion is delayed with respect to aggregation because the initial rate of vesicle aggregation would be expected to be second order with respect to vesicle concentration (Wilschut *et al.*, 1981). The ratio of the rate of aggregation to the rate of fusion is relatively constant over a wide range of Ca^{2+} concentrations, but this ratio decreases as the Mg^{2+} concentration is increased. Thus, aggregation and fusion appear to be different functions of the Mg^{2+} concentration. Mg^{2+}-induced fusion of small PS vesicles is associated with a higher degree of leakiness per fusion event compared to Ca^{2+}-induced fusion. Even at low Mg^{2+} concentrations the relative rate of release of contents during fusion is considerable; the vesicles, therefore, appear to be aggregating, becoming destabilized (as indicated by the leakage of contents), and also fusing at a much slower rate (Wilschut *et al.*, 1981).

The geometry of Mg^{2+} binding to PS molecules in the highly curved bilayer of small vesicles could be different than its binding to those in unstrained large vesicles. Mg^{2+} binding could cause packing defects in the vesicle membrane, and because aggregated small vesicles would be expected to be in the primary minimum (Nir and Bentz, 1978; Nir *et al.*, 1981) and hence in close contact, the membranes could intermix at these defect points. The delay in fusion with respect to aggregation could be attributed to the time of lateral diffusion of defects along the bilayer until their alignment at the point of contact (Wilschut *et al.*, 1981).

With respect to the possible involvement of ion-induced isothermal phase transitions in membrane fusion (Papahadjopoulos *et al.*, 1977; Papahadjopoulos, 1978), the correlation certainly holds for large PS vesicles; at 25°C, Ca^{2+} induces the transformation of the initially fluid bilayer to the crystalline phase, whereas Mg^{2+} does not. The Mg^{2+}-induced fusion of small PS vesicles may be attributed to factors other than an isothermal phase transition as discussed above. The observation that Ca^{2+} induces a phase transition in an equilibrium structure obtained after a long-term incubation does not make it necessary, however, that such a phase transition to a crystalline Ca/PS complex should be involved in the membrane fusion process at the point of contact. The induction of fusion by Sr^{2+} or Ba^{2+} in large PS vesicles both above and below the T_c of the metal ion/PS complex and the increased initial rate of fusion above the T_c (Düzgüneş *et al.*, 1983c) lends support to the hypothesis that such ionotropic phase transitions may not be strictly necessary for membrane fusion. It should be noted that more extensive fusion of small PS vesicles by La^{3+} is observed at the T_c of the La/PS complex than at temperatures above and below the T_c (Hammoudah *et al.*, 1981). Similarly, when small PS vesicles are incubated in the presence of Mg^{2+} at 12°C, large multilamellar structures are formed in contrast to the aggregated vesicular structures, which are only slightly larger than the original vesicles, obtained at 37°C. This observation indicates that the isothermal phase change from the liquid–crystalline state to gel state, which occurs when the experiment is carried out at 12°C (the T_c of the Mg/PS complex is 18°C), may be related to the extensive fusion of the bilayers at this temperature (Papahadjopoulos *et al.*, 1977). Recent experiments on the temperature dependence of Mg^{2+}-induced fusion of small PS vesicles, as monitored by the intermixing of internal contents of the vesicles, show that at lower temperatures fusion is slower than at temperatures above the T_c of the complex although it is accompanied by extensive release of the contents (Wilschut *et *al.*, 1983), reflecting a more drastic structural transformation and corroborating the findings with freeze-fracture electron microscopy mentioned earlier (Papahadjopoulos *et al.*, 1977).

Calcium ion specificity over Mg^{2+} in membrane fusion is observed in several phospholipid vesicle systems other than large PS vesicles. PS/PE (1:1) vesicles undergo fusion in the presence of either Ca^{2+} or Mg^{2+}, with Ca^{2+} requiring a lower threshold concentration than Mg^{2+}. If 10% of the PE is replaced by PC, however, the fusogenic capacity of Mg^{2+} is abolished (Düzgüneş *et al.*, 1981d). It appears that the fusion-facilitating effect of PE is readily countered by PC, which is more hydrated than PE (Jendrasiak and Hasty, 1974; Hauser, 1975) and presumably less susceptible to dehydration by divalent cations.

Small unilamellar vesicles composed of mixtures of PS and PC in which the PS component is greater than 50% also exhibit some Ca^{2+}/Mg^{2+} specificity which can be explained by the differential binding capacity of the ions onto the PS molecules as predicted by a modified Gouy-Chapman formalism (Düzgüneş *et al.*, 1981a). As in the pure PS vesicles, the Mg^{2+}-induced aggregation and release of contents are faster than the intermixing of internal contents in these PS/PC mixtures (N. Düzgüneş, unpublished data).

Another example of absolute specificity of Ca^{2+} over Mg^{2+} in membrane fusion is to be found in the enhancement of the rate and extent of membrane fusion by synexin, a water-soluble protein isolated from the cytosol of a large number of cell types (Creutz *et al.*, 1978; Düzgüneş *et al.*, 1980; Hong *et al.*, 1981, 1982a,b). Synexin lowers the threshold concentration of Ca^{2+} required to induce the fusion of large PS or PS/PE vesicles and increases the initial rate of fusion, whereas Mg^{2+} is not effective (Hong *et al.*, 1981). Although Mg^{2+} itself does not activate synexin, it nevertheless enables Ca^{2+} to induce the fusion of PA/PE vesicles at a threshold concentration about one order of magnitude lower than that required in the absence of synexin (Hong *et al.*, 1982b). This effect of Mg^{2+} could be a consequence of the reduction of the electrostatic repulsive forces between the membranes, or of the direct participation of Mg^{2+} in fusion after the membranes have been brought to close proximity by the combined action of Mg^{2+} and the Ca^{2+}-activated synexin. It is known that Mg^{2+} itself can induce the fusion of PA/PE vesicles at higher concentrations than those used in the synexin experiments (Sundler *et al.*, 1981). It is unclear at present how synexin enhances the Ca^{2+}-induced fusion of certain types of phospholipid vesicles. It has been proposed that tight anhydrous synexin–Ca^{2+}–phospholipid complexes between two interacting vesicles could reduce the activation energy for the intermixing of the phospholipids between the apposed membranes adjacent to these complexes (Hong *et al.*, 1982a).

Implications for Biological Membranes

The function of membrane proteins, some of which exhibit enzymatic activity, may be modulated by the physical state and composition of the adjacent lipid bilayer (Kimelberg, 1976; Cullis and de Kruijff, 1979). Several membrane proteins appear to have an affinity for negatively charged phospholipids. For example, cytochrome *c* oxidase requires tightly bound CL for maximal activity (Robinson *et al.*, 1980) and segregates this lipid in the boundary layer (Cable and Powell, 1980). The hydrophobic myelin protein, lipophilin, associates preferentially with PS in PC/PS mixed membranes (Boggs *et al.*, 1982). It has been proposed that lactate dehydrogenase from *Escherichia coli* plasma membranes is surrounded predominantly by PG or PS (Kovatchev *et al.*, 1981). Similarly, the Na, K-ATPase from *Electrophorus electricus* has a selectivity for negatively charged spin-labeled lipids (Brotherus *et al.*, 1980). Thus, the phase changes induced by divalent cations in negatively charged phospholipids would be expected to alter the interaction of these phospholipids with the membrane proteins which they surround. The constraints imposed upon the configuration and molecular motion of the phospholipids by their complexing with Ca^{2+} or Mg^{2+} would affect the molecular organization of the protein–phospholipid interface and could lead to configurational changes in the protein, or lateral redistribution or clustering of the proteins in the plane of the bilayer. Thus, membrane organization could be altered when the membrane components seek the most favorable thermodynamic state after the alteration of the lipid packing by divalent cations (Israelachvili, 1977). The lateral distribution of intramembranous particles in biological or reconstituted membranes observed in freeze-fracture electron microscopy depends on the physical state of the phospholipids in the membrane (Verkleij *et al.*, 1972; Speth and Wunderlich, 1973; Wunderlich *et al.*, 1973; Chen and Hubbell, 1973; Kleeman and McConnell, 1974). Phase separations in phospholipids induced by divalent cations would be expected to segregate the proteins laterally into more fluid phospholipid domains. If the proteins require negatively charged phospholipids for activity (cf. Kimelberg and Papahadjopoulos, 1972; Niggli *et al.*, 1981) and if the fluid domains consist of neutral phospholipids which do not interact appreciably with divalent cations, the activity of the protein could be drastically diminished. On the other hand, if the proteins are active only in liquid–crystalline phase lipids (Silvius and McElhaney, 1980), lateral segregation would result in increased activity.

The binding of Ca^{2+} to negatively charged phospholipids in membranes would decrease the surface potential and hence the ion concentrations near the surface of the membrane; this in turn may affect ionic conductances across the membrane, for example in excitable membranes (Ohki, 1978). Ca^{2+}-binding to PS in rod outer segment disk membranes would be expected to reduce the diffusion coefficient of this ion in the cytoplasm (McLaughlin and Brown, 1981).

Studies on the interaction of divalent cations with pure acidic phospholipid vesicles as a model membrane system constitute a basis for the understanding of their effects in more complex biological membranes. Most relevant in this regard is the possibility of lipid domains in membranes (Klausner *et al.*, 1980; Bearer and Friend, 1980; Karnovsky *et al.*, 1982). Thus, although negatively charged phospholipids are a minority among total phospholipids in cellular membranes, they could be segregated in their own domains and constitute high local concentrations that could behave in ways similar to the model phospholipid membranes. The asymmetric distribution of acidic phospholipids across certain membranes (Bergelson and Barsukov, 1977; Rothman and Lenard, 1977; Op den Kamp, 1979) is another factor which would concentrate these phospholipids in particular domains. For example, PS and PE are predominately located in the inner monolayer of erythrocyte membranes (Zwaal *et al.*, 1973; Op den Kamp, 1979), and synaptosome plasma membranes (Fontaine *et al.*, 1980). However, the localization of aminophospholipids in subcellular membranes such as microsomes and Golgi-derived secretory vesicles (Sundler *et al.*, 1977; Higgins and Dawson, 1977; Nilsson and Dallner, 1977) and in platelet plasma membranes (Schick *et al.*, 1976; Otnaess and Holm, 1976) is controversial at present (Op den Kamp, 1979).

The localization as well as the mole fraction of fusogenic phospholipids in lateral domains or on one half of the lipid bilayer may be a factor which regulates the fusion susceptibility of a membrane. If the inner monolayer of the plasma membrane is rich in acidic phospholipids and PE, it would be prone to fuse with secretory vesicles, which may also have an asymmetric distribution of phospholipids when the ionic environment is suitable. Ca^{2+} is known to be involved in stimulus-secretion coupling in many secretory cells (Poste and Allison, 1973; Rubin, 1974; Douglas, 1974). The entry of Ca^{2+} into the cell would increase the Ca^{2+} concentration transiently near the plasma membrane, and if the concentration is above the appropriate threshold for the particular membrane, secretory vesicles close to or already apposed to the plasma membrane would undergo fusion with the latter. Intracellular concentrations of Mg^{2+} could contribute to the close apposition of the vesicles to the plasma membrane by reducing the surface charge density of the two membranes (Portis *et al.*, 1979). Millimolar concentrations of Mg^{2+} could reduce appreciably the Ca^{2+} concentration required for fusion, especially if the membranes or the domains within the membranes contain significant amounts of PE (Düzgüneş *et al.*, 1981d). Concerning the question of whether phospholipid vesicle fusion is fast enough to account for the rapid release of neurotransmitter at synapses (Heuser, 1978; Parsegian, 1977), we should note the recent kinetic analysis by Nir *et al.* (1982), which indicates that the Ca^{2+}-induced fusion of small PS vesicles can take place within a time scale of 10 msec, and the electron microscopic observations of Miller and Dahl (1982), which essentially confirm this prediction.

In the previous section we have emphasized the role of phospholipids and their metal ion complexes in membrane fusion. We do not underestimate, however, the involvement of proteins, particularly Ca^{2+}-binding proteins, in membrane fusion in biological systems. Proteins could participate in defining the specificity of attachment of particular membranes to one another before fusion. Intramembranous particles have been observed in regions of the plasma membrane where fusion activity occurs (Satir, 1974; Heuser, 1978; Dreyer *et al.*, 1973; Venzin *et al.*, 1977). Some of these proteins may be Ca^{2+} gates for the local entry of Ca^{2+} during stimulation (Venzin *et al.*, 1977; Satir and Oberg, 1978). We have shown recently that synexin enhances the rate of Ca^{2+}-induced fusion of phospholipid vesicles (Düzgüneş *et al.*, 1980; Hong *et al.*, 1981, 1982a,b). This enhancement is specific for Ca^{2+} and for certain phospholipid compositions; for example, the fusion of PS/PC membranes is not affected by synexin although the fusion rate of PS/PE membranes is considerably increased (Hong *et al.*, 1981). In addition, the threshold concentration of Ca^{2+} required to induce the fusion of PA/PE membranes is reduced to about 10 μM in the presence of synexin and 1–1.5 mM Mg^{2+} (Hong *et al.*, 1982a,b). It is thus possible that synexin is the intracellular receptor for Ca^{2+} in certain cells which perform Ca^{2+}-induced secretion. The involvement in membrane fusion of other proteins, such as intrinsic membrane glycoproteins, is another intriguing possibility.

In this review, we have discussed the effects of Ca^{2+} and Mg^{2+} on phospholipid membranes with the objective of understanding their effects on cellular membranes. These studies constitute the *synthetic* approach to the elucidation of the function and underlying structural components of biological membranes whereby the molecular components and their interaction with their aqueous environment are studied in isolation and in simple sequential recombinaton. Even these simple systems have proved to be difficult to analyze (exemplified by the controversy over the binding constant of Ca^{2+} to PS described briefly by McLaughlin *et al.*, 1981). Nevertheless, studies on well-defined phospholipid membrane systems are invaluable as a complement to the *analytical* approach in which cellular membranes are fractionated, enzymatically modified, and the individual protein and lipid components extracted and analyzed. Our understanding of the behavior of simple model membrane systems in the presence of divalent cations should also facilitate questions on possible molecular mechanisms of protein–lipid interactions and membrane fusion in biological systems not raised until now.

Acknowledgments

We thank Drs. J. Bentz and K. Hong (University of California, San Francisco) for discussions, and Ms. J. Swallow for the preparation of the manuscript.

References

Bearer, E. L., and Friend, D. S. (1980). *Proc. Natl. Acad. Sci. U.S.A.* **77,** 6601–6605.
Bentz, J., and Nir, S. (1980). *Bull. Math. Biol.* **42,** 191–220.
Bentz, J., Düzgüneş, N., and Nir, S. (1983). *Biochemistry* **22,** 3320–3330.
Bergelson, L. D., and Barsukov, L. I. (1977). *Science* **197,** 224–230.
Boggs, J. M. (1980). *Can. J. Biochem.* **58,** 755–770.
Boggs, J. M., Moscarello, M. A., and Papahadjopoulos, D. (1982). *In* "The Molecular Biology of Lipid–Protein Interactions" (O. H. Griffith and P. Jost, eds.), Vol. 2, pp. 1–51. Wiley, New York.
Brotherus, J. R., Jost, P. C., Griffith, O. H., Keana, J. F. W., and Hokin, L. E. (1980). *Proc. Natl. Acad. Sci. U.S.A.* **77,** 272–276.
Cable, M. B., and Powell, G. L. (1980). *Biochemistry* **19,** 5679–5686.
Chapman, D. (1975). *Q. Rev. Biophys.* **8,** 185–235.
Chapman, D., Peel, W. E., Kingston, B., and Lilley, T. H. (1977). *Biochim. Biophys. Acta* **464,** 260–275.
Chapman, D., Williams, R. M., and Ladbrooke, B. D. (1967). *Chem. Phys. Lipids* **1,** 445–475.
Chen, Y. S., and Hubbell, W. L. (1973). *Exp. Eye Res.* **17,** 517–532.
Cohen, J. A., and Cohen, M. (1981). *Biophys. J.* **36,** 623–651.
Cowley, A. C., Fuller, N. L., Rand, R. P., and Parsegian, V. A. (1978). *Biochemistry* **17,** 3163–3168.
Creutz, C. E., Pazoles, C. J., and Pollard, H. B. (1978). *J. Biol. Chem.* **253,** 2858–2866.
Cullis, P. R., and deKruijff, B. (1976). *Biochim. Biophys. Acta* **436,** 523–540.
Cullis, P. R., and deKruijff, B. (1979). *Biochim. Biophys. Acta* **559,** 399–420.
Cullis, P. R., Verkleij, A. J., and Ververgaert, P.H.J.Th. (1978). *Biochim. Biophys. Acta* **513,** 11–20.
Daemen, F. J. M. (1973). *Biochim. Biophys. Acta* **300,** 255–288.
Douglas, W. W. (1974). *Biochem. Soc. Symp.* **39,** 1–28.
Dreyer, F., Peper, K., Akert, K., Sandric, C., and Moor, H. (1973). *Brain Res.* **62,** 373–380.
Düzgüneş, N., and Ohki, S. (1977). *Biochim. Biophys. Acta* **467,** 301–308.
Düzgüneş, N., and Ohki, S. (1981). *Biochim. Biophys. Acta* **64,** 734–747.
Düzgüneş, N., Hong, K., and Papahadjopoulos, D. (1980). *In* "Calcium-Binding Proteins: Structure and Function" (F. L. Siegel, E. Carafoli, R. H. Kretsinger, D. H. MacLennan, and R. H. Wasserman, eds.), pp. 17–22. Elseiver-North Holland, New York.

Düzgüneş, N., Nir, S., Wilschut, J., Bentz, J., Newton, C. Portis, A., and Papahadjopoulos, D. (1981a). *J. Membr. Biol.* **59,** 115–125.
Düzgüneş, N., Rehfeld, S. J., Freeman, K. B., Newton, C., Eatough, D. J., and Papahadjopoulos, D. (1981b). *Intl. Biophys. Congr., VII, Abstr.*, p. 226.
Düzgüneş, N., Hong, K., Wilschut, J., Lopez, N., and Papahadjopoulos, D. (1981c). *Intl. Biophys. Congr., VII, Abstr.*, p. 105.
Düzgüneş, N., Wilschut, J., Fraley, R., and Papahadjopoulos, D. (1981d). *Biochim. Biophys. Acta* **642,** 182–195.
Düzgüneş, N., Wilschut, J., Hong, K., Fraley, R., Perry, C., Friend, D. S., James, T. L., and Papahadjopoulos, D. (1983a). *Biochim. Biophys. Acta* (in press).
Düzgüneş, N., Wilschut, J., and Papahadjopoulos, D. (1983b). *In* "Physical Methods on Biological Membranes and their Model Systems" (F. Conti, ed.). Plenum, New York (in press).
Düzgüneş, N., Paiement, J., Freeman, K., Lopez, L., Wilschut, J., and Papahadjopoulos, D. (1983c). *Biophys. J.* **41,** 30a.
Eisenberg, M., Gresalfi, T., Riccio, T., and McLaughlin, S. (1979). *Biochemistry* **18,** 5213–5223.
Ekerdt, R., and Papahadjopoulos, D. (1982). *Proc. Natl. Acad. Sci. U.S.A.* **79,** 2273–2277.
Findlay, E. J., and Barton, P. G. (1978). *Biochemistry* **17,** 2400–2405.
Fontaine, R. N., Harris, R. A., and Schroeder, F. (1980). *J. Neurochem.* **34,** 209–277.
Gaber, B. P., Yager, P., and Peticolas, W. L. (1978). *Biophys. J.* **21,** 161–176.
Galla, H. J., and Sackmann, E. (1975). *Biochim. Biophys. Acta* **401,** 509–529.
Gresh, N. (1980). *Biochim. Biophys. Acta* **597,** 345–357.
Hammoudah, M. M., Nir, S., Bentz, J., Mayhew, E., Stewart, T. P., Hui, S. W., and Kurland, S. J. (1981). *Biochim. Biophys. Acta* **645,** 102–114.
Hark, S. K., and Ho, J. T. (1979). *Biochem. Biophys. Res. Commun.* **91,** 665–670.
Hark, S. K., and Ho, J. T. (1980). *Biochim. Biophys. Acta* **601,** 54–62.
Harlos, K., and Eibl, H. (1980). *Biochim. Biophys. Acta* **601,** 113–122.
Hauser, H. (1975). *In* "Water: A Comprehensive Treatise" (F. Franks, ed.), Vol. 4, pp. 209–303. Plenum, New York.
Hauser, H., and Dawson, R. M. C. (1967). *Eur. J. Biochem.* **1,** 61–69.
Hauser, H., Phillips, M. C., and Barratt, M. D. (1975). *Biochim. Biophys. Acta* **413,** 341–353.
Hauser, H., Levine, B. A., and Williams, R. J. P. (1976a). *Trends Biochem. Sci. (Pers. Ed.)* **1,** 278–281.
Hauser, H., Darke, A., and Phillips, M. C. (1976b). *Eur. J. Biochem.* **62,** 335–344.
Hauser, H., Finer, E. G., and Darke, A. (1977). *Biochem. Biophys. Res. Commun.* **76,** 267–274.
Heuser, J. E. (1978). *In* "Transport of Macromolecules in Cellular Systems" (S. C. Silverstein, ed.), pp. 445–464. Dahlem Konferenzen, Berlin.
Higgins, J. A., and Dawson, R. M. C. (1977). *Biochim. Biophys. Acta* **470,** 342–356.
Hinz, H.-J., and Sturtevant, J. M. (1972). *J. Biol. Chem.* **247,** 6071–6075.
Hoekstra, D. (1982). *Biochemistry* **21,** 2833–2840.
Hong, K., Düzgüneş, N., and Papahadjopoulos, D. (1981). *J. Biol. Chem.* **256,** 3641–3644.
Hong, K., Düzgüneş, N., and Papahadjopoulos, D. (1982a). *Biophys. J.* **37,** 297–305.
Hong, K., Düzgüneş, N., Ekerdt, R., and Papahadjopoulos, D. (1982b). *Proc. Natl. Acad. Sci. U.S.A.* **79,** 4642–4644.
Israelachvili, J. N. (1977). *Biochim. Biophys. Acta* **469,** 221–225.
Ito, T., and Ohnishi, S. I. (1974). *Biochim. Biophys. Acta* **352,** 29–37.
Ito, T., Ohnishi, S. I., Isinaga, M., and Kito, M. (1975). *Biochemistry* **14,** 3064–3069.
Jacobson, K., and Papahadjopoulos, D. (1975). *Biochemistry* **14,** 152–161.
Janiak, M. J., Small, D. M., and Shipley, G. G. (1976). *Biochemistry* **15,** 4575–4580.

Jendrasiak, G. L., and Hasty, J. H. (1974). *Biochim. Biophys. Acta* **337,** 79–91.
Karnovsky, M. J., Kleinfeld, A. M., Hoover, R. L., and Klausner, R. D. (1982). *J. Cell Biol.* **94,** 1–6.
Kimelberg, H. K. (1976). *Mol. Cell. Biochem.* **10,** 171–190.
Kimelberg, H. K., and Papahadjopoulos, D. (1972). *Biochim. Biophys. Acta* **282,** 277–292.
Klausner, R. D., Kleinfeld, A. M., Hoover, R. L., and Karnovsky, M. J. (1980). *J. Biol. Chem.* **255,** 1286–1295.
Kleeman, W., and McConnell, H. M. (1974). *Biochim. Biophys. Acta* **345,** 220–230.
Kovatchev, S., Vaz, W. L. C., and Eibl, H. (1981). *J. Biol. Chem.* **256,** 10369–10374.
Kurland, R. J., Hammoudah, M., Nir, S., and Papahadjopoulos, D. (1979a). *Biochem. Biophys. Res. Commun.* **88,** 927–932.
Kurland, R., Newton, C., Nir, S., and Papahadjopoulos, D. (1979b). *Biochim. Biophys. Acta* **551,** 137–147.
Ladbrooke, B. D., and Chapman, D. (1969). *Chem. Phys. Lipids* **3,** 304–367.
Lansman, J., and Haynes, D. H. (1975). *Biochim. Biophys. Acta* **394,** 335–347.
Lau, A., McLaughlin, A., and McLaughlin, S. (1981). *Biochim. Biophys. Acta* **645,** 279–292.
Lee, A. G. (1977). *Biochim. Biophys. Acta* **472,** 237–281.
Lentz, B. R., Barenholz, Y., and Thompson, T. E. (1976). *Biochemistry* **15,** 4521–4528.
Lis, L. J., Kauffman, J. W., and Shriver, D. F. (1975). *Biochim. Biophys. Acta* **406,** 453–464.
Luna, E. J., and McConnell, H. M. (1977). *Biochim. Biophys. Acta* **470,** 303–316.
Luzzati, V., and Tardieu, A. (1974). *Annu. Rev. Phys. Chem.* **25,** 79–94.
MacDonald, R. C., Simon, S. A., and Baer, E. (1976). *Biochemistry* **15,** 885–891.
McIver, D. J. L. (1979). *Physiol. Chem. Phys.* **11,** 289–302.
McLaughlin, S., and Brown, J. (1981). *J. Gen. Physiol.* **77,** 475–487.
McLaughlin, S. G. A., Szabo, G., and Eisenman, G. (1971). *J. Gen. Physiol.* **58,** 667–687.
McLaughlin, S., Mulrine, N., Gresalfi, T., Vaio, G., and McLaughlin, A. (1981). *J. Gen. Physiol.* **77,** 445–473.
Miller, D. C., and Dahl, G. P. (1982). *Biochim. Biophys. Acta* **689,** 165–169.
Nagle, J. F., and Wilkinson, D. A. (1978). *Biophys. J.* **23,** 159–175.
Nancollas, G. H. (1966). "Interactions In Electrolyte Solutions." Elsevier, Amsterdam.
Newton, C. (1978). Ph.D. Thesis, State Univ. of New York, Buffalo, New York.
Newton, C., Pangborn, W., Nir, S., and Papahadjopoulos, D. (1978). *Biochim. Biophys. Acta* **506,** 281–287.
Niggli, V., Adumyah, E. S., and Carafoli, E. (1981). *J. Biol. Chem.* **256,** 8588–8592.
Nilsson, O. S., and Dallner, G. (1977). *Biochim. Biophys. Acta* **464,** 453–458.
Nir, S., and Bentz, J. (1978). *J. Coll. Interface Sci.* **65,** 399–414.
Nir, S., Newton, C., and Papahadjopoulos, D. (1978). *Bioelectrochem. Bioenerg.* **5,** 116–133.
Nir, S., Bentz, J., and Portis, Jr., A. R. (1980a). *Adv. Chem. Ser.* **188,** 75–106.
Nir, S., Bentz, J., and Wilschut, J. (1980b). *Biochemistry* **19,** 6030–6036.
Nir, S., Bentz, J., and Düzgüneş, N. (1981). *J. Coll. Interface Sci.* **84,** 266–269.
Nir, S., Wilschut, J., and Bentz, J. (1982). *Biochim. Biophys. Acta* **688,** 275–278.
Nir, S., Bentz, J., Wilschut, J., and Düzgüneş, N. (1983). *Prog. Surface Sci.* **13,** 1–124.
Ohki, S. (1978). *Bioelectrochem. Bioenerg.* **5,** 204–214.
Ohki, S. (1982). *Biochim. Biophys. Acta* **689,** 1–11.
Ohki, S., and Düzgüneş, N. (1979). *Biochim. Biophys. Acta* **552,** 438–449.
Ohki, S., and Kurland, R. (1981). *Biochim. Biophys. Acta* **645,** 170–176.
Ohki, S., and Papahadjopoulos, D. (1970). *In* "Surface Chemistry of Biological Systems" (M. Blank, ed.), pp. 155–174. Plenum, New York.
Ohki, S., and Sauvé, R. (1978). *Biochim. Biophys. Acta* **511,** 377–387.
Ohki, S., Düzgüneş, N., and Leonards, K. (1982). *Biochemistry* **21,** 2127–2133.

Ohnishi, S.-I. (1975). *Adv. Biophys.* **8,** 35–82.
Ohnishi, S.-I., and Ito, T. (1973). *Biochem. Biophys. Res. Commun.* **51,** 132–138.
Ohnishi, S.-I., and Ito, T. (1974). *Biochemistry* **13,** 881–887.
Op den Kamp, J. A. F. (1979). *Ann. Rev. Biochem.* **48,** 47–71.
Otnaess, A.-B., and Holm, T. (1976). *J. Clin. Invest.* **57,** 1419–1425.
Papahadjopoulos, D. (1968). *Biochim. Biophys. Acta* **163,** 240–254.
Papahadjopoulos, D. (1977). *J. Coll. Interface Sci.* **58,** 459–470.
Papahadjopoulos, D. (1978). *In* "Membrane Fusion" (G. Poste, and G. L. Nicolson, eds.), pp. 765–790. Elsevier, Amsterdam.
Papahadjopoulos, D., and Miller, N. (1967). *Biochim. Biophys. Acta* **135,** 624–630.
Papahadjopoulos, D., Poste, G., Schaeffer, B. E., and Vail, W. J. (1974). *Biochim. Biophys. Acta* **352,** 10–28.
Papahadjopoulos, D., Vail, W. J., Jacobson, K., and Poste, G. (1975). *Biochim. Biophys. Acta* **394,** 483–491.
Papahadjopoulos, D., Vail, W. J., Pangborn, W. A., and Poste, G. (1976). *Biochim. Biophys. Acta* **448,** 265–283.
Papahadjopoulos, D., Vail, W. J., Newton, C., Nir, S., Jacobson, K., Poste, G., and Lazo, R. (1977). *Biochim. Biophys. Acta* **465,** 579–598.
Papahadjopoulos, D., Portis, A., and Pangborn, W. (1978). *Ann. N.Y. Acad. Sci.* **308,** 50–63.
Papahadjopoulos, D., Poste, G., and Vail, W. J. (1979). *Methods Membr. Biol.* **10,** 1–121.
Parsegian, V. A. (1977). *In* "Society for Neuroscience Symposia" (W. M. Cowan and J. A. Ferrendelli, eds.), Vol. 2, pp. 161–171. Society for Neuroscience, Bethesda, Maryland.
Phillips, M. C., Hauser, H., and Paltauf, F. (1972). *Chem. Phys. Lipids* **8,** 127–133.
Portis, A., Newton, C., Pangborn, W., and Papahadjopoulos, D. (1979). *Biochemistry* **18,** 780–790.
Poste, G., and Allison, A. C. (1973). *Biochim. Biophys. Acta* **300,** 421–465.
Rainier, S., Jain, M. K., Ramirez, F., Ioannou, P. V., Marecek, J. F., and Wagner, R. (1979). *Biochim. Biophys. Acta* **558,** 187–198.
Ranck, J. L., Mateu, L., Sadler, D. M., Tardieu, A., Gulik-Krzywicki, T., and Luzzati, V. (1974). *J. Mol. Biol.* **85,** 249–277.
Rand, R. P. (1981). *Annu. Rev. Biophys. Bioeng.* **10,** 277–314.
Rand, R. P., and Sengupta, S. (1972). *Biochim. Biophys. Acta* **255,** 484–492.
Rehfeld, S. J., Düzgüneş, N., Newton, C., Papahadjopoulos, D., and Eatough, D. J. (1981a). *FEBS Lett.* **123,** 249–251.
Rehfeld, S. J., Düzgüneş, N., Papahadjopoulos, D., and Eatough, D. J. (1981b). *Biophys. J.* **33,** 111a.
Robinson, N. C., Strey, F., and Talbert, L. (1980). *Biochemistry* **19,** 3656–3661.
Rojas, E., and Tobias, J. M. (1965). *Biochim. Biophys. Acta* **94,** 394–404.
Rothman, J. E., and Lenard, J. (1977). *Science* **195,** 743–753.
Rubin, R. P. (1974). "Calcium and the Secretory Process," pp. 189. Plenum, New York.
Sacré, M.-M., Hoffman, W., Turner, M., Tocanne, J.-F., and Chapman, D. (1979). *Chem. Phys. Lipids* **69,** 69–83.
Satir, B. H. (1974). *Symp. Soc. Exp. Biol.* **38,** 399–418.
Satir, B. H., and Oberg, S. G. (1978). *Science* **199,** 536–538.
Schick, P. K., Kurica, K. B., and Chacko, G. K. (1976). *J. Clin. Invest.* **57,** 1221–1226.
Seimiya, T., and Ohki, S. (1972). *Nature (London) New Biol.* **239,** 26–27.
Seimiya, T., and Ohki, S. (1973). *Biochim. Biophys. Acta* **298,** 546–561.
Sheetz, M. P., and Chan, S. I. (1972). *Biochemistry* **11,** 4573–4581.
Shimshick, E. J., and McConnell, H. M. (1973). *Biochemistry* **12,** 2351–2360.
Shinitzky, M., and Barenholz, Y. (1978). *Biochim. Biophys. Acta* **515,** 367–394.

Shipley, G. G. (1973). *In* "Biological Membranes" (D. Chapman and D. F. H. Wallach, eds.), Vol. 3, pp. 1–89. Academic Press, London.
Silvius, J. R., and McElhaney, R. N. (1980). *Proc. Natl. Acad. Sci. U.S.A.* **77,** 1255–1259.
Simon, S. A., Lis, L. J., Kauffman, J. W., and MacDonald, R. C. (1975). *Biochim. Biophys. Acta* **375,** 317–326.
Sklar, L. A., Miljanich, G. P., and Dratz, E. A. (1979). *J. Biol. Chem.* **254,** 9592–9597.
Speth, V., and Wunderlich, F. (1973). *Biochim. Biophys. Acta* **291,** 621–628.
Spiker, R. C., Jr., and Lewin, I. W. (1976). *Biochim. Biophys. Acta* **433,** 457–468.
Stewart, T. P., Hui, S. W., Portis, A. R., and Papahadjopoulos, D. (1979). *Biochim. Biophys. Acta* **556,** 1–16.
Strehlow, V., and Jähnig, F. (1981). *Biochim. Biophys. Acta* **641,** 301–310.
Sundler, R., Sarcione, S. L., Alberts, A. W., and Vagelos, P. R. (1977). *Proc. Natl. Acad. Sci. U.S.A.* **74,** 3350–3354.
Sundler, R., Düzgüneş, N., and Papahadjopoulos, D. (1981). *Biochim. Biophys. Acta* **649,** 751–758.
Suurkuusk, J., Lentz, B. R., Barenholz, Y., Biltonen, R. L., and Thompson, T. E. (1976). *Biochemistry* **15,** 1393–1401.
Szoka, F., Jr., and Papahadjopoulos, D. (1980). *Annu. Rev. Biophys. Bioeng.* **9,** 467–508.
Thompson, T. E., Lentz, B. R., and Barenholz, Y. (1977). *In* "Biochemistry of Membrane Transport. FEBS Symposium No. 42" (G. Semenza and E. Carafoli, eds.), pp. 47–71. Springer-Verlag, Berlin.
Tocanne, J. F., Ververgaert, P.H.J.Th., Verkleij, A. J., and van Deenen, L. L. M. (1974). *Chem. Phys. Lipids* **12,** 201–219.
Tokutomi, S., Eguchi, G., and Ohnishi, S.-I. (1979). *Biochim. Biophys. Acta* **551,** 78–88.
Tokutomi, S., Lew, R., and Ohnishi, S.-I. (1981). *Biochim. Biophys. Acta* **643,** 276–282.
Tokutomi, S., Ohki, K., and Ohnishi, S.-I. (1980). *Biochim. Biophys. Acta* **596,** 192–200.
Träuble, H., and Eibl, H. (1974). *Proc. Natl. Acad. Sci. U.S.A.* **71,** 214–219.
Träuble, H. (1977). *In* "Structure of Biological Membranes: 34th Nobel Symposium" (S. Abrahamsson and I. Pascher, eds.), pp. 509–550. Plenum, New York.
Vail, W. J., and Stollery, J. G. (1979). *Biochim. Biophys. Acta* **551,** 174–184.
van Dijck, P. W. M., deKruijff, B., Aarts, P. A. M. M., Verkleij, A. J., and de Gier, J. (1978a). *Biochim. Biophys. Acta* **506,** 183–191.
van Dijck, P. W. M., deKruijff, B., Verkleij, A. J., van Deenen, L. L. M., and de Gier, J. (1978b). *Biochim. Biophys. Acta* **512,** 84–96.
Venzin, M., Sandric, C., Akert, K., and Wyss, V. R. (1977). *Brain Res.* **130,** 393–404.
Verkleij, A. J., Ververgaert, P. H. J., van Deenen, L. L. M., and Elbers, P. F. (1972). *Biochim. Biophys. Acta* **288,** 326–332.
Watts, A., Harlos, K., Maschke, W., and Marsh, D. (1978). *Biochim. Biophys. Acta* **510,** 63–74.
Williams, R. J. P. (1976). *Symp. Soc. Exp. Biol.* **30,** 1–17.
Wilschut, J., Düzgüneş, N., Fraley, R., and Papahadjopoulos, D. (1980). *Biochemistry* **19,** 6011–6021.
Wilschut, J., Düzgüneş, N., and Papahadjopoulos, D. (1981). *Biochemistry* **20,** 3126–3133.
Wilschut, J., Düzgüneş, N., and Papahadjopoulos, D. (1983). (In preparation.)
Wu, S. H., and McConnell, H. M. (1975). *Biochemistry* **14,** 847–854.
Wunderlich, F., Speth, V., Batz, W., and Kleinig, H. (1973). *Biochim. Biophys. Acta* **298,** 39–49.
Zwaal, R. F. A., Roelofsen, B., and Colley, C. M. (1973). *Biochim. Biophys. Acta* **300,** 159–182.

Chapter 8

The Effect of the Proton and of Monovalent Cations on Membrane Fluidity[1]

Hansjörg Eibl

Introduction

Membranes and phospholipid bilayers are held together by hydrophobic bonding of their nonpolar hydrocarbon chains. However, both the apolar and the polar regions of the molecule contribute to the characteristic properties of phospholipid–water dispersions, which represent simple model systems for biological cells. Because only complex mixtures of naturally occurring phospholipids are obtained by extraction from biomembranes, it is advantageous to study the principal effects of ionic interactions on structurally defined synthetic phospholipids. This approach has been very helpful in gaining an understanding of the properties of bilayer membranes on a molecular level.

Throughout this chapter, the term *fluidity* is used to describe the physical

[1]This work was supported by the Deutsche Forschungsgemeinschaft through SFB 33 and by the Stiftung Volkswagenwerk.

ISBN 0-12-053002-3

state of a bilayer membrane with respect to the packing of the acyl chains, including a transition region (the so-called main transition) where abrupt changes of the fluidity have been observed. The main transition is a result of a phase change. The physical states of the phospholipid molecule below the transition (the gel phase) and above the transition (the liquid-crystalline phase) are different. The main transition of a phospholipid dispersion may be triggered, and thus changes in the fluidity may be induced, by alteration of physical parameters such as the temperature, the surface pressure, or the ionic environment.

For ionic interactions, the strongest effects were observed with negatively charged phospholipids and cations. Therefore, the following discussion will concentrate on this aspect, including the influence of structural modifications within the phospholipid molecule on the physical behavior of bilayer systems in water dispersions. First, this chapter is concerned with the classical bilayer-forming molecules, the phosphatidylcholines, and their structural modifications within the apolar and the polar region. Second, it concentrates on the effect of protons; third, it evaluates the influence of monovalent cations on phospholipid bilayer structures.

The effect of divalent cations, which is closely related to that of the proton, will be discussed separately. Of course, trivalent cations bind strongly to charged membrane surfaces, but their contribution to the functional properties of biomembranes other than analytical aspects can be neglected. It is the intent, in this chapter, to discuss the more general aspects of ionic interactions within the membrane surface rather than to present a comprehensive review of relevant publications.

The Effect of Structural Variation within the Phospholipid Molecule

The lipid most thoroughly investigated in model systems is 1,2-dipalmitoyl-*sn*-glycero-3-phosphocholine, which attracted interest because it is the main phospholipid in lung alveolar surfactant (King and Clements, 1972). The two identical fatty acid chains within one molecule may be of functional importance in the lung (Träuble *et al.*, 1974). The first calorimetric heating curves for 1,2-dipalmitoyl-*sn*-glycero-3-phosphocholine were published by Phillips *et al.* (1969), and as a result of the original work of Chapman, different physical techniques were introduced to study the physical properties of phospholipids (Chapman and Wallach, 1968). According to recent publications (Chen *et al.*, 1980; Füldner, 1981; Ruocco and Shipley, 1982; Stümpel *et al.*, 1981) 1,2-dipalmitoyl-*sn*-glycero-3-phosphocholine–water dispersions show three thermal transitions on heating from 0 to 60°C; the subtransition I (transition temperature T_{S_I}), the pretransition (transition temperature T_P), and the main transition (transition temperature T_M). By definition, the state below the temperature of the main transition is called the gel or ordered phase, and that above the main transition is the liquid–crystalline or fluid phase. The two lipid phases differ in packing and density of the hydrocarbon chains. The gel state is densely packed (all trans conformation of the hydrocarbon chains) and the liquid-crystalline state is more losely packed (trans and gauche conformations).

The subtransition and the pretransition are of only marginal influence on the fluidity in bilayer membranes. Therefore, the following discussion will concentrate on the main transition T_M, and a change in fluidity will be considered to involve a phase change, the transition from the gel to the

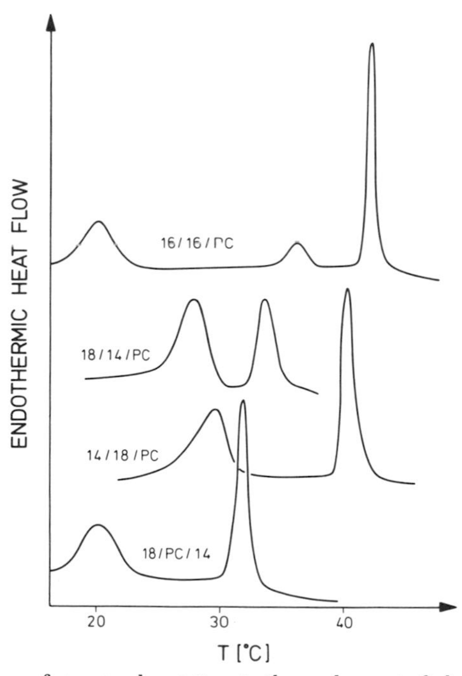

Fig. 1. The influence of structural variation in the apolar part of phosphatidylcholines; calorimetric heating curves of saturated phosphatidylcholines in excess water. Heating rate, 1.25°C/min; lipid concentration, 3 g/50 ml distilled water; sensitivity range, 1 mcal/sec. 16/16/PC indicates palmitoyl residues in the 1- and 2-positions and phosphocholine in the 3-position of glycerol (natural configuration).

liquid–crystalline state or vice versa. This has considerable advantages from an experimental point of view, because such phase changes may be reliably detected by such physical techniques as calorimetry, X-ray analysis, and various spectroscopic methods. The dramatic changes in the fluidity of the system at T_M are accompanied by abrupt changes of the structural and functional properties of the membranes, including alterations of the molecular volume (Träuble and Haynes, 1971), of conductivity and carrier-induced ion transport (Boheim *et al.*, 1980), and of the activity of membrane-bound and phospholipid-dependent enzymes (Eibl *et al.*, 1982).

The effect of structural variation in the apolar region of phosphatidylcholine on the transition temperature T_M is discussed first. The modifications involve varying the chain length of the fatty acid residues, and varying the specific distribution of the fatty acids (Keough and Davis, 1979; Stümpel *et al.*, 1981) and of the phosphocholine residue (Stümpel *et al.*, 1983) over the 1, 2, and 3 positions of the glycerol molecule. As shown in Fig. 1, the calorimetric heating scans of the various phosphatidylcholines differ greatly in the number of phase transitions observed and also in the value of the transition enthalpies (cf. Table I). The comparison includes 1,2-dipalmitoyl-*sn*-glycero-3-phosphocholine, the respective 1,3-isomer, and 1,2- or 1,3-isomers of phosphatidylcholines that contain stearoyl and myristoyl chains within one molecule. Based on the number of thermal transitions and on the X-ray diffraction patterns of the respective lipid phases, the phosphatidylcholines may be subdivided into two groups.

The 1,2-diacyl-*sn*-glycero-3-phosphocholines with two identical fatty acid chains form one group (data are shown for 1,2-dipalmitoyl-*sn*-glycero-3-phosphocholine as a typical example); they are characterized by the appearance of three transitions (a subtransition I, a pretransition, and a main transition), by their respective transition enthalpies, and by the X-ray diffraction patterns of the four lipid phases: L_ϵ (L_ϵ'), L_β (L_β'), P_β(P_β'), and L_α[2].

All of the other structural isomers of phosphatidylcholines studied in this laboratory fall in the other group according to their physical properties, for example, mixed chain 1,2-diacyl-*sn*-glycero-3-phosphocholines, mixed chain 1,3-isomers, and 1,3-isomers with identical fatty acid chains. These lipids represent the general case of phosphatidylcholine behavior and show only two transitions; a subtransition II and a main transition. The three lipid phases involved, L_ϵ (L_ϵ'), P_β (P_β'), and L_α, are characterized by their X-ray diffraction patterns. An important difference in this group is the absence of a pretransition and consequently of the L_β (L_β') phase. A more detailed dis-

[2]Throughout this chapter, the tilt angle of the acyl or alkyl chains of the phospholipids to the membrane normal was not determined. Consequently, primed and unprimed lipid phases were not distinguished.

TABLE I

Influence of Structural Variation in the Apolar Part of Phosphatidylcholines: Physical Properties of Phosphatidylcholine Bilayer Structures[a]

Phosphatidylcholines	T(°C)	Phase	d(Å)	s(Å)		
1,2-Dipalmitoyl-*sn*-glycero-3-PC	5	L_ε	58.70	10.00	6.78	4.40
	15 (T_{S_I})	↓ 4.1			4.20	3.88
	20	L_β	63.50	4.20	4.05	
	35 (T_P)	↓ 1.6				
	38	P_β	70.00	4.18		
	41 (T_M)	↓ 8.7				
	45	L_α	68.00	4.50		
1,3-Dipalmitoyl-glycero-2-PC	5	L_ε	57.04	9.63	6.70	4.14
	25 ($T_{S_{II}}$)	↓ 8.0				
	35	P_β	47.10	4.12		
	39 (T_M)	↓ 9.4				
	45	L_α	62.16	4.55		
1-Stearoyl-2-myristoyl-*sn*-glycero-3-PC	4	L_ε	67.30	9.55	4.49	3.72
	24 ($T_{S_{II}}$)	↓ 7.9				
	30	P_β	—	4.16		
	33 (T_M)	↓ 6.0				
	39	L_α	68.40	4.55		
1-Myristoyl-2-stearoyl-*sn*-glycero-3-PC	4	L_ε	60.40	9.98	6.75	4.34
	26 ($T_{S_{II}}$)	↓ 6.6			4.15	3.86
	31	P_β	69.90	4.15		
	42 (T_M)	↓ 8.2				
	46	L_α	63.80	4.55		
1-Stearoyl-3-myristoyl-*sn*-glycero-2-PC	5	L_ε	58.30	9.82	6.69	4.33
	16 ($T_{S_{II}}$)	↓ 5.9			4.15	3.76
	20	P_β	68.30	4.13		
	30 (T_M)	↓ 7.1				
	35	L_α	63.40	4.55		

[a] PC, phosphatidylcholine; T, temperature; T_{S_I} and $T_{S_{II}}$, temperature of subtransition I and subtransition II; $L_\varepsilon(L_\varepsilon')$, $L_\beta(L_\beta')$, $P_\beta(P_\beta')$ and L_α, different lipid phases characterized by their X-ray diffraction patterns; d, long spacings; s, short spacings. The arrow (e.g., L_ε to L_β) indicates a phase change and the number indicates the respective transition enthalpy (kcal/mol).

cussion of this is given by Stümpel *et al.* (1983). These studies will now be extended to phosphatidylethanolamines, -serines, and -glycerols.

Structural variation in the polar region of the phospholipid molecule will cause sensitivity in protons and monovalent and divalent cations and may allow variation of bilayer fluidity by alteration of the ionic environment. A comparison of the transition behavior of different phospholipids that have a

TABLE II

The Influence of Structural Variation in the Polar Region of Phospholipids[a]

$$\begin{array}{l} \qquad\qquad\qquad\qquad\qquad CH_2{-}O{-}CO{-}(CH_2)_{14}{-}CH_3 \\ \qquad\qquad\qquad\qquad\qquad | \\ CH_3{-}(CH_2)_{14}{-}CO{-}O{-}CH \\ \qquad\qquad\qquad\qquad\qquad | \\ \qquad\qquad\qquad\qquad\qquad CH_2{-}X \end{array}$$

Apolar region X	Charge per molecule (pH 7)	T_P (°C)	T_M (°C)
$PO_4^-{-}(CH_2)_2{-}N^+(CH_3)_3$	0	34	41
$PO_4^-{-}(CH_2)_4{-}N^+(CH_3)_3$	0	34	40
$PO_4^-{-}CH_2{-}CHOH{-}CH_2{-}OH$; Na^+	−1	36	41
$PO_4^-{-}CH_2{-}CH_3$; Na^+	−1	35	42
$N^+(CH_3)_3$; Cl^-	+1	36	42
$PO_4^-{-}(CH_2)_2{-}N^+(CH_3)_2{-}(CH_2)_2 - N^+(CH_3)_3$; Cl^-	+1	33	40
$PO_4^-{-}(CH_2)_2{-}N^+H_3$	0	—	65
$PO_4^-{-}H$; Na^+	−1	—	64

[a]T_P, pretransition temperature; T_M, main transition temperature.

constant apolar region, 1,2-dipalmitoyl-*sn*-glycerol, and modification of the polar part of the molecule, is presented in Table II.

In one group of compounds, neither large variation in structure nor in charge (neutral, negatively, and positively charged molecules) is reflected in the phase transition behavior; the typical properties of 1,2-dipalmitoyl-*sn*-glycero-3-phosphocholine, the pretransition and the main transition, are retained in many of the structural analogs. Obviously, these transitions are not sensitive to these modifications. However, there is a second group of molecules in which small variations of structure and introduction of charge will have a significant influence on the physical properties of the respective bilayer membranes. These lipids do not show a pretransition, and the main transition is shifted to higher temperatures as has been demonstrated for phosphatidylethanolamines and phosphatidic acids. The difference in the transition temperature between 1,2-dipalmitoyl-*sn*-glycero-3-phosphocholine and the respective phosphoethanolamine or phosphate is about 24°C at pH values near 7 (Eibl, 1977; Eibl and Woolley, 1979).

The phospholipid structures presented in Table II differ from each other only in the polar region of the molecules. A search for structural analogies that could explain the large shift in the phase transition temperature at pH 7 between the two groups of phospholipid molecules emphasizes that the interplay of forces in the polar region of these molecules is strongly influenced by the presence or absence of protons.

It is generally known that at pH 7 the lipids of group I are in the deprotonated state whereas the lipids of group II are in a partially protonated state. This observation opens the discussion of whether the strong differences in the physical behavior of phospholipids with respect to the polar region can be related to the degree of protonation and thus are critically dependent on the p*K* values of their phosphate and ammonium groups. If this is correct, the behavior of the phospholipids presented in Table II is representative for this class of lipids, and other members may be grouped in the same way according to the degree of protonation. This is indeed possible for many other structures already studied in our laboratory. Therefore, these general observations will necessarily lead to the prediction that phosphatidic acids and phosphatidylethanolamines must adopt a lecithin-like structure at high pH values if the protons are removed from the lipid structure. This should result in a drop of the main transition temperature accompanied by the appearance of a pretransition in those cases where the phospholipids do contain two identical fatty acids within one molecule.

The Effect of the Proton

A comprehensive understanding of the effect of protons on phospholipid bilayer membranes may be derived from a comparison of the different states of ionization of phosphatidic acid with the respective phase transition temperature. A detailed discussion of the phosphatidic acid system may then allow predictions for the behavior of other phospholipids at different states of protonation and may lead to a general description of the protonation effect in membrane surfaces.

The first attempt to demonstrate the influence of charge alteration on the properties of bilayer systems was undertaken by Träuble and Eibl (1974). Phosphatidic acid was chosen because it allows the dissociation of two protons from one phospholipid molecule by variation of the pH from 2 to 12. The initial experiments were performed with 1,2-dimyristoyl-*sn*-glycero-3-phosphate, but later the work was continued on chemically more stable ether analogs that confirmed the principal results of the earlier studies (Eibl and Blume, 1979).

The state of protonation is described by the degree of dissociation α. By definition, a value of $\alpha_1 = 0$ to 1 for the first deprotonation step in phosphatidic acid corresponds to zero–100% dissociation respectively, pK_1 being the pH at which $\alpha_1 = 0.5$. Then, $\alpha_2 = 0$ to 1 describes the second deprotonation step, with pK_2 at $\alpha_2 = 0.5$ ($\alpha_1 = 1$ corresponding to $\alpha_2 = 0$).

In Table III, the melting points of different solid, water-free phosphatidic acids in the various ionization states (free acid, monosodium, and disodium salt) are compared with their aqueous bilayer phase transition temperatures T_M at different degrees of dissociation α. In the case of 1,2-dihexadecyl-*sn*-glycero-3-phosphate, the melting points of the solid lipid and of T_M of the lipid dispersion in water for $\alpha_1 = 0$ are nearly equal, with a value of about 62°C. Thus, the presence or absence of water does not influence the state of the fully protonated lipid, and the dispersion seems to represent a distribution of solid lipid in water with negligible interaction between lipid and water molecules.

The presence of excess water with phosphatidic acids containing ester bonds, as shown for 1,2-dimyristoyl-, 1,3-dimyristoyl- and 1,2-dipalmitoyl-*sn*-glycero-3-phosphoric acid, lowers T_M in comparison to the melting points of the fully protonated lipids. Nevertheless, the apolar parts of these molecules, the chains, dominate the physical properties. Then, with increasing charge at $\alpha_1 = 1$, the solid sodium salts of phosphatidic acid show a biphasic behavior. A change in the optical density is observed at 60–80°C

TABLE III

A Comparison of the Melting Points (m.p.) of Phosphatidic acid, Mono- and Disodium Salt, with the Phase Transition Temperature (T_M) of Water Dispersions at Distinct Values of the Degree of Dissociation α

		Degree of dissociation		
		α_1		α_2
sn-Glycero-3-phosphoric acid		0	1 0	1
1,2-Dimyristoyl-	m.p.	59	165	260
	T_M[a]	45	52	28
1,3-Dimyristoyl-	m.p.	47	158	254
	T_M	38	42	23
1,2-Dipalmitoyl-	m.p.	69	160	253
	T_M	62	71	46
1,2-Dihexadecyl-	m.p.	62	166	257
	T_M	62	75	50

[a]Temperatures in degree Centigrade.

that lies in the range of the main transition temperature T_M of water dispersions of the lipids and indicates chain melting (sinter point). The macroscopic ordering of the lipid is maintained until the true melting point is reached, which occurs at much higher temperatures, about 150°C for the monosodium salts and about 250°C for the disodium salts of the phosphatidic acids. The appearance of two melting regions, chain melting and disruption of the headgroup lattice, is typical for phospholipid molecules. It demonstrates that the polar and the apolar regions in one and the same molecule can interact specifically and independently with foreign molecules or with cations of the ionic environment. However, in most cases a change in the polar region will also be reflected in the apolar part of the molecule, and vice versa (Eibl and Blume, 1979).

Biological membranes are usually exposed to water and regulated by specific properties of biomembranes that separate inner (cell interior) and outer (cell exterior) aqueous regions. It is important to understand how the properties of membrane-forming molecules such as phospholipids may change in an aqueous environment because of the presence or absence of protons within the membrane surface. Complete titration curves of different phosphatidic acid–water dispersions were first described by Eibl and Blume (1979), and the transition enthalpies were compared at different degrees of dissociation (Blume and Eibl, 1979). A stable plateau region of the phase transition temperature is observed over a wide range of pH values. For 1,2-dihexadecyl-*sn*-glycero-3-phosphate, the transition temperature T_M of the plateau region is more than 20°C above the T_M of the fully protonated state (see Fig. 2). The points of inflexion near pH 4 and 10 indicate that the upper and lower limits of the plateau coincide with the p*K* values of phosphatidic acid, with a pK_1 of about 3.5 and a pK_2 of about 9.5. These p*K* values were confirmed independently by acid base titration of 1,2-dimyristoyl-*sn*-glycero-3-phosphate, in which a pK_1 of 3.2 and a pK_2 of 9.5 were determined (Träuble and Eibl, 1974).

If we compare the T_M values of 1,2-dihexadecyl-*sn*-glycero-3-phosphoric acid at different values of α_1, it is rather surprising that variation of α_1 from 0 to 0.5 causes an increase in T_M of 13°C. It indicates a denser packing of the chains with increasing surface charge and demonstrates that the lipid structure is particularly stable when negatively charged phosphate groups and protons are located simultaneously within the membrane surface (p*K* regions). Therefore, the upper value in the T_M/pH diagram of Fig. 2 does not represent the fully protonated state of the lipid as predicted from the electrostatic theory; it corresponds to the degree of protonation $\alpha_1 = 0.5$, the pK_1 of the phosphatidic acid system. A further increase of charge from $\alpha_1 = 0.5$ to $\alpha_2 = 0.5$ only slightly lowers T_M despite the fact that the charge per phospholipid molecule is increased by one unit from pK_1 to pK_2. The slope

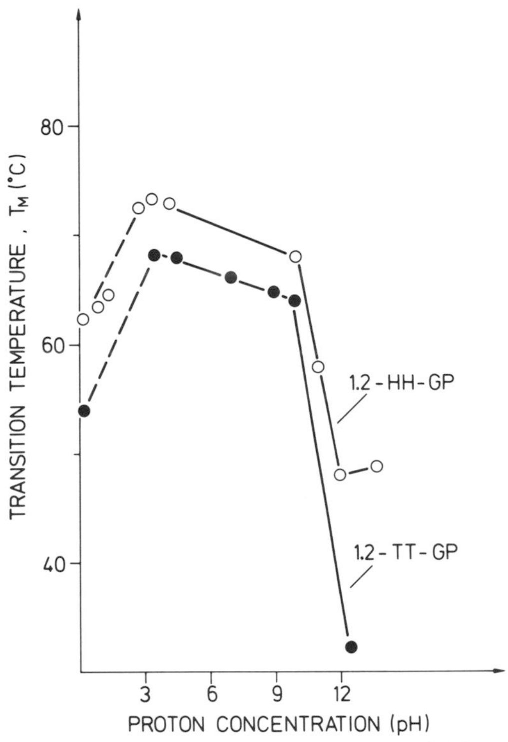

Fig. 2. The effect of proton concentration on the temperature of the main transition T_M of 1,2-dihexadecyl- and 1,2-ditetradecyl-*sn*-glycero-3-phosphoric acid. Dispersions contained 20 mg of lipid in 20 ml of bidistilled water. The pH was adjusted by the addition of dilute NaOH. The dotted lines indicate regions where two transitions were sometimes observed, a lower and an upper transition.

of the plateau region, $dT_M/d\alpha$, is about −6°C. For $\alpha_2 > .5$, however, a dramatic change in T_M, a sudden drop of about 25°C, is observed. The slope $dT_M/d\alpha$ in this region is about −150°C. It is the result of a sudden destabilization of the lipid surface as a consequence of a complete loss of interfacial hydrogen bonds. Simultaneously, water can penetrate into the polar lipid structure. The hydration leads to a molecular packing that closely resembles that of the respective phosphatidylcholines. This is also indicated by the very similar T_M values for phosphatidylcholines and phosphatidic acids at pH 12.

The large differences between the transition temperatures of phosphatidic acid, T_M, at $\alpha_2 = 0.5$ and at $\alpha_2 = 1.0$, can be explained simply by the strong structuring effect of protons within the membrane surface. For $\alpha_2 = 1.0$, the protons are completely extracted from the lipid surface, resulting in the observed drop of T_M. If we compare the fully protonated state, $\alpha_1 = 0$ and $T_M = 62°C$, with that of the fully deprotonated state, $\alpha_2 = 1$ and $T_M = 50°C$ in the case of 1,2-dihexadecyl-*sn*-glycero-3-phosphoric acid, the difference in T_M of about 12°C represents the electrostatic effect with a slope $dT_M/d\alpha$ of about $-6°C$.

The extended plateau for T_M, far above the T_M values for the fully proto-

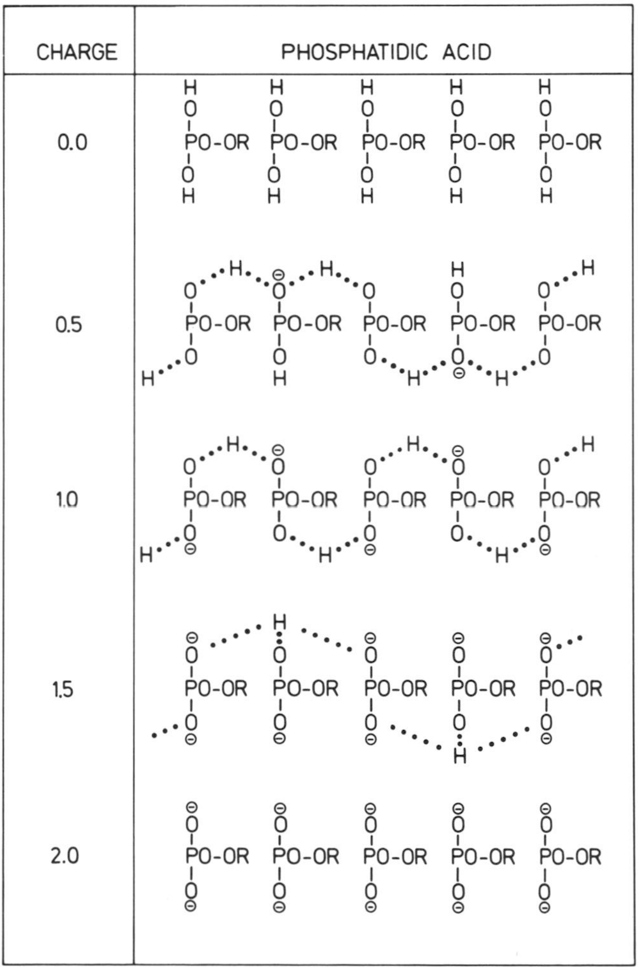

CHARGE	PHOSPHATIDIC ACID
0.0	PO-OR PO-OR PO-OR PO-OR PO-OR
0.5	PO-OR PO-OR PO-OR PO-OR PO-OR
1.0	PO-OR PO-OR PO-OR PO-OR PO-OR
1.5	PO-OR PO-OR PO-OR PO-OR PO-OR
2.0	PO-OR PO-OR PO-OR PO-OR PO-OR

Fig. 3. The polar region of phosphatidic acid molecules. With increasing charge, a continuous lattice of hydrogen bonds is formed. At full charge, the hydrogen bonds are broken and stabilization is lost.

nated and the fully deprotonated states, is the result of strong hydrogen bonding between protons in the membrane surface and negatively charged phosphate groups. The lipid matrix is stabilized between pK_1 and pK_2 by a continuous net of hydrogen bonds interconnecting the nearest neighbor molecules; this is shown schematically in Fig. 3. Hydrogen bond stabilization requires the presence of both proton-donating and proton-accepting groups, for instance, protonated phosphate or ammonium groups, and negatively charged phosphate residues, as demonstrated for $\alpha_1 = 0$ and $\alpha_2 = 1$ where a proton donor or a proton acceptor alone will not result in a stabilization of the membrane system. The strong influence of protons in the phosphatidic acid structure is also documented by the first single crystal analysis of a phosphatidic acid molecule, the monosodium salt of 1,2-dimyristoyl-*sn*-glycero-3-phosphoric acid (Harlos *et al.*, 1983).

The detailed discussion of the phosphatidic acid system should allow predictions for other phospholipid–water dispersions. Thus, the increased difference between the pK values of phosphatidylethanolamines, p$K_1 < 2$ and p$K_2 > 11$, should result in an extended plateau for the phase transition temperature T_M in comparison with phosphatidic acids (Stümpel *et al.*, 1980). This has indeed been observed; see Fig. 4. Again the partially protonated state has a T_M of about 25°C above T_M of the fully deprotonated state. The slope of the plateau region, $dT_M/d\alpha$, is only -1°C.

If we now consider phospholipid systems with only one dissociable proton, for instance, methylphosphatidic acid (Eibl and Woolley, 1979) or phosphatidylglycerol (Watts *et al.*, 1981), the prediction is that the plateau region bordered by pK_1 and pK_2 in phosphatidic acids and phosphatidylethanolamines will disappear, and a maximum value for T_M is expected at the pK value of the phospholipid, at $\alpha = 0.5$, (see Fig. 4). Two regions in the T_M/pH diagram, which are separated by the point of inflexion, are important. A change in α from 0 to 0.5 will increase T_M with increasing charge. In this region, the network of hydrogen bonds is formed by partial deprotonation of the phosphate groups, because stabilization by hydrogen bonding depends on both the simultaneous presence of proton donator and proton acceptor molecules. A change from $\alpha = 0.5$ to 1.0 then will decrease T_M with increasing charge, because the continuous band of hydrogen bonds between the phosphate groups is broken by the complete extraction of protons from the membrane surface. The results shown by the behavior of more complex phospholipid structures with polyproton systems, such as phosphatidylserines (Cevc *et al.*, 1981), are in good agreement with these predictions.

Lipids without protons bound to the membrane surface, in the fully deprotonated state, will arrange in a lamellar phase; an example is the phosphatidylcholines, which form planar bilayers or vesicles. Most previous work

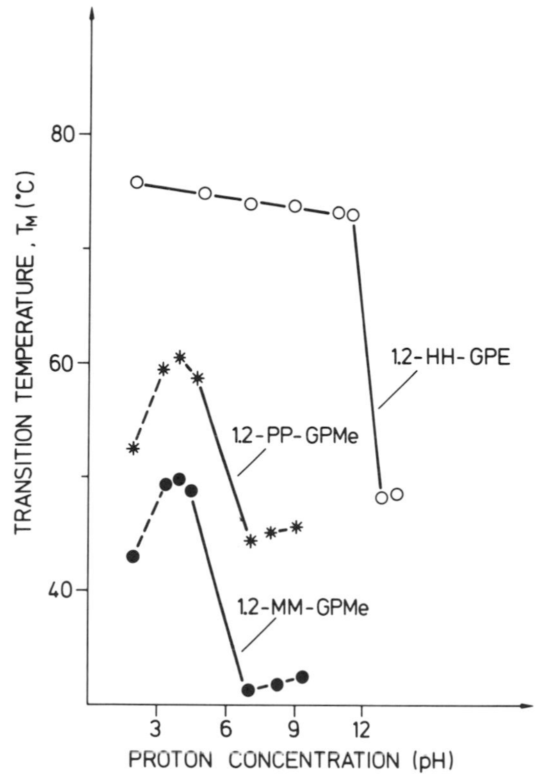

Fig. 4. The effect of proton concentration on the temperature of the main transition T_M. 1,2-Dihexadecyl-*sn*-glycero-3-phosphoethanolamine, a two-proton system, is compared with 1,2-dipalmitoyl- and 1,2-dimyristoyl-*sn*-glycero-3-phosphoric acid methyl esters as one-proton systems.

has dealt with lamellar structures because they are the dominant phases in biological membranes. However, information about phospholipid molecules which arrange in hexagonal phases or inverted micelles is increasing. This has been reported for phospholipids with unsaturated chains, such as 1,2-dioleoyl-*sn*-glycero-3-phosphoethanolamine (Cullis and de Kruijff, (1976), egg phosphatidylserine (Hope and Cullis, 1980), and negatively charged phospholipids in the presence of divalent cations (Deamer *et al.*, 1970). Also defined synthetic phospholipids with saturated chains may adopt hexagonal structures by interaction with calcium ions (Harlos and Eibl, 1980a,b).

In addition, phase changes from the lamellar phase in the liquid–crystalline state to the hexagonal phase even in the absence of divalent cations were described for synthetic phosphatidylethanolamines and phosphatidic acids with saturated chains. It was demonstrated for saturated chains that, deviating from the temperature of the main transition which increases with increasing chain length a phase change from the lamellar to the hexagonal phase has a reversed temperature dependence. The temperature of the transition to the hexagonal phase decreases with increasing chain length of the saturated chains (Harlos and Eibl, 1981).

In the light of the present discussion, it is important to recall that phosphatidylethanolamines and phosphatidic acids at high pH in the fully deprotonated state form lamellar structures in the presence of monovalent cations. After protonation of the ammonium group, hexagonal phases are observed for phosphatidylethanolamines at pH values near 8. Protonation of the phosphate group to pK_2 will increase the main transition temperature of phosphatidic acid dispersions but will not allow the transition from the lamellar to a hexagonal phase at temperatures below 100°C. This transforma-

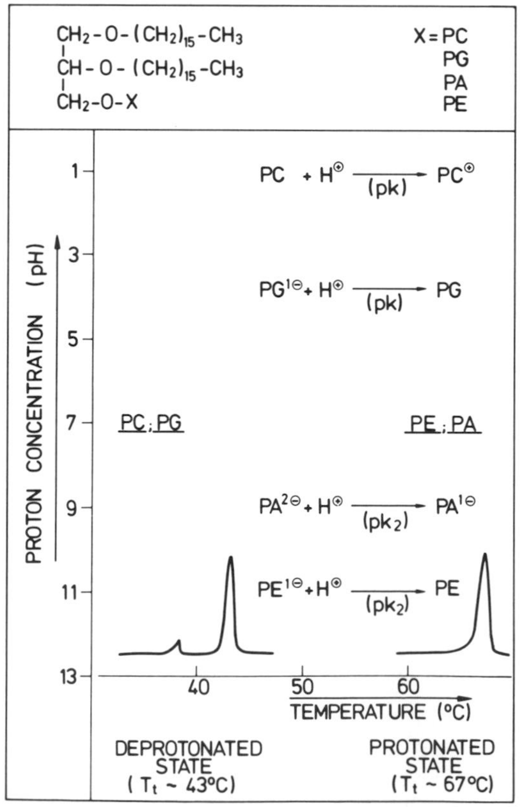

Fig. 5. A comparison of the deprotonated and protonated states of phospholipid molecules. At pH 13, all lipids are in the deprotonated (low temperature) state which is characterized by the appearance of a pretransition. The protonated (high temperature) state is obtained by protonation.

tion is then possible at pH values below 5 near the pK_1 of the phosphatidic acid.

The effect of protons on the transition from lamellar gel to liquid crystalline is summarized in Fig. 5. Both, the phase changes lamellar gel to liquid crystalline and lamellar liquid crystalline to hexagonal do influence the fluidity of apolar acyl or alkyl chains. The phospholipids included in the discussion are those that frequently are active constituents of biomembranes; phosphatidylethanolamine, phosphatidic acid, phosphatidylglycerol, and phosphatidylcholine (Jain, 1972). Water dispersions of these lipids at pH 13 do not contain protons in the membrane surface, and they form stable lamellar structures between 10 and 80°C. With 1,2-dihexadecyl-*sn*-glycerol as the constant apolar part of the molecule, the lipids in the deprotonated state have a very similar transition behavior with a pretransition at about 39°C (which may depend on the concentration of monovalent cations as will be discussed later) and a main transition near 44°C. The slightly higher phase transitions of these lipids in comparison to the ester analogs, for instance, 1,2-dipalmitoyl-*sn*-glycero-3-phosphocholine with $T_p \sim 36°$ and $T_M \sim 41°C$, are caused by ether linkage of the long chain residues.

If we now increase the proton concentration from pH 13 to pH 11, the amino group of phosphatidylethanolamine is protonated, and the temperature of the main transition is increased from 44 to 69°C. The lamellar structure is retained. At pH 7, an additional transition which was characterized by X-ray analysis appears at 87°C as a transition from the lamellar to the hexagonal phase. In phosphatidic acid, protonation of the phosphate group, pK_2 with $\alpha_2 = 0.5$, occurs at pH 9; this induces the lamellar gel to liquid–crystalline transition. A transition from the lamellar to the hexagonal phase is only observed at pH values below 5 in the region of the first p*K* of the phosphate group, pK_1 with $\alpha_1 = 0.5$. If we now further increase the proton concentration, the phosphate group of phosphatidylglycerol is protonated at pH 3.5 and that of phosphatidylcholine at pH values less than 2. A sudden increase of the temperature of the main transition is again observed, the result of the formation of a continuous band of hydrogen bonds between the negatively charged phosphate groups.

In general, natural membranes function at pH values near 7. The differentiated interaction of membrane-forming phospholipids with protons results in the large difference of the p*K* values and allows a classification of the lipids at pH 7 on the basis of their physical state as a result of the respective degree of dissociation α. At pH 7, the negatively charged phosphate in phosphatidylcholine is shielded from divalent cations by intramolecular compensation through the positive charge of the choline group. Therefore, phosphatidylcholines are in the deprotonated, low temperature state (see Fig. 5) and exclusively occur in the lamellar phase. Phosphatidylglycerols and phosphatidic acids carry one negative charge per molecule, which introduces sensitivity against divalent cations. Phosphatidylglycerols occupy the deprotonated, low temperature state like phosphatidylcholines, however phosphatidic acids the protonated high temperature state like phosphatidylethanolamines. Both lipids prefer the lamellar structure but divalent cations can induce a phase change lamellar liquid crystalline to hexagonal. Phosphatidylethanolamines do not interact with divalent cations and are in the protonated, high temperature state. Independent of the presence of divalent cations they can occur in the lamellar phase but may arrange in a hexagonal structure at temperatures above the main transition. From phosphatidylcholine to phosphatidylethanolamine there is a stepwise increase in the capacity to undergo a phase change from the lamellar structure to the hexagonal phase. Phosphatidylcholines are retained in the lamellar structure even in the presence of divalent cations, whereas phosphatidylethanolamines can occur in the hexagonal phase in the absence of divalent cations. The lamellar and the hexagonal phases are biologically important because they are stable in biologically relevant ranges of temperature.

We have discussed at length that a change in the degree of dissociation α can influence the temperature of the main transition and induce a phase change from a lamellar to a hexagonal structure. We now may ask the complementary question: whether the main transition can alter α as a result of the considerable expansion of the bilayer area at T_M. Indeed, a phase change from the gel to the liquid-crystalline state will shift α to larger values as protons are released from the membrane surface (Träuble and Eibl, 1975), and, vice versa, a phase change from the liquid-crystalline to the gel phase will change α to smaller values. The lipid will bind protons and thus reduce the proton concentration (pH) of the aqueous environment. The phospholipid membrane represents a reservoir of hydrogen ions; their activation by a gel to liquid–crystalline transition can be reversed by a transition back to the gel phase.

To summarize, we have discussed that the degree of dissociation α can strongly influence the main transition temperature and thus the fluidity of phospholipid bilayer systems. For a given phospholipid molecule, three states of the bilayer membrane can be distinguished. First and second, the fully protonated and the fully deprotonated lipids represent bilayer systems that differ in the temperature of the main transition by about 6°C for one-proton systems like phosphatidylglycerol, and by about 12°C for two-proton systems such as phosphatidic acid. The transition temperature of the fully protonated state is always higher than that of the deprotonated one, which corresponds to the electrostatic theory (Träuble *et al.*, 1976). Third, the partly deprotonated or partly protonated state is characterized by a transition temperature far above the two other states. This contradicts the electrostatic theory and must be explained by hydrogen bond stabilization.

The Effect of Monovalent Cations

In their interaction with negatively charged phospholipid–water dispersions, monovalent cations interfere with proton binding in those pH regions where small variations in the proton concentration result in large changes of the temperature of the main transition. These pH regions are characterized by the degree of dissociation α. They are sensitive to monovalent cations for $\alpha \sim 0.5$ in a one-proton system such as phosphatidylglycerol, for $\alpha_2 \sim 0.5$ in a two-proton system (phosphatidic acid), and for $\alpha_n \sim 0.5$ for a n proton system. Monovalent cations at physiological concentrations scarcely bind to negatively charged lipids. However, they can strongly influence the binding characteristic of protons to lipid membranes. This can be demonstrated for 1,2-dimyristoyl-*sn*-glycerol-3-phosphoric acid methylester by comparison of the p*K* values of the phosphate group in the presence of sodium chloride at different molar concentrations (Träuble *et al.*, 1976). p*K* values about 5.6 were found for 3×10^{-4} *M* and about 2.4 for 2×10^{-1} *M* salt solutions, showing that with increasing salt concentration the apparent p*K* of the lipid–water dispersion is shifted towards the true pK_1 of phosphoric acid of about 2.0. The qualitative explanation is that additional electrostatic work is needed to dissociate protons from a negatively charged surface. An important consequence of this effect is that the lamellar gel to liquid–crystalline transition in negatively charged phospholipid membranes may be triggered by an increase in salt concentration, when $\alpha_n \sim 0.5$. The partly protonated state, for instance, 1,2-dihexadecyl-*sn*-glycero-3-phosphate at $\alpha_2 = 0.5$ (pH 10, cf. Fig. 2), is destabilized at constant pH and constant temperature by an increase of sodium chloride concentration, because of a shift of α, from $\alpha_2 = 0.5$ to $\alpha_2 = 1.0$. The release of protons from the membrane surface leads to the expected decrease in the temperature of the main transition, $\Delta T_M = 25°C$, as described by Träuble and Eibl (1975).

A comparison of the phase behavior of 1,2-dihexadecyl-*sn*-glycero-3-phosphoethanolamine with the respective phosphocholine at pH 13 shows certain differences. Both lipids have a main transition temperature of about 44°C, but the phosphatidylethanolamine lacks the pretransition at 39°C. This may be explained by the difference in surface charge; phosphatidylcholines have a zwitterionic structure caused by the presence of one negatively charged phosphate and one positively charged choline group. However, phosphatidylethanolamine at high pH is negatively charged, and the membrane surface is surrounded by a diffuse layer of monovalent cations that are independent of the lipid polar group. If the sodium ion concentration is increased to about 1 *M*, one sodium ion is found in a volume of 1400 $Å^3$, the approximate volume of the lipid molecule (Jähnig *et al.*, 1979).

Under these conditions, a pretransition is then detected within the phosphatidylethanolamine system (Stümpel *et al.*, 1980) as well as in the case of phosphatidic acid at pH 13 (Harlos *et al.*, 1979).

A further increase of the concentration of monovalent cations up to 5–10 *M* results in a strong increase in the temperature of the main transition; this resembles the proton effect and probably indicates binding of monovalent cations (Harlos *et al.*, 1979; Stümpel *et al.*, 1980). An analysis of the properties of negatively charged lipid membranes at high ionic strength is important for practical reasons as well as from a theoretical point of view because halophilic bacteria do grow at high salt concentrations. However, systematic studies of negatively charged phospholipids at 3 or 5 *M* concentrations of NaCl are not available at this time.

In summary, monovalent cations can influence the phase behavior of negatively charged phospholipid membranes in two ways. First, they can interact with partly protonated systems at $\alpha_n \sim 0.5$, which results in an increase of the degree of dissociation α and in the expected decrease of the temperature of the main transition. Second, they can interact with negatively charged phospholipids in the deprotonated state at $\alpha_n = 1$; at low salt, a small increase of the temperature of transition is observed with increasing salt concentration; at molar concentrations of salt, a pretransition is induced; and at high salt, a strong increase in the temperature of the main transition indicates binding of monovalent cations.

Concluding Remarks

A detailed understanding of the specific interactions between phospholipids and proteins, the main constituents of biomembranes, requires simple model systems, because biological membranes are complex in structure and function. Phospholipid-dependent membrane proteins can be extracted from natural sources and, after purification, may be reconstituted within a defined phospholipid environment. Questions may then be asked, such as whether an alteration of charge or a phase transformation of the lipid matrix can regulate biologically active proteins.

For instance, the influence of different lipid states on carrier-induced ion transport was studied in black film experiments (Boheim *et al.*, 1980). Sodium ion transport in phosphatidylcholine membranes was active only in the liquid–crystalline but not in the gel state. Spontaneous current fluctuations were also recorded in the region of the gel to liquid–crystalline transition in the absence of the carrier.

Mixtures of phospholipids are necessary for the proper functioning of certain phospholipid-dependent enzymes. D-β-Hydroxybutyrate apodehydrogenase has a specific requirement for phosphatidylcholine (Sekuzu *et al.*, 1963), but the natural mixture of phospholipids of the mitochondrial inner membrane gives better maximal activity than does phosphatidylcholine alone (Nielsen and Fleischer, 1973). The optimal reactivation of the enzyme activity was mimicked by synthetic mixtures of phosphatidylcholine, phosphatidylethanolamine, and negatively charged phospholipids, even without the use of unsaturation in the fatty acid chains (Eibl *et al.*, 1982). However, the liquid–crystalline state of the lipid mixture was required for reactivation, the gel state being ineffective (Churchill *et al.*, 1983).

The results on the influence of protons on the phase behavior of phospholipids indicate that the aggregation and fusion events of phosphatidic acid dispersions triggered by calcium ions (Verkleij and de Gier, 1981) may also be induced in the absence of divalent cations by small variations of the proton concentration.

The amphiphilic structure of phospholipid molecules, with apolar and polar elements, endows them with ideal properties for the formation of biological interfaces. Structural variation of the apolar part will influence the pretransition, which is only found in 1,2-diacyl systems, as well as the temperature of the main transition and the temperature of the transformation from the lamellar to the hexagonal phase. Modification of the polar region will not effectively change the physical properties of bilayer systems with the exception that protons become structural elements of the membrane surface. Then, the degree of dissociation α, the state of protonation, determines the lipid structure. The resulting partly protonated states of the membrane differ from the unprotonated ones by the temperatures of the transitions and by the ability to form hexagonal phases.

The experimental results on phospholipid systems collected in the last decade have provided an understanding of the physical properties of phospholipid–water systems on a molecular level. The different phases observed—lamellar structures in their different states of fluidity, hexagonal, and other nonlamellar phases—are the results of a sensitive interplay between internal structural elements, the apolar and polar regions, and their interactions with monovalent and divalent cations in the aqueous environment.

Acknowledgments

Many thanks to Dr. John M. Seddon for helpful discussions and for carefully reading the manuscript. The author is grateful to Dr. Roland C. Aloia for his continuous stimulation.

References

Blume, A., and Eibl, H. (1979). *Biochim. Biophys. Acta* **558,** 13–21.

Boheim, G., Hanke, W., and Eibl, H. (1980). *Proc. Natl. Acad. Sci. U.S.A.* **77,** 3403–3407.

Cevc, G., Watts, A., and Marsh, D. (1981). *Biochemistry* **20,** 4955–4965.

Chapman, D., and Wallach, D. F. H. (1968). *In* "Biological Membranes" (D. Chapman, ed.), pp. 125–202. Academic Press, New York.

Chen, S. C., Sturtevant, J. M., and Gaffney, B. J. (1980). *Proc. Natl. Acad. Sci. U.S.A.* **77,** 5060–5063.

Churchill, P., McIntyre, J. O., Eibl, H., and Fleischer, S. (1983). *J. Biol. Chem.* **258,** 208–214.

Cullis, P. R., and de Kruijff, B. (1976). *Biochim. Biophys. Acta* **436,** 523–540.

Deamer, D. W., Leonard, R., Tardieu, A., and Branton, D. (1970). *Biochim. Biophys. Acta* **219,** 47–60.

Eibl, H. (1977). *In* "Polyunsaturated Fatty Acids" (W. Kunau and R. T. Holman, eds.), pp. 229–244. Am. Oil Chem. Soc., Champaign, Illinois.

Eibl, H., and Blume, A. (1979). *Biochim. Biophys. Acta* **553,** 476–488.

Eibl, H., and Woolley, P. (1979). *Biophys. Chem.* **10,** 261–271.

Eibl, H., Churchill, P., McIntyre, J. O., and Fleischer, S. (1982). *Biochem. Int.* **4,** 551–557.

Füldner, H. H. (1981). *Biochemistry* **20,** 5707–5710.

Harlos, K., and Eibl, H. (1980a). *Biochemistry* **19,** 895–899.

Harlos, K., and Eibl, H. (1980b). *Biochim. Biophys. Acta* **601,** 113–122.

Harlos, K., and Eibl, H. (1981). *Biochemistry* **20,** 2888–2892.

Harlos, K., Stümpel, J., and Eibl, H. (1979). *Biochim. Biophys. Acta* **555,** 409–416.

Harlos, K., Eibl, H., Pascher, J., and Sundell, L. (1983). *Chem. Phys. Lipids* (in press).

Hope, M. J., and Cullis, P. R. (1980). *Biochem. Biophys. Res. Commun.* **92,** 846–852.

Jähnig, F., Harlos, K., Vogel, H., and Eibl, H. (1979). *Biochemistry* **18,** 1459–1468.

Jain, M. K. (1972). "The Biomolecular Lipid Membrane: A System." Van Nostrand-Reinhold, Princeton, New Jersey.

Keough, K. M. W., and Davis, P. J. (1979). *Biochemistry* **18,** 1453–1459.

King, R. J., and Clements, J. A. (1972). *Am. J. Physiol.* **223,** 715–726.

Nielsen, N. C., and Fleischer, S. (1973). *J. Biol. Chem.* **248,** 2549–2555.

Phillips, M. C., Williams, R. M., and Chapman, D. (1969). *Chem. Phys. Lipids* **3,** 234–244.

Ruocco, M. J., and Shipley, G. G. (1982). *Biochim. Biophys. Acta* **684,** 59–66.

Sekuzu, I., Jurtshik, P., Jr., and Green, D. E. (1963). *J. Biol. Chem.* **238,** 975–982.

Stümpel, J., Harlos, K., and Eibl, H. (1980). *Biochim. Biophys. Acta* **599,** 464–472.

Stümpel, J., Nicksch, A., and Eibl, H. (1981). *Biochemistry* **20,** 662–665.

Stümpel, J., Eibl, H., and Nicksch, A. (1983). *Biochim. Biophys. Acta* **727,** 246–254.

Träuble, H., and Eibl, H. (1974). *Proc. Natl. Acad. Sci. U.S.A.* **71,** 214–219.

Träuble, H., and Eibl, H. (1975). *In* "Functional Linkage in Biomolecular Systems" (F. O. Schmitt, D. M. Schneider, and D. M. Crothers, eds.), pp. 59–101. Raven Press, New York.

Träuble, H., and Haynes, D. H. (1971). *Chem. Phys. Lipids* **7,** 324–355.

Träuble, H., Eibl, H., and Sawada, H. (1974). *Naturwissenschaften* **61,** 344–354.

Träuble, H., Teubner, M., Woolley, P., and Eibl, H. (1976). *Biophys. Chem.* **4,** 319–342.

Verkleij, A. J., and de Gier, J. (1981). *In* "Liposomes from Physical Structure to Therapeutic Applications" (C. G. Knight, ed.), pp. 83–100. Elsevier/North-Holland Biomedical Press, Amsterdam.

Watts, A., Harlos, K., and Marsh, D. (1981). *Biochim. Biophys. Acta* **645,** 91–96.

Chapter **9**

Membrane Fluidity and Cytoplasmic Viscosity

Alec D. Keith and Andrea M. Mastro

Introduction

Some knowledge and viewpoints about the membrane boundary of cells date back to the late nineteenth century. The work of Overton (1899) is probably the most clearly defined of the early investigators. Overton qualified the cell boundary as having thin film properties and selective permeability, and showed that plant root cells would maintain their size and shape properties in solutions of about 7% sucrose concentration although higher concentrations resulted in cellular plasmolysis. By using 7% sucrose plus a percentage of some additional compound, he could measure the time dependency of recovery from plasmolysis and in this way establish a rank order of permeability for a wide variety of molecular species, which he did for several thousand.

Membrane Fluidity in Biology, Vol. 2
General Principles

ISBN 0-12-053002-3

Another aspect of his work implied the interaction between the cell boundary and the cytoplasmic structure. It showed the elasticity of the cell cytoplasm by the discovery that during plasmolysis, the cell boundary separates from the cell wall although visible cytoplasmic structures are maintained at approximately the same relationship to each other even when they are forced into a smaller volume.

Development of the Cytoplasmic Theory

The concept of the fluidity of the cytoplasm was implicit in the earliest development of the general cell theory. The discoverers of protozoans witnessed cytoplasmic streaming, the movement of small particles through protozoans, and also saw formation of pseudopodia. The latter implied a connective network between the cell membrane and cytoplasm, because deformation of either the cytoplasm or membrane surface caused overall changes in cell shape. They also noted that the material near the cell wall appeared to flow more slowly than that nearer the cell interior, although no visible membrane separated them; this also suggested a physical connection between membranes and cytoplasm. Elementary biology texts still refer to these two areas of cytoplasm as the ectoplasm and the endoplasm.

Some of the comments of these early cell biologists were collected by Conklin (1940) in the history of "The Cell and Protoplasm" and are repeated here. Wolff (in 1759) wrote that "Every organ is composed at first of a little mass of clear, viscous, nutritive fluid, which possesses no [visible] organization of any kind but is at most composed of globules." The protoplasm was described by Schleiden as slime, mucous, mucilage; Nägelé described it as a mixture of gum, sugar, and proteid. Felix Dujardin coined the term "sarcode" for the "living jelly" of animal cells. Sarcode was "substance glutinous, perfectly homogeneous, elastic, contractile, without any trace of organization, fibrous, membranes or appearance of cellulosity." Purkinge used the word "protoplasm" to describe the "jelly" of animal cells.

Protoplasmic streaming was first seen in and described in plant cells by Corti; Treviranus again described it in 1807. Meyen thought the fluid moved similarly to the planets around the sun. A scientific breakthrough occurred when Cohn (1848) and DeBarry (1849) proposed that animal cell sarcode and plant cell protoplasm were the same. In 1892 Hertwig formulated a "protoplasm theory" which went so far as to describe the protoplasm as "life" itself.

Finally, as the century moved on, scientists realized that protoplasm had a nucleus distinct from the rest of the cytoplasm. Many scientists of the second half of the nineteenth century studied fertilized eggs and described changes in the nucleus and in the cytoplasm. Gradually, separate structures were identified in the nucleus and cytoplasm: mitotic spindles, chromosomes, mitochondria, and so on. Altmann (1890) regarded mitochondria as similar to bacteria. Some organelles were not seen but were believed to exist "logically" as a smallest unit of cytoplasm: . . . "gemmules, plastidules, pangenes, plasomes, ideoblasts." As Conklin (1940) says,

> gradually the view developed that protoplasm is not merely a combination of many of the most complex chemical compounds that are known, but that it is a complicated organization. . . . In this protoplasmic organization the constituent parts that were first recognized were the nucleus and its surrounding material; the latter was named "cytoplasm" by Kölliker, in 1862.

As microscopic techniques improved it was clear that the cell "protoplasm" contained organelles. The largest was the nucleus; the rest was cytoplasm or protoplasm. With advances in light and electronmicroscopy many more organelles in the cytoplasm, such as mitochondria, lysozomes, and vacuoles, became visible.

Advances in biochemistry allowed organelles to be separated from the rest of the cell. However, these fractionation techniques still left the amorphous ground substance undefined and unfractionable. The 100,000 × g supernatant of a cell homogenate was probably the closest thing to the cytoplasm. Little was really known about the "substance of life."

Recently, a major symposium "Organization of the Cytoplasm" was held with the purpose of addressing the topic of cytoplasm as a holistic entity (Brinkley, 1981a). It is interesting to compare this symposium with one held in 1939 to commemorate the knowledge of the protoplasmic unit of living things collected during the 100 years since the formulation of the cell theory by Schleiden and Schwann (Moulton, 1940). Sixteen papers were presented in 1939; their titles covered such areas as "Molecular Structure in Protoplasm," "Protoplasm and Colloids," and "Cellular Differentiation and Internal Environment." In contrast, although the 1981 conference had 93 presentations, some of the titles were relatively unchanged: "Structural Organization of the Cytoplasm," "Organizing the Cytoplasm for Motility," and "Cellular and Molecular Aspects of the Amoeboid Movement." Many of the 1981 papers contained the names of organelles and structures unnamed, although not totally unknown, in 1939. Structures such as *microtubules, microfilaments*, and *intermediate filaments* were beyond the limits of the techniques but not the scientific imagination of the scientists of the 1930s. For example, in the 1939 symposium a very interesting discussion of cytoplasm by Sponsler (1940) was entitled "Molecular Structure in the Protoplasm." He combined what was known at the time about molecular and microscopic properties of cytoplasm into a workable model. He felt that the protoplasm could only function if "it were organized structurally along some defined plan. . . . The conception of activities demands a dynamic, a changing, moving, active structure." He knew that the cytoplasm was mostly water and protein, and he also knew that small particles existed in the cytoplasm which were below the limits of microscopic resolution. Combining this information with the then recent discovery regarding protein structure, he "indulged in wild speculation" and concluded that "the particles which may be seen as diffractions spots or visible light in the ultramicroscope are composed actually of these protein chains in various arrangements." He described how the protein chains could join with each other and with organelles in a loosely organized network or aggregation. This network, likened to a sponge, allowed for both the organization and movement required by the cell cytoplasm.

As pointed out in the 1981 conference (Brinkley, 1981b), Porter, a pioneer cell biologist–electronmicroscopist, predicted in 1961 that the "structureless medium in which are suspended all the resolvable elements of the cytoplasm will, in time, be shown to contain complex organizations of macromolecules."

Here lies the crux of the issue. Fluidity of the cytoplasm cannot be understood without understanding the cystoplasmic organization. One cannot describe viscosity and fusion within the cytoplasm well without knowledge of size and spacing of barriers (various organelles) or without an understanding of restrictions on movement placed by affinities of particles for organelles. It is also obvious that movements depend on the intrinsic property of the particle in question, such as its size, shape, and charge.

Development of Membrane Theory

As stated earlier, organized studies on biological membranes began just before the turn of the century. Most of the landmarks of the first half of the twentieth century concerned membrane structure and conveyed to most readers the concept of a static structure. Gorter and Grendel (1925) presented the first evidence of a phospholipid bilayer, for red blood cell membranes, implying that all membranes may have the same general structure. Soon after this, Plowe (1931) showed the mechanical elasticity of plant membrane–cytoplasm complexes. Danielli and Davson (1935) presented electrophysiology work that allowed construction of a membrane model. This

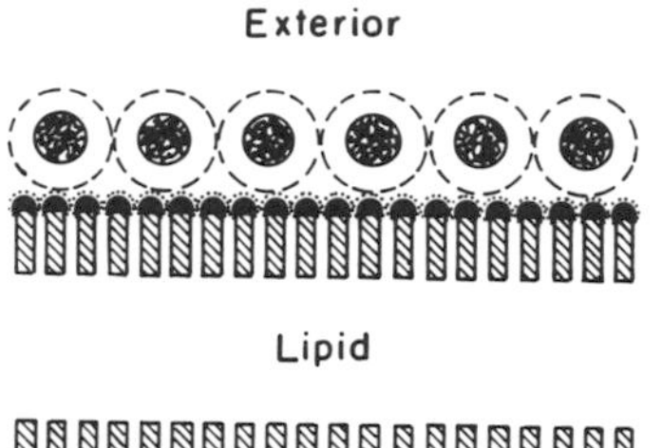
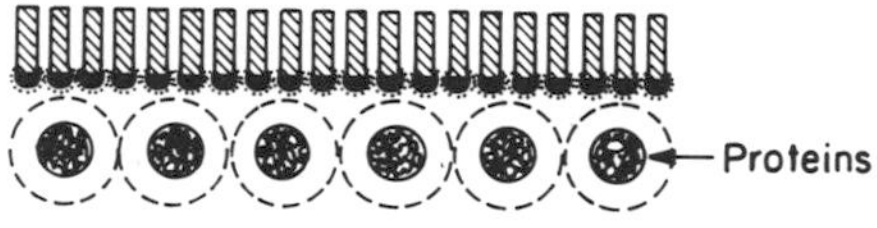

Fig. 1. The Danielli–Davson membrane model. This membrane model shows a lipid bilayer of undefined dimensions because these authors felt that the phospholipid bilayer interior contained an unknown thickness of nonpolar lipids. The phospholipid-to-protein size relation is rather small for the proteins, but the general features are similar to those of the Singer–Nicholson model currently accepted.

model, shown in Fig. 1, incorporated the Gorter–Grendel bilayer (although no credit was given) in the general model shown. Also in the 1930s, the first major papers employing the technique of X-ray diffraction were presented. Schmitt *et al.* (1935) worked with myelinated nerves from frogs and presented evidence to suggest a structure of two membranes pressed tightly up against each other. We now recognize their presentation to represent two membranes in close apposition.

Some years later, Robertson (1959) and others using electronmicroscopy established the general dimensions of biological membranes and the generalization that all biological membranes are probably similar.

The papers presented up to this point in history concentrated on a static membrane structure. Essentially nothing was known or had been contributed about the structure of membranes with regard to time, the movement of membrane molecules, and movement of molecules through membranes. It was not until the late 1960s that the concept of viscosity or fluidity was brought into the membrane literature (Keith *et al.*, 1968; Hubbell and McConnell, 1968). From that time, the membrane was regarded as a dynamic structure with considerable molecular movement, molecular exchange, permeability based on membrane dynamics and the dynamics of lipid–protein interactions. Dynamics were brought into the consideration of functional biological membranes. These concepts were collected together in a general model known as the Singer–Nicholson "Fluid Mosaic" Membrane Model (Singer and Nicolson, 1972).

Cytoplasm

Components of the Cytoplasm

Diffusion of a material (solute) refers to its movement through a solvent. In the case of cells, the solvent is water, which comprises more than 90% of the cell and is the major component of cytoplasm. Dissolved in this aqueous medium are numerous molecules and organelles of various sizes; these are all contained in the cellular cytoplasm. Visible at the light or electronmicroscope level are large organelles such as mitochondria, lysosomes, ribosomes on the endoplasmic reticulum or other membranes, and secretory granules as well as various other small organelles. More recent technical improvements have revealed that the remaining ground substance itself has many structural elements; the cytoplasm has a cystoskeleton.

One component of the cytoskeleton is the microtubular network. Microtubules, primarily bundles of tubulin subunits and associated proteins, have been found in the cytoplasm of almost all cells (Goldman *et al.*, 1976). Furthermore, these structures are not restricted to the mitotic spindle. Microtubules seem to play a role in motility and also shape, secretion, and other specialized functions of cells. They exist as bundles of tubulin in a hollow tubule about 250 Å outer diameter and 150 Å inner diameter. Their appearance and disappearance in cells seem to be controlled by microtubule organizing centers, which apparenlty provide direction as well as rate of growth of the polymers. Microtubules also have associated proteins (MAPS) which copurify with them. These proteins may be important in microtubule association with other parts of the cytoskeleton or with other organelles such as plasma membrane. For example, the protein "ankyrin" has recently been described by Branton *et al.* (1981), as a good candidate for a microtubleassociated protein that could link the microtuble system with the proteins of the membranes of red blood cells.

The second class of prominent cytoskeletal proteins is the actin system. Known as the major protein of muscle cells, actin has also been found in virtually all nonmuscle cells (Goldman *et al.*, 1976). In these cells it is usually seen as fibrillar actin (F actin) about 50 Å in diameter although the globular (G actin) is also found. The actin fibers are often seen in bundles arranged parallel with the plane of the plasma membrane. It is apparently the ability of actin to undergo sol–gel transformation that accounts for the increased viscosity of this subcortical cytoplasm (Pollard *et al.*, 1981). Actin is also the main component of the stress fibers laid down by fibroblasts cells in culture; it comprises the core of microvilli and filopodia.

Assembly of F-actin filaments, like that of microtubules, appears to be directed. There are several models to account for the polarity of organization of unit addition and of control size, but it is still unclear exactly how filamentation proceeds or is controlled *in vivo*.

The second-order assembly of F actin into bundles is not totally understood. Various actin networks such as those found in microvilli have actin associated proteins. How or if these control assembly is not known. Actin has been described as having at least 20 actin-associated proteins. Some may be important for the actin-bundle assembly–disassembly. Others may themselves use actin as a framework for organization in a cell.

The third major group of fibers in the cytoskeleton is the intermediate filaments. They are 100 Å in diameter and are found in a variety of cell types. At the protein level they are a heterogeneous family and have been subclassified on the basis of electronmicroscopic and immunofluorescent studies with various cell types (Osborn *et al.*, 1981). The general classes of intermediate filaments are α-keratins (epithelial cells), vimentins (mesenchyme cells), desmins (muscle cells), neurofilaments (neurons), and glial filaments (glial cells). Little is known about the control assembly of these fibers but there is some evidence that, as do the other fibers, they have organizing centers for polymerization. The function of the intermediate filament network is also unclear at this time. However, it is often found in association with the microtubular network and it has been suggested that their interactions may be an important control mechanism *in vivo* (Liem *et al.*, 1981; Runge and Williams, 1981; Wiche *et al.*, 1981).

In fact, it would not seem unreasonable that all cytoskeletal elements interact in such a way as to provide organization to the cytoplasm as well as to provide the bones and muscle for movement and shape changes. One can imagine that this network serves as a focus for "cytoplasmic" proteins, especially those which are part of one metabolic pathway.

More recently Porter and his collaborators have called to our attention just such a structure of the cytoplasm (Wolosewick and Porter, 1979a,b). With high voltage electronmicroscopy they have visualized a fine, cytoplasmic, three-dimensional network of fibers which they have labelled microtrabeculae. This microtrabecular lattice (MTL) consists of protein-rich fibers ranging from 2 to 3 nm to as much as 10 nm in diameter. The fibers are not actin, microtubules, or intermediate filaments, but they can cross-link or associate with these other fibers. The fibers appear to vary from a few to more than 100 nm in length. Accordingly, the interstitial spaces vary from 50 to 150 nm across. The authors suggest that this lattice can deform and reform at different places in the cell as the cell moves and divides. Small molecules presumably could move easily within the water-rich spaces of the lattice. Large organelles appear to be associated with the network. Large molecules would be restrained from total random diffusion by the presence of the network. The lattice is reminiscent of the framework hypothesis of cytoplasmic ground substance as described by Frey-Wyssling (1953). Schliwa *et al.* (1981a,b) have used a detergent extraction technique to obtain high voltage electronmicrographs which support this picture. However, these results are not accepted by everyone (Heuser and Kirschner, 1980). Although artifacts of preparation have been suggested as the origin of the microtrabecular lattice, Wolosewik and Porter (1979b) have presented convincing evidence for its existence. Its dynamic picture is, of course, revealed in a photograph that is fixed in time.

Physical State of Cytoplasm

The general consideration of sol–gel states of cytoplasm imposed by the cytoskeletal structure has been discussed for many years. It is interesting that the significance of a sol or a gel state is not always obvious, although it is very important to bulk phase movement. Convection processes proceed well in the sol state but not in the gel state. Diffusional processes occur at approximately the same rates in both gels and sols; however, if convection occurs in sol states, it will overwhelm the diffusional processes so that diffusion will not readily be detectable by conventional methods.

Cell size is an important consideration. Cells of the size of bacteria or viruses have molecular diffusion processes that occur rapidly with respect to cell size because of the small cell radii. Higher organisms may have cell structures that would result in longer times for diffusion to occur across a cell from boundary to boundary or boundary to center. For example, when comparing a 1-μm diameter bacterium to a 20-μm diameter mammalian cell, the relative diffusion time from boundary to center will be extraordinarily different.

Over several years the view of the cytoplasm has ranged from one extreme to the other. It has been viewed as a structureless ground substance, or a broth containing all of the enzymes of intermediary metabolism apparently meeting by chance, or a network in which all components are stabilized into a framework with a high degree of organization. Notably the first view now lacks explanation for organization; the second provides little sense of flexibility for division, movement, shape changes, etc. The real structure probably lies somewhere in between. Any model must allow for structure–function relationship.

Approaches other than microscopy have been used to examine the relationship between cytoplasmic structure and function. For example, Clegg

(1981) applied centrifugal stratification to artemia cell systems and examined the metabolic activity of enzymes associated with the soluble cytoplasm (100,000 × *g* supernatant of homogenate). Although this treatment changed the gross ultrastructural arrangements within the cells, the enzymes required for the metabolic activity seemed not to have been changed in functional association. Thus, he concluded that the enzymes are associated in a fixed way which allows them to withstand centrifugal forces.

Lasek and Brady (1981) chose to examine a specialized cell, the neuron, which has a natural separation of cytoplasm. Cytoplasm in the cell body is where protein synthesis occurs. There is also cytoplasm in the axon where no protein synthesis occurs. Thus, proteins and enzymes must be transported from the body to the axoplasm. The investigators suggest that part of the segregation and movement of components results from a selective affinity of the proteins to a cytoskeletal or filamentous network.

It is clear that attempts to measure cytoplasmic fluidity or diffusion of molecules in the cytoplasm must allow for structural impediments in the cell. These properties of cells are important in interpreting the results of measurements of movement in the cytoplasm. The numerous approaches of attempting to define cell fine structure can lead to predictions as to how things will move in the cell. On the other hand, measurement of diffusion in cytoplasm can help determine the correct structural models.

Measurements of Diffusion in Cells

During the past 50 or 60 years, several laboratories have performed experiments to directly measure the diffusion of particles in the cell cytoplasm. Basically, particles were placed in the cytoplasm and their movements were microscopically observed. All of these measurements were at the macroscopic level, but various workers (of the past decade) have attempted limited deductions about the physical properties of the cytoplasm from the movements of particles. In a relatively thorough study, Crick and Hughes (1950) used a magnetic particle method to measure the viscosity of chick embryo fibroblasts grown in culture. In their experiments, cells were allowed to phagocytose particles of magnetite or finely divided iron particles and were then subjected to a magnetic field. Twisting or dragging of the particles was observed under a phase contrast microscope. They summarized their findings by describing the protoplasm as not only viscous but also elastic. (The modulus of rigidity is 10^2 dynes/cm^2.) They suggested that there was evidence of thixotropic behavior.

Crick and Hughes were well aware of the limitations of their work.

". . . Our technique is not sufficiently precise to enable us to say whether the inside of the chick cell is uniform in its physical properties." They also discussed whether their studies revealed information about the structure of the cytoplasm and considered examining movement of particles in model solutions.

A discussion of the relation of their findings to the main theories of cytoplasmic structure of the 1930s and 1940s is very interesting. The two main theories, "bush heap" and "framework," suggested an underlying structure in the cytoplasm that was beyond the limits of microscopic detection. The main difference appeared to be the order and permanence of the association of the components of these structures. Frey-Wyssling (1953) described the cytoplasmic structure as one of ". . . wonderful coordination. The framework cannot represent an unordered pile but must possess an organized and well defined structure". Crick and Hughes (1950) felt that structure need not imply a "cytoskeleton" in terms of a fixed and rigid structure; they likened the cytoplasm to "mother's work basket—a jumble of beads and buttons of all shapes and sizes, with pins and threads for good measure, all jostling about and held together by 'colloidial forces.'" They go on to say that the structure may be "a rather nonspecific, transient affair for a normal cell". Thus, this aspect of the general discussion has not changed very much over the past 35 years.

Fluorescence Polarization

Fluorescence polarization was used by Cercek *et al.* (1978) to look at the rotational motion of fluorescein molecules in the cytoplasm of numerous cells. In this technique, the state of the fluorescein molecule is determined by the properties of its microenvironment. The local pH, dielectric constant, polarity, and viscosity can determine the exitation and emission spectra of each molecule, and the average signal from a cell gives an indication of the heterogeneity of the cytoplasm. The viscosity of the cytoplasm is estimated from the amount of rotation of a fluorescein molecule between the time of absorption and emission of light. Fluorescence polarization is inverse to the rotational rate of a probe molecule. The greater the viscosity, the less the rotation and the greater the fluorescence. The motion of the molecule being determined is relevant only to the local viscosity within a few angstroms of the rotating probe molecule.

Using this approach, Cercek *et al.* measured numerous cells in various physiological states (i.e., growth versus resting) and found that there was a change in the average fluorescence polarization of the molecules as the cells moved through various phases of their cycle. Obviously the picture is an average of all the probe molecules in the cell.

In another experiment, yeast cells were centrifuged to change the cytoplasmic matrix before addition of the fluorescein probe (Cercek and Cercek, 1974). In this case the fluorescence was also changed. These studies suggest that the complexity of the cytoplasm is greater than a homogeneous aqueous solution; however, they give no quantitative numbers for viscosity.

Ultralow Temperature Autoradiography

The microinjection of radioactive probes and low temperature autoradiography were combined in a study designed to determine the cytoplasmic diffusion coefficient (Paine *et al.*, 1975). ^{3}H-dextrans with hydrodynamic radii of 12, 23, and 36 Å were found to have D_c of 3.5×10^{-7}, 2.5×10^{-7}, and 1.5×10^{-7} cm^2/sec, respectively, in the cytoplasm of oocytes. Inulin, a protein of 5,500 MW and sucrose (342 MW) were found to diffuse at 2×10^6 and 3×10^7 cm^2/sec. In general, tracers of these sizes moved at approximately 0.1 to 0.4 of the diffusion coefficient found in bulk phase water.

Self-Diffusion Coefficient of Water

Although diffusion usually refers to the movement of particles in a cytoplasmic solvent (i.e., water) the self-diffusion of water is also the subject of much study. The diffusion properties of water are intimately associated with the physical properties of water. This topic is beyond the scope of this discussion but is discussed elsewhere (Beall and Hazlewood, 1983; Beall *et al.*, 1981; Ling, 1979). However, the application of post field gradient nuclear magnetic resonance (NMR) techniques to the study of water in living cells has led many investigators to report that the self-diffusion coefficient of water in cytoplasm is only about one-half that found in bulk phase water (Beall and Hazlewood, 1983). Furthermore, this diffusion coefficient varies within the same cells as they traverse the cell cycle (Beall *et al.*, 1976) and among different lines of normal and neoplastically transformed cells (Beall *et al.*, 1981).

Fluorescence Recovery after Bleaching

Wojcieszyn *et al.* (1981) applied the method of fluorescence recovery after photo bleaching (FRAP) to measure translational diffusion of relatively large proteins (IgG and bovine serum albumin) in the cytoplasm of cultured human fibroblasts. They microinjected the molecules into cells via a red blood cell ghost microinjection procedure, and then followed the return of fluorescence in the defined area of the cytoplasm which had been bleached by a laser beam. They obtained diffusion coefficients (D) approximately 10^{-8} cm^2/sec at 22°C. Bovine serum albumin behaved as if it were in a solution with a viscosity of 61% sucrose. It diffused about 70-fold slower than in a simple buffer. At 5°C, the diffusion was only 20% of that observed at 37°C. Interestingly, treatment of cells with colchicine reversed the reduction in viscosity seen at 5°C, which suggests that cell cytoskeletal barriers impede diffusion and contribute to apparent cytoplasmic viscosity.

Electron Spin Resonance (ESR)

Spin-labeled molecules have been used to characterize the dynamics of membranes (Wu and McConnell, 1975; Smith, 1972). More recently ESR techniques have been used to examine cytoplasmic viscosity, and can be used to measure diffusion for approximately 20–500 Å. These measurements are independent of bulk flow. Furthermore the sensitivity of the technique allows determination of concentrations as low as 10^{-7}–10^{-4} M of the probe, depending on the sample. The probes are small molecules of about 300 MW; therefore, it is possible to follow the movement of these compounds in the aqueous cytoplasm. Two examples of commonly used probes are shown in Fig. 2. Charged spin-labeled molecules can also be used to estimate sequestering within a local environment or charge–charge interaction with cytoskeletal elements.

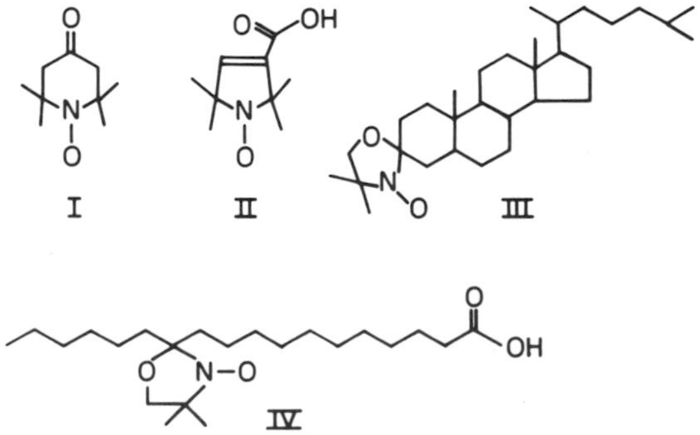

Fig. 2. Spin labels treated in the present monograph. I. 2,2,6,6-tetramethylpiperidone (TEMPONE); II. 2,2,5,5-tetramethylpyrrolinecarboxylic acid (PCAOL); III. the dimethyloxazolidine nitroxide of cholestane located at the 3 position of the A ring; IV. the 12-dimethyloxazolidine derivative of stearic acid.

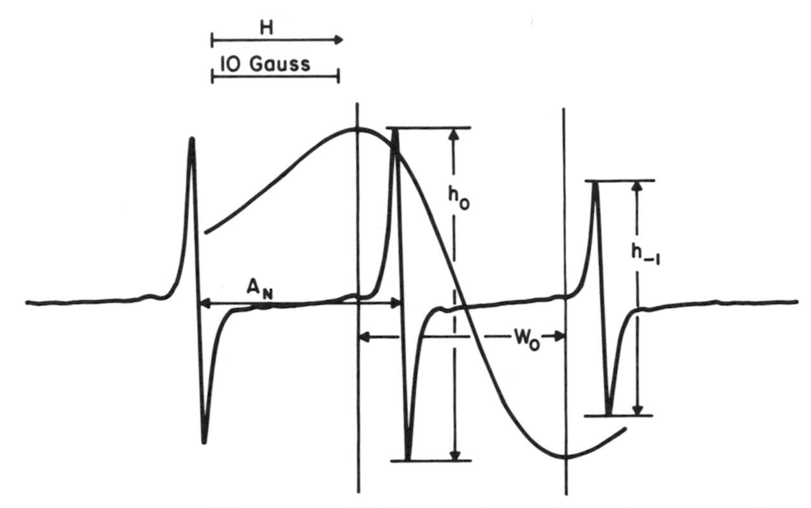

Fig. 3. An EPR signal from a spin label in aqueous medium. This spectrum shows the general spectral features and the measurements that are routinely taken to calculate rotational correlation time, diffusion coefficient, and hyperfine splitting. The midline first-derivative line height is h_0, h_{-1} is the high-field first-derivative line height, and W_0 is the midfield line width taken from an expanded spectrum. The nitrogen hyperfine coupling constant (A_N) is a direct measure of the local polarity experienced by the nitroxide molecule.

This laboratory has used spin label applications to estimate cytoplasmic diffusion of various cellular protoplasms (Keith and Snipes, 1974; Keith *et al.*, 1977a; Mastro and Keith, 1981). Various parameters of the signal obtained from spin label probes relate to the physical properties of the microenvironment of the spin label, as is shown in Fig. 3. Specifically, one constant, the hyperfine coupling constant (A_N), shown in the same figure, is sensitive to solvent polarity and is an effective measurement of the local dielectric constant within a few angstroms of the spin label molecule. By comparing the A_N of spin labels in known solvents, one can estimate the polarity of the environment of the probe in small local environments of heterogeneous media.

The rotational motion of spin-labeled molecules about the sum of rotational axes is measured by a derived parameter, rotational correlation time (τ_c). Local viscosity in aqueous media and other restrictions to rotational motion within a few angstroms of the probe can increase the τ_c values in comparison with τ_c values in water. By these approaches, translational motion of spin labels can also be determined.

Rotational correlation time τ_c is defined in bulk phase forms by the Stoke's Einstein equation

$$\tau_c = \frac{4\pi r^3 \eta}{3kT}$$

where η is the viscosity of the solvent, k is Boltzman's constant, and T is in degrees Kelvin. One of the forms for τ_c that is convertible to the Stoke's Einstein form is

$$\tau_c = KW_0\left[\left(\frac{h_0}{h_{-1}}\right)^{1/2} - 1\right]$$

where K is a proportionality constant, W_0 is the midline width, h_0 is the midline first derivative line height, and h_{-1} is the high field first derivative line height.

Translational motion of molecules is similarly treated in a bulk phase equation

$$D = \frac{3kT}{6\pi r \eta}$$

D is a rate constant (coefficient) if the same terms are defined as above. $1/\tau_c$ can be handled in a similar manner, as a rate constant. The equivalent equation for the application of spin label collision frequency to diffusion is

$$D = K_1\frac{\omega_{ex}}{M} = K_2\left[\frac{\Delta H}{M}\right]$$

where K_1 and K_2 are proportionality constants, ω_{ex} is the electron spin exchange frequency, M is the molarity of the probe molecule, and ΔH is the line broadening in frequency units caused by increased collision frequency resulting from electron spin exchange. The useful parameters for taking measurements of ESR spectra are also shown in Fig. 3.

The diffusion coefficient parameter is of value in estimating movement and thereby diffusion of the probe in the cytoplasm over a range of about 20–500 Å average probe spacing. Translational diffusion of small molecules employing spin exchange frequency in intact cells have been presented by Mastro and Keith (1981). When spin labels collide with each other or with other paramagnetic species, there is a loss of signal interest from line broadening, and consequently significant or drastic loss of first derivative line height is observed. The collision frequency of spin labels with each other in isotropic media is directly proportional to the concentration of spin label and the viscosity of the solvent. The addition of barriers to that space limits spin–spin collision frequency and therefore limits spin exchange. Thus the motion of a spin label molecule about allowable rotational axes (τ_c) and translational movement over a few hundred angstroms (D) can be influenced independently but measured simultaneously. For example, in a very viscous solution (e.g., 70% sucrose) both the τ_c and D may be significantly modified compared to these parameters in water, indicating that the diffusion coefficient has decreased and the rotational rate has decreased. In an aqueous solution with rigid barriers, such as polyacrylamide beads with small mesh sizes, the translational motion (D) can be hindered significantly at the same time that the τ_c is relatively unaffected (Keith *et al.*, 1977a). Changing the concentration of the spin label molecule provides additional evidence for the role of local viscosity as being strongly influenced by barriers in the local environment. Measurements of spin–spin exchanges in model environments such as viscous solutions or in confining space provide data for interpretation of diffusion in the cytoplasm. We have used the spin label approach to measure diffusion in mammalian cells (Keith and Snipes, 1974; Mastro and Keith, 1981; Hammerstedt *et al.*, 1979).

In a series of experiments the spin label probes *TEMPONE* or *PCAOL* (Fig. 2) were added to various mammalian cells in culture. Spin labels were also added to model systems such as solutions of various sucrose concentrations or to polyacrylamide beads with different internal pore spaces. The spectra were recorded and the effect of spin label concentrations on line width were calculated. The change in line width (ΔH) is directly proportional to the translational diffusion (Mastro and Keith, 1981). The rotational diffusion coefficients were calculated directly from the ESR spectra.

In the model systems in which space was the confining parameter, the translational diffusion of the molecules was changed without affecting the rotational motion (Keith *et al.*, 1977a). That is, the molecules were free to rotate about their own axes but were limited in translational movement through the beads over the general dimension of on-center cubic lattice spacing between probe molecules. In the high viscosity sucrose solutions, both the rotational and translation motions were decreased. In mammalian cells, the spin labels were reduced in motion for both rotational and translational motion compared to that of bulk water. In most cell lines, spin-label molecules behaved as if they were in 15% sucrose solution. The rotational correlation time was about twice that observed for the same probe in bulk phase water. The diffusion coefficient ranged from 1×10^{-6} to 5×10^{-6} cm^2/sec, depending on the cell line, cell state of growth, and concentration of spin label. Taking into account the molecular weight differences between bovine serum albumin, BSA (68,000 MW), and PCAOL (172 MW), the diffusion coefficients estimated with the ESR technique agree reasonably well with the fluorescein photobleaching method. The apparent diffusion of a molecule such as BSA is somewhat slower than a small probe because of the relative sizes of internal cellular barriers and intrabarrier spaces.

Membranes

Movement of Proteins on Cell Membranes

A couple of years after spectroscopic studies had shown that the phospholipid environment of biological membranes was fundamentally fluid in nature, Frye and Edidin (1970) showed, in an elegant experiment, that proteins could migrate in the plasma membrane of mouse–human heterokaryons. By using fluorescence-labeled antibodies specific for cell surface antigens, it was possible to observe the change from a nonmosaic pattern of labeling shortly after cell fusion to a mosaic or mixed fluorescent labeling after 40 min of incubation. The authors concluded that the surface antigens probably moved in the plasma membrane by a diffusion process indicating a fairly fluid membrane. They assumed that an average fluorescent-labeled antigen had a radius of about 100 Å and the membrane matrix had approximately a 1–2 (average 1.5) poise viscosity. Using these values produces a calculation of a diffusion coefficient of about 3×10^{-9} cm^2/sec. This value is substantially lower than that given for the work (cited earlier) on serum albumin in cell cytoplasm.

The comparison of comparative viscosities of membranes and cytoplasm is far from simple and straightforward. Membrane proteins, including membrane proteins bearing complexed antibodies, may have their movement modified or directed by attachments to the cytoskeleton. Such attachments could readily change an apparent mixing time either to a shorter or to a longer time interval. A complex such as antigen–antibody has part of its surface in contact with membrane lipids and part of its surface at the aqueous interface; therefore, the viscosity experienced by the complex is a composite of both environments. Interactions with neighboring proteins may also modify the migration rate. The potential for exchanging boundary lipids reported by Jost and others (Jost *et al.*, 1973) would also change the migration rate. Certainly, the general consideration that the membrane matrix is adequately fluid to allow lateral protein movement is true and has been borne out in a large number of subsequent experiments (Albrecht-Bühler, 1973; Albrecht-Bühler and Yarnell, 1973) and subsequent work. The movement of surface proteins sometimes is observed to be orderly and perhaps directed, as in the capping phenomenon where surface immunoglobulin molecules gather over the pole of the cell (Taylor *et al.*, 1971).

A review article by Cherry (1979) brought the knowledge and considerations of protein movement in membranes to a sophisticated state; he treats perpendicular and parallel rotational axes as well as translational motion of proteins and membranes.

It was indicated earlier in this chapter that diffusion time into or across higher organism cells would require a much longer time than bacterial cells because of the greater distances involved. A potential compensating mechanism utilizing membranes for two-dimensional diffusion into mammalian cells might shorten the time required to traverse a cell, a condition normally expected of three-dimensional diffusion (Adam and Delbrück, 1968). For example, a medium with a diffusion constant of 10^{-8} cm^2/sec would require about 200 seconds to traverse a sphere of 20 μm, three-dimensional diffusion would require about 6,000 sec and would also have a much lower concentration at the end point because of concentration gradient effects. Solubility and membrane compatibility are important considerations in regard to translocation of solutes within cells.

Spectroscopic Studies of Membranes

The scientific literature during the late 60s and 70s flourished, with numerous papers treating the dynamic nature of phospholipid models of membranes and membrane or cellular preparations directly. During this time the terms *fluidity, dynamic, molecular tumbling, membrane phase transition, membrane phase separation,* and others became well known. The general methods of differential thermal analysis (DTA), differential scanning calorimetry (DSC), electron spin resonance (ESR), spin labels, nuclear magnetic resonance (NMR), and fluorescence spectroscopy were often used to establish these dynamic features.

The general concept of membrane fluidity was established by a large number of qualitative experiments showing that movement (mainly rotational motion) of probe molecules took place in the phospholipid-rich zones of membranes. Documentation of most of the studies employing ESR can be found in reviews (Keith *et al.*, 1973; Wu and McConnell, 1975; Smith, 1972; Griffith and Waggoner, 1969). Rotational motion of molecules or partitioning between membrane hydrocarbon rich zones and aqueous zones only implys that translational motion takes place in membranes.

Very important and sophisticated treatments dealing with translational motion of spin label probes located in biological membranes were published from two separate laboratories in 1971 (Sackmann and Trauble, 1972a,b; Trauble and Sackmann, 1972; Hubbell and McConnell, 1971). These studies, using spin-labeled fatty acids and steroids, showed that such molecules located in membranes laterally diffuse and have molecular collisions with each other. Assignment of diffusion constants to a given membrane or phospholipid preparation from the data obtained from these studies requires the effective rotational radius of the probes used. As the measurements were recorded in terms of line broadening resulting from close encounters between paramagnetic species, an effective radius is the distance over which an electron spin exchange occurs. Errors made in the estimation of this effective radius could possibly modify some of the diffusion constants, but the more important aspect, that of translational diffusion of small molecules and membranes, was effectively established.

Fluorescence spectroscopy has been used extensively to establish the fluidity of the hydrocarbon phase of membranes (Hubbell and McConnell, 1971; Steim, 1965; Tourtellotte *et al.*, 1970; McElhaney, 1982). Implicit in the observation of phase transitions by a biophysical technique is the consideration that the state of freedom of molecular motion is different at temperatures above and below the phase transition. By this implication, thermal analyses have been instrumental in establishing the nature and interactions involved in membrane lipid behavior (Hubbell and McConnell, 1971; Steim, 1965; Tourtellotte *et al.*, 1970; McElhaney, 1982).

Modifying Membrane Fluidity

One realistic and straightforward way of modifying membrane properties of aqueous phospholipid dispersions is to control the cholesterol content of a given preparation. Both the membrane (Demel and de Kruijff, 1976) and the associated interfacial water with phospholipid dispersions is modified by the addition of cholesterol. The general effect on phase transitions with increasing cholesterol content is to obscure phase transition behavior. Probe molecules generally have more motion at lower temperatures with cholesterol added and less motion at temperatures above an expected phase transition; the net effect is to obscure the detection of a phase transition. In preparations of dimyristoyl lecithin containing 40 moles of water per mole of phospholipid, the aqueous regions increased in viscosity as cholesterol was increased (Keith *et al.*, 1977b). An explanation of these effects has been well presented by Demel and de Kruijff (1976), where they suggest that dimyristoyl lecithin ordinarily undergoes a prominent phase transition at about 23°C, from a tightly packed crystalline to a more loosely packed liquid–crystalline state at temperatures above the transition. These authors offer the explanation that intercallation of cholesterol into the phospholipid bilayers causes an intermediate gel state that has no marked temperature sensitivity to form. Their model appears to be consistent with published data.

A second way to modify membrane or phospholipid fluidity is to add lysophosphatides. In one study, the hibernation of ground squirrels was associated with the loss of a phase transition and the concomitant appearance of about 7% lysophosphatides in liver mitochondria membranes (Keith *et al.*, 1975). Subsequent *in vitro* studies showed that the addition of lysophosphatides could obscure observable phase transitions in the same preparations. In other work many of the perturbing influences of lysophosphatides such as lysis of red blood cells, induction of cell fusion, increased permeability, and general effects on membrane bound enzymes were shown (Weltzien, 1979). Lysophosphatides apparently have their main effects on membranes because of their general geometry and detergent-like properties. Lysophosphatidyl choline is capable of forming micelles in aqueous medium.

Amphiphilic molecules that have geometry very different from that of phospholipids may act as membrane or phospholipid perturbers. Molecules with quasispherical geometry possessing a polar group and a bulky nonpolar moiety have been shown by both biological and physical methods of detection to be capable of lowering membrane phase transitions (Eletr and Keith, 1972; Eletr *et al.*, 1974). Other studies have shown that some of these molecules are antiviral for several enveloped viruses (Snipes *et al.*, 1975, 1977, 1979; Snipes and Keith, 1978a,b). The antiviral activity apparently is mediated by effects on membrane phospholipids or hydrophobic binding sites on viral membrane proteins.

References

Adam, G., and Delbrück, M. (1968). *In* "Structural Chemistry and Molecular Biology" (A. Rich and N. Davidson, eds.), pp. 198–215. Wiley, New York.

Albrecht-Bühler, G. (1973). *Exp. Cell Res.* **78,** 67–70.

Albrecht-Bühler, G., and Yarnell, M. M. (1973). *Exp. Cell Res.* **78,** 59–66.

Beall, P. T., and Hazlewood, C. F., eds. (1983). "Biophysics and Physiology of Water in Living Cells." Academic Press, New York (in press).

Beall, P. T., Hazlewood, C. F., and Rao, P. N. (1976). *Science* **192,** 904–907.

Beall, P. T., Asch, B. B., Medina, D., and Hazlewood, C. F. (1981). *In* "The Transformed Cell" (I. L. Cameron and T. B. Pool, eds.), pp. 294–325. Academic Press, New York.

Branton, D., Cohen, C. M., and Tyler, J. M. (1981). *Cell* **24,** 24–32.

Brinkley, B. R., ed. (1981a). *Cold Spring Harbor Symp. Quant. Biol.* **46,** 1.

Brinkley, B. R. (1981b). *Cold Spring Harbor Symp. Quant. Biol.* **46,** 1029–1040.

Cercek, L., and Cercek, B. (1974). *Radiat. Environ. Biophys.* **11,** 209–219.

Cercek, L., Cercek, B., and Ockey, C. H. (1978). *Biophys. J.* **23,** 395–405.

Cherry, R. J. (1979). *Biochim. Biophys. Acta* **559,** 289–327.

Clegg, J. S. (1981). *Cold Spring Harbor Symp. Quant. Biol.* **46,** 23–37.

Conklin, E. G. (1940). *In* "The Cell and Protoplasm" (F. R. Moulton, ed.), pp. 6–19. The Science Press, Washington, D.C.

Crick, F. H. C., and Hughes, A. F. W. (1950). *Exp. Cell Res.* **1,** 37–80.

Danielli, J. F., and Davson, H. (1935). *J. Cell. Physiol.* **5,** 495–508.

Demel, R. A., and de Kruijff, B. (1976). *Biochim. Biophys. Acta* **457,** 109–132.

Eletr, S., and Keith, A. D. (1972). *Proc. Natl. Acad. Sci. U.S.A.* **69,** 1353–1357.

Eletr, S., Williams, M. A., Watkins, T., and Keith, A. D. (1974). *Biochim. Biophys. Acta* **339,** 190–201.

Frey-Wyssling, A. (1953). "Submicroscopic Morphology of Protoplasm." Am. Elsevier, New York.

Frye, L. D., and Edidin, M. (1970). *J. Cell Sci.* **7,** 319–335.

Goldman, R., Pollard, T., and Rosenbaum, J., eds. (1976). "Cold Spring Harbor Conferences on Cell Proliferation," Cell Motil., Vol. 3. Cold Spring Harbor Lab., Cold Spring Harbor, New York.

Gorter, E., and Grendel, F. (1925). *J. Exp. Med.* **41,** 439–443.

Griffith, O. H., and Waggoner, A. S. (1969). *Acc. Chem. Res.* **2,** 17–24.

Hammerstedt, R. H., Keith, A. D., Boltz, R. C., and Todd, P. W. (1979). *Arch. Biochem. Biophys.* **194,** 565–580.

Heuser, J. E., and Kirschner, M. W. (1980). *J. Cell Biol.* **86,** 212–234.

Hubbell, W. L., and McConnell, H. M. (1968). *Proc. Natl. Acad. Sci. U.S.A.* **61,** 12–16.

Hubbell, W. L., and McConnell, H. M. (1971). *J. Am. Chem. Soc.* **93,** 314–326.

Jost, P. C., Brooks, U. J., and Griffith, O. H. (1973). *J. Mol. Biol.* **76,** 313–318.

Keith, A. D., and Snipes, W. (1974). *Science* **183,** 666–668.

Keith, A. D., Waggoner, A. S., and Griffith, O. H. (1968). *Proc. Natl. Acad. Sci. U.S.A.* **61,** 819–826.

Keith, A. D., Sharnoff, M., and Cohn, G. E. (1973). *Biochim. Biophys. Acta* **300,** 379–419.

Keith, A. D., Aloia, R. C., Lyons, J., Snipes, W., and Pengelley, E. T. (1975). *Biochim. Biophys. Acta* **394,** 204–210.

Keith, A. D., Snipes, W. C., Mehlhorn, R. J., and Gunter, T. (1977a). *Biophys. J.* **19,** 205–218.

Keith, A. D., Snipes, W., and Chapman, D. (1977b). *Biochemistry* **16,** 634–641.

Lasek, R. J., and Brady, S. T. (1981). *Cold Spring Harbor Symp. Quant. Biol.* **46,** 113–124.

Liem, R. K. H., Keith, C. H., Leterrier, J. F., Trenkner, E., and Shelanski, M. L. (1981). *Cold Spring Harbor Symp. Quant. Biol.* **46,** 341–350.

Ling, G. N. (1979). *In* "The Aqueous Cytoplasm" (A. D. Keith, ed.), pp. 23–60. Dekker, New York.

McElhaney, R. N. (1982). *Chem. Phys. Lipids* **30,** 229–259.

Mastro, A. M., and Keith, A. D. (1981). *In* "The Transformed Cell" (I. L. Cameron and T. B. Pool, eds.), pp. 327–345. Academic Press, New York.

Moulton, F. R., ed. (1940). "The Cell and Protoplasm." The Science Press, Washington, D.C.

Osborn, M., Geisler, N., Shaw, G., Sharp, G., and Weber, K. (1981). *Cold Spring Harbor Symp. Quant. Biol.* **46,** 413–429.

Overton, E. (1899). *Vierteljahrsschr. Naturforsch. Ges. Zuerich* **44,** 88–135.

Paine, D. L., Moore, L. C., and Horowitz, S. B. (1975). *Nature (London)* **254,** 109–119.

Plowe, J. Q. (1931). *Protoplasma* **12,** 196–221.

Pollard, T. D., Aebi, U., Cooper, J. A., Fowler, W. E., and Tseng, P. (1981). *Cold Spring Harbor Symp. Quant. Biol.* **46,** 513–524.

Robertson, J. D. (1959). *Biochem. Symp.* **16,** 3–43.

Runge, M. S., and Williams, R. C. (1981). *Cold Spring Harbor Symp. Quant. Biol.* **46,** 483–493.

Sackmann, E., and Trauble, H. (1972a). *J. Am. Chem. Soc.* **94,** 4482–4491.

Sackmann, E., and Trauble, H. (1972b). *J. Am. Chem. Soc.* **94,** 4492–4499.
Schliwa, M., van Blerkom, J., and Porter, K. R. (1981a). *Proc. Natl. Acad. Sci. U.S.A.* **78,** 4329–4333.
Schliwa, M., van Blerkom, J., and Pryzwansky, K. B. (1981b). *Cold Spring Harbor Symp. Quant. Biol.* **46,** 51–67.
Schmitt, F. O., Bear, R. S., and Clark, G. L. (1935). *Radiology* **25,** 131–151.
Singer, S. J., and Nicolson, G. L. (1972). *Science* **175,** 720–731.
Smith, I. C. P. (1972). *In* "Biological Applications of Electron Spin Resonance" (H. M. Swartz, J. R. Bolton, and D. C. Borg, eds.), pp. 484–539. Wiley, New York.
Snipes, W., and Keith, A. D. (1978a). *In* "Light Transducing Membranes" (D. W. Deamer, ed.), pp. 109–126. Academic Press, New York.
Snipes, W., and Keith, A. D. (1978b). *Symp. Pharmacol. Eff. Lipids [Pap.], 1978* Monograph No. 5, pp. 63–74.
Snipes, W., Person, S., Keith, A., and Cupp, J. (1975). *Science* **187,** 64–66.
Snipes, W., Person, S., Keller, G., Taylor, W., and Keith, A. (1977). *Antimicrob. Agents Chemother.* **11,** 98–104.
Snipes, W., Keller, G., Woog, J., Vickroy, T., Deering, R., and Keith, A. (1979). *Photochem. Photobiol.* **29,** 785–790.
Sponsler, O. L. (1940). *In* "The Cell and Protoplasm" (F. R. Moulton, ed.), pp. 166–187. The Science Press, Washington, D.C.
Steim, J. M. (1965). *Arch. Biochem. Biophys.* **112,** 599–605.
Taylor, R. B., Duffus, W. P. H., Raff, M. C., and de Petris, S. (1971). *Nature (London), New Biol.* **233,** 225–229.
Tourtellotte, M. E., Branton, D., and Keith, A. (1970). *Proc. Natl. Acad. Sci. U.S.A.* **66,** 909–915.
Trauble, H., and Sackmann, E. (1972). *J. Am. Chem. Soc.* **94,** 4499–4510.
Weltzien, H. U. (1979). *Biochim. Biophys. Acta (Reviews on Biomembranes)* **559,** 259–287.
Wiche, G., Herrmann, H., Leichtfried, F., and Pytlela, R. (1981). *Cold Spring Harbor Symp. Quant. Biol.* **46,** 475–482.
Wojcieszyn, J. W., Schlegel, R. A., Wu, E. S., and Jacobson, K. A. (1981). *Proc. Natl. Acad. Sci. U.S.A.* **78,** 4407–4410.
Wolosewick, J. J., and Porter, K. R. (1979a). *Am. J. Anat.* **147,** 303–324.
Wolosewick, J. J., and Porter, K. R. (1979b). *J. Cell Biol.* **82,** 114–139.
Wu, H., and McConnell, H. M. (1975). *Biochemistry* **14,** 847–854.

Index

D

E

M

N

O

U

V

W

X

Z